D0651281

MICROCOMPARTMENTATION AND PHASE SEPARATION IN CYTOPLASM

VOLUME 192

International Review of **Cytology** A Survey of **Cell Biology**

Guest Edited by

Harry Walter
Aqueous Phase Systems
Washington, D.C.

Donald E. Brooks
Department of Pathology and Laboratory Medicine
University of British Columbia
Vancouver, British Columbia, Canada

Paul A. Srere
Department of Biochemistry
Veterans Affairs Medical Center
Dallas, Texas

MICROCOMPARTMENTATION AND PHASE SEPARATION IN CYTOPLASM

VOLUME 192

ACADEMIC PRESS
San Diego London Boston New York Sydney Tokyo Toronto

This book is printed on acid-free paper.

Academic Press
A Division of Harcourt Brace & Company
525 B Street, Suite 1900, San Diego, CA 92101-4495
http://www.apnet.com

Academic Press
24-28 Oval Road, London NW1 7DX
http://www.hbuk.co.uk/ap/

International Standard Serial Number: 0074-7696

International Standard Book Number: 0-12-364596-4

PRINTED IN THE UNITED STATES OF AMERICA
00 01 02 03 04 05 EB 9 8 7 6 5 4 3 2 1

Paul A. Srere

Paul A. Srere, our colleague and coeditor, died unexpectedly on July 11, 1999 shortly after finishing his work on this book.

This book provided a medium for his long promulgated view that the description in standard texts of biochemical reactions taking place in cytoplasm is misleading. He believed that the high macromolecular concentration in cytoplasm and the microcompartmentation and channeling of macromolecules which point to a complex cytoplasmic organization preclude analogies to reactions of dilute chemicals as studied *in vitro.*

Paul was thoughtful, enthusiastic, and untiring in his work. His insights and comments greatly improved this volume.

We will miss him sorely.

The Editors

CONTENTS

Effects of Specific Binding Reactions on the Partitioning Behavior of Biomaterials

Gerhard Kopperschläger

Properties of Interfaces and Transport across Them

Heriberto Cabezas

Compartmentalization of Enzymes and Distribution of Products in Aqueous Two-Phase Systems

Folke Tjerneld and Hans-Olof Johansson

PART II. PHYSICOCHEMICAL PROPERTIES OF CYTOPLASM

Macromolecular Crowding and Its Consequences

H.-O. Johansson, D. E. Brooks, and C. A. Haynes

Lens Cytoplasmic Phase Separation

John I. Clark and Judy M. Clark

Cytoarchitecture and Physical Properties of Cytoplasm: Volume, Viscosity, Diffusion, Intracellular Surface Area

Katherine Luby-Phelps

Intracellular Compartmentation of Organelles and Gradients of Low Molecular Weight Species

Tak Yee Aw

Macromolecular Compartmentation and Channeling

Judit Ovádi and Paul A. Srere

The State of Water in Biological Systems

Keith D. Garlid

Mechanisms for Cytoplasmic Organization: An Overview

Len Pagliaro

PART III. CYTOPLASM AND PHASE SEPARATION

Can Cytoplasm Exist without Undergoing Phase Separation?

D. E. Brooks

Consequences of Phase Separation in Cytoplasm

Harry Walter

CONTRIBUTORS

Numbers in parentheses indicate the pages on which the authors' contributions begin.

Tak Yee Aw (223), *Department of Molecular and Cellular Physiology, Louisiana State University Medical Center, Shreveport, Louisiana 71130*

D. E. Brooks (155, 321), *Department of Pathology and Laboratory Medicine, University of British Columbia, Vancouver, B.C., Canada V6T 2B5*

Heriberto Cabezas (99), *U. S. Environmental Protection Agency, Office of Research and Development, National Risk Management Research Laboratory, Sustainable Technology Division, Systems Analysis Branch, Cincinnati, Ohio 45268*

John I. Clark (171), *University of Washington, School of Medicine, Seattle, Washington 98195*

Judy M. Clark (171), *University of Washington, School of Medicine, Seattle, Washington 98195*

Keith D. Garlid (281), *Department of Biochemistry and Molecular Biology, Oregon Graduate Institute of Science and Technology, Portland, Oregon 97291*

C. A. Haynes (155), *Biotechnology Laboratory and Department of Chemical and Bioresource Engineering, University of British Columbia, Vancouver, B.C. Canada V6T 1Z3*

Hans-Olof Johansson (137, 155), *Department of Biochemistry, University of Lund, S-22100, Lund, Sweden and Biotechnology Laboratory, University of British Columbia, Vancouver, B.C., Canada V6T 1Z3*

Göte Johansson (33), *Department of Biochemistry, University of Lund, S-22100, Lund, Sweden*

Gerhard Kopperschläger (61), *Institute of Biochemistry, Medical School, University of Leipzig, Liebigstrsse 16, Germany*

Katherine Luby-Phelps (189), *Department of Physiology, The University of Texas Southwestern Medical Center, Dallas, Texas 75235*

Judit Ovádi (255), *Institute of Enzymology, Biological Research Center, Hungarian Academy of Sciences, Budapest, Hungary*

Len Pagliaro (303), *CEREP, Inc., Redmond, Washington 98052*

Paul A. Srere (255),* *Department of Biochemistry, Veterans Affairs Medical Center, Dallas, Texas 75216*

Folke Tjerneld (137), *Department of Biochemistry, University of Lund, S-22100, Lund, Sweden*

Vladimir Tolstoguzov (3), *Nestlé Research Centre, CH-1000 Lausanne 26, Switzerland*

Harry Walter (33, 331), *Aqueous Phase Systems, Washington, DC 20008*

*Deceased.

PREFACE

Phase separation is a common occurrence when two or more structurally distinct macromolecules are dissolved, above certain concentrations, in water. One can obtain as many immiscible aqueous phases as the number of macromolecular species present. This well-known phenomenon has led us to examine in this volume the possibility, on theoretical and practical grounds, that phase separation occurs in the liquid phase of cytoplasm since it contains a high concentration of a highly diverse protein population. Phase separation would impose on cytoplasm an organization comprised of many phase compartments and interfaces that would produce partitioning of biomaterials among phases. It would provide a basis for the known microcompartmentation of cytoplasm and related phenomena.

This volume has been divided into three parts which deal, respectively, with aqueous phase systems and their properties, the physicochemical properties of cytoplasm, and a consideration, based on the book's first two parts, of the likelihood of phase separation in cytoplasm and of its consequences.

Thus, Part I reviews aqueous phase separation *in vitro*, the partitioning of biomaterials in selected phase systems, and the nature of the interface. The chapters include those on the physicochemical properties of protein–protein and protein–polysaccharide phase systems (V. Tolstoguzov); the parameters contributing to the partitioning behavior of biomaterials in two–polymer aqueous phase systems (G. Johansson and H. Walter), including how phase systems can, by specific molecular binding, yield bioaffinity partitioning (G. Kopperschläger); and the restriction (i.e., compartmentalization) of enzymes to a given phase (without binding to a solid support) with enzyme products partitioning to other phases (F. Tjerneld and H.-O. Johansson). The chapter by H. Cabezas provides a detailed analysis of the properties of interfaces (a kind of simple membrane) and the parameters which determine molecular transport across them.

Part II scrutinizes the effects of macromolecular crowding on aqueous phase separation and on the relative volumes of the phases and the size of the interface obtained (H.-O. Johansson, D. E. Brooks, and C. A. Haynes); the cytoarchitecture of cytoplasm (K. Luby-Phelps); intracellular compartmentation and channeling of molecules (J. Ovádi and P. A. Srere) as well as formation of molecular gradients and compartmentation of organelles (T. Y. Aw). A phase separation noted in lens cytoplasm is described (J. I. Clark and J. M. Clark); and observations on the behavior of water in biological systems (K. D. Garlid) are presented. Mechanisms that contribute to cytoplasmic organization other than and in addition to phase separation are briefly summarized in L. Pagliaro's chapter.

In Part III, D. E. Brook's chapter examines, on theoretical grounds, the likely nature of aqueous phases in cytoplasm if present and considers some potential experimental approaches to determine the occurrence of aqueous phase separation in cells. Based on the known physicochemical properties of aqueous phases and on the partitioning of biomaterials in them, H. Walter's chapter outlines anticipated consequences of phase separation in cytoplasm and juxtaposes some of these with reported observations in cytoplasm.

It is the hope of the editors that this volume will reenforce an appreciation of the complexity of the liquid phase of cytoplasm and tempt readers to devise experimental paths to test the model presented.

Harry Walter
Donald E. Brooks
Paul A. Srere

Part I

Physicochemical Properties of Aqueous Phase Systems and Partitioning Behavior of Biomaterials

Compositions and Phase Diagrams for Aqueous Systems Based on Proteins and Polysaccharides

Vladimir Tolstoguzov
Nestlé Research Centre, CH-1000 Lausanne 26, Switzerland

Limited thermodynamic compatibility of proteins with other proteins and proteins with polysaccharides is a fundamental phenomenon that has been demonstrated in more than 200 biopolymer pairs. These systems can undergo a liquid–liquid phase separation resulting in the different macromolecular components primarily concentrated in the different phases. This occurs under conditions (pH values and ionic strengths) inhibiting attraction between nonidentical biopolymers, i.e., the formation of interbiopolymer complexes. Generally, phase separation takes place when the total concentration of the macromolecular components exceeds a certain critical value. The excluded volume of the macromolecules determines both their thermodynamic activity and phase separation threshold. Phase diagrams of biopolymer mixtures and physicochemical features of biphasic systems are considered here. Attention is centered on the limited compatibility of the main classes of proteins and various polysaccharides and on the effects of variables such as pH, ionic strength, temperature and shear forces on the phase state, equilibrum and structure of these two-phase liquid systems. The general nature of the phenomenon of thermodynamic incompatibility of biopolymers accounts for its importance in structure formation in cytoplasm.

KEY WORDS: Biopolymer incompatibility, Excluded volume, Phase behavior of biopolymer solutions, Phase diagram, Protein-salt-water systems, Protein-protein-water systems, Protein-polysaccharide-water systems, Membraneless osmosis, Biphasic system features.

I. Introduction: Limited Compatibility of Macromolecular Compounds

The history of the experimental study of polymer incompatibility, like the history of the whole polymer science, started with natural macromolecules.

0074-7696/00 $30.00

The first observation of a limited thermodynamic compatibility (or a limited cosolubility) of biopolymers (by which is meant proteins or polysaccharides) was published more than 100 years ago. Professor W. Beijerinck (1896, 1910) discovered the impossibility of mixing biopolymers in the common solvent, water. The two solutions of gelatin and starch formed a water-in-water emulsion instead of a homogeneous mixture. This emulsion settled into two liquid layers. One of the layers was the transparent gelatin solution containing a small quantity of starch while the other layer contained starch and a small amount of gelatin. Ostwald and Hertel (1929) then found a difference in phase behavior between cereal and potato starches in mixtures with gelatin. They showed that, after phase separation, the liquid phase rich in starch turns into a powder-like precipitate. Doi and Nikuni (1962) discovered the crystallization of amylopectin in mixtures with gelatin. They hypothesized that this phenomenon could be a model for starch deposition in the plant cell. Using the cloud point method, Doi (1965) constructed first phase diagrams for the gelatin-amylopectin-water system using both temperature-gelatin content and gelatin-amylopectin contents as coordinates. Bungenberg de Jong (1936) showed that on mixing aqueous solutions of gelatin and gum arabic two different types of phase separation could occur. He determined the first phase diagrams for both phase separation types and gave them the names "simple" and "complex coacervation." Simple coacervation results in concentration of the biopolymers in different phases, i.e., it corresponds to biopolymer incompatibility. It takes place at a sufficiently high ionic strength and pH values above the isoelectric point (IEP) of gelatin where the biopolymer macro-ions have like net charges. Complex coacervation is the formation of interbiopolymer complexes, i.e., phase separation results in concentration of biopolymers into a single liquid, highly hydrated, concentrated phase with a solution of either an excess of one of the macro-ions or a nearly pure solvent as the other phase. For instance, at pHs below gelatin's IEP and at a low ionic strength electrostatic complexes of the positively charged macro-ions of gelatin and the negatively charged macro-ions of gum arabic are formed. Dobry and Boyer-Kawenoki (1948) showed that phase separation occurs in mixed solutions of many water-soluble polymers such as gelatin, serum albumin, gum arabic, glycogen, polyvinyl alcohol, polyvinylpyrrolidone, methylcellulose, polyacrylic acid, etc. Later, phase behavior of mixtures of gelatin with polysaccharides were studied by Grinberg *et al.* (1970, 1971), Grinberg and Tolstoguzov (1972, 1997), Tolstoguzov *et al.* (1974a,b, 1985), Tolstoguzov (1990, 1991, 1992), Kasapis *et al.* (1993), and Clewlow *et al.* (1995).

During the last 50 years the major efforts in the field of polymer compatibility were stimulated by practical problems related to mixtures of the following macromolecular compounds.

(i) Polymer blends are of great applied importance for controlling mechanical, optical, and other physical properties of materials such as plastics, rubbers, films, fibers, glues, etc. Normally, chemically or structurally dissimilar polymers have limited thermodynamic compatibility or limited miscibility, i.e., for thermodynamic reasons (low mixing entropy of macromolecules), these polymers are immiscible with each other on a molecular level. The Flory-Huggins (Flory, 1953; Tanford, 1961) theory of polymer solutions, developed for flexible linear and neutral polymers, predicts their phase behavior. Polymers form single-phase mixed solutions only when their mixing process is exothermic. This type of polymer pair is rarely encountered (Krause, 1978; Kwei and Wang, 1978). It was shown, theoretically and experimentally, that the transition from polymer mixtures to polyelectrolyte mixtures increases compatibility. This enhancement of the cosolubility of polyelectrolytes reflects an increase in mixing entropy due to contributions of low molecular weight counterions under electrically neutral conditions (Khokhlov and Nyrkova, 1992).

(ii) Mixtures of polysaccharides and synthetic water-soluble polymers are of practical significance in the pharmaceutical industry and biotechnology. Albertsson (1958, 1972) was a pioneer in the systematic study of phase behavior of aqueous mixed solutions of polysaccharides, modified polysaccharides, and synthetic polymers. Applications of these systems for fractionation and purification of biomaterials, such as cells, cell organelles, and biopolymers, have been developed and discussed by Albertsson (1995), Walter *et al.* (1985, 1991), and Walter and Brooks (1995). In the pioneering works of Ogston (1937) and Laurent and Ogston (1963), an original approach for estimating thermodynamic interactions and compatibility of water-soluble polymers was developed. This approach was based on the excess of osmotic pressure in mixed solutions of incompatible polymers and used for the study of biphasic systems, as proposed by Albertsson. The authors assumed that phase separation of biopolymers occurs in biological systems (Edmond and Ogston, 1968).

(iii) Protein-salt-water systems are of interest in the medical (ophthalmology) and food technologies although they may be less relevant to the possibility of phase separation in native cytoplasm. Pioneering investigations on protein-salt-water systems were carried out on lactoglobulin and seed storage globulins (including arachin and soybean globulins) by Tombs at Unilever Research in the 1960s-1970s (Tombs, 1970a,b, 1972, 1975, 1985; Tombs *et al.,* 1974) and then on lens crystallins by Benedek *et al.* at M.I.T. (Asherie *et al.,* 1996; Lomakin *et al.,* 1996; Liu *et al.,* 1995; Broide *et al.,* 1991). Phase separation in salt solutions of seed storage globulins was systematically studied in several other laboratories (Ismond *et al.,* 1988, 1990; Popello *et al.,* 1990, 1991, 1992; Suchkov *et al.,* 1990, 1997; Tolstoguzov,

1988a, 1991, 1992). The formation of highly concentrated noncrystalline phases is of significance for both lens crystallins and seed storage globulins.

(iv) Biopolymer mixtures are of importance for controlling the composition-structure-property relationship in formulated foods (Grinberg and Tolstoguzov, 1972, 1997; Kasapis *et al.,* 1993; Morris, 1990; Polyakov *et al.,* 1997; Tolstoguzov *et al.,* 1985; Tolstoguzov, 1978, 1986, 1988b, 1991, 1993b, 1996a,b, 1997a,b). Proteins and polysaccharides are the main structure forming food materials. Each food contains biopolymer mixtures. Liquid aqueous biphasic systems, i.e., water-in-water emulsions, are typical of foods. The term a "water-in-water emulsion" was invented (Tolstoguzov, 1986, 1988a,b) to distinguish them from oil-in-water and water-in-oil emulsions (Tolstoguzov *et al.,* 1974b).

Systematic investigations in the areas of protein-polysaccharide mixtures and protein mixtures started in the laboratory of novel food forms of the USSR Academy of Sciences about 30 and 20 years ago, respectively. These works were reviewed by Grinberg and Tolstoguzov (1997), Ledward (1993), Polyakov *et al.* (1997), Samant *et al.* (1993), Tolstoguzov *et al.* (1985), and Tolstoguzov (1978, 1986, 1988c, 1991, 1997a). Until that time information on the thermodynamic compatibility of biopolymers had been confined to gelatin-polysaccharide mixtures. Gelatin, however, behaves like classical flexible linear chain polymers, i.e., unlike both compact globular proteins and rigid polysaccharides. The next investigations were into the polyelectrolyte nature of biopolymers and the ability of neutral polysaccharides to bind low molecular weight ions in aqueous solutions. However, biopolymer mixtures remained experimentally unstudied for a long time. For instance, it was not clear at all whether aqueous solutions of globular proteins and their mixtures with polysaccharides could undergo phase separation. By contrast, the formation of interbiopolymer complexes, e.g., by oppositely charged globular proteins with each other and with polysaccharides, i.e., complex coacervation, was well known and attracted more attention (Ledward, 1993; Poole *et al.,* 1984; Tolstoguzov, 1978, 1986, 1990, 1994b).

Phase state analysis of more than 200 biopolymer pairs showed the general nature of their thermodynamic incompatibility in a common aqueous solvent. Experimental studies have shown that incompatibility occurs in the following biopolymer mixtures: (i) proteins and polysaccharides; (ii) proteins belonging to different classes according to Osborne's classification (i.e., albumins, globulins, glutelines, and prolamines); (iii) native and denatured proteins; (iv) aggregated and nonaggregated forms of the same protein; and (v) structurally dissimilar polysaccharides (Antonov *et al.,* 1975, 1979, 1980, 1982, 1987; Grinberg *et al.,* 1970, 1971; Grinberg and Tolstoguzov, 1972, 1997; Kalichevsky *et al.,* 1986, Kalichevsky and Ring, 1987; Polyakov *et al.,* 1979, 1980, 1985a,b, 1986a,b, 1997; Tolstoguzov, 1978, 1992, 1997a). Incompatibility also probably occurs between dissolved and adsorbed (at the oil/water interface) forms of the same protein. Denaturation and partial hydrolysis of proteins oppositely influence their incompatibility with other biopolymers (Tolstoguzov, 1991).

Since both biological and food systems generally contain the same proportions of macromolecular components, information about phase behavior of foods is of great interest for understanding that in cytoplasm. However, phase separation probably occurs more often in food than in biological systems. The reason seems to be that the biological systems are mainly based on specific interactions (mainly attraction) of biopolymers, while food is based on nonspecific interactions (both attraction and repulsion) of macromolecular compounds.

The aim of this chapter is to consider phase behavior of biopolymer mixtures, since this information is of importance in the new concept of phase separation in cytoplasm proposed by Walter and Brooks (1995).

II. Phase Diagrams

The phase behavior of a mixed biopolymer solution is quantitatively characterized by a phase diagram describing the boundary conditions of phase separation and the partitioning of the components (i.e., the water and the biopolymers) between the phases. Phase diagrams are used for graphically presenting effects of temperature, pH, salt concentration, and other variables on phase behavior of biopolymers. Two types of liquid-liquid phase separation, namely, in salt solutions of proteins and in mixed biopolymer solutions, will be considered.

A. Phase Behavior of Protein-Salt-Water Systems

Figure 1 gives the typical phase diagrams of the 11S broad bean globulin (legumin) in an aqueous salt solution. These phase diagrams plot: (a) salt concentration versus protein concentration; (b and c) temperature versus protein concentration. The latter resemble the behavior observed for lens crystallin described in the chapter by Clark and Clark in this volume. They show a reversible transition of salt legumin solutions from the single-phase to the two-phase state on diluting with water (a), cooling (b) and shearing (c) (Popello *et al.,* 1990, 1991, 1992; Suchkov *et al.,* 1990, 1997; Tolstoguzov, 1988b, 1992, 1996b). The solid lines are the binodal (coexistence) curves and present compositions of the coexisting phases. The regions lying above and below the binodal curves represent biphasic and single-phase systems, respectively. The right and left branch of each binodal correspond to the

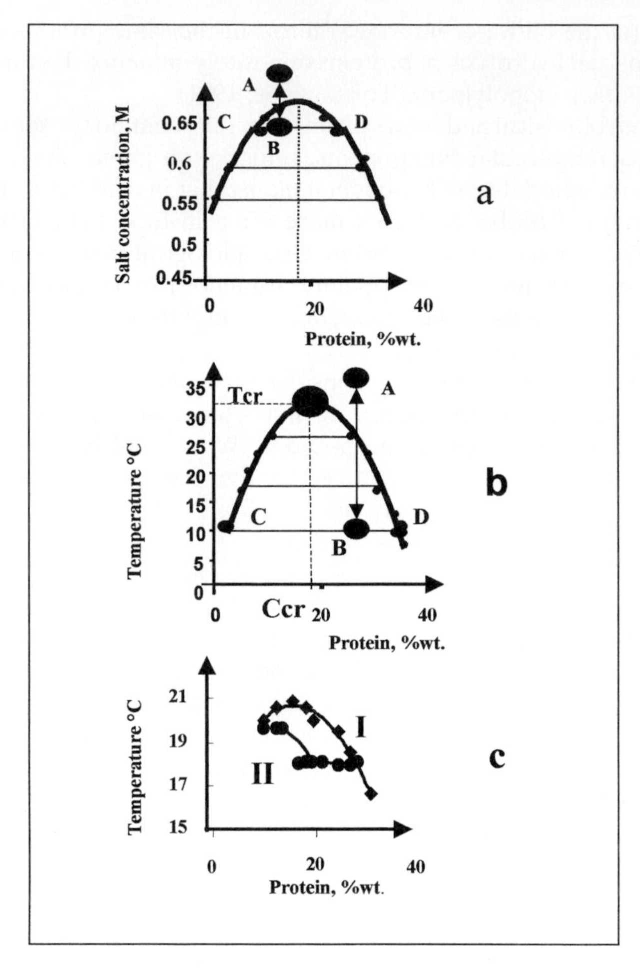

FIG. 1 Phase diagrams of the 11S globulin (legumin) of broad bean-sodium chloride-water system: (a) pH 4.8, T = 8°C; (b) pH 4.8, 0.6 M NaCl; (c) pH 4.8, 0.6 M NaCl; the binodals (I) and (II) are determined by turbidimetric and rheological techniques, respectively.

composition of the concentrated bottom phase (called mesophase) and the less concentrated top phase. The thin horizontal lines are the tie-lines. Each tie-line connects three points representing the compositions of the system as a whole and of both equilibrium phases. The relative amount of the coexisting phases is estimated by the inverse lever rule. The phase volume ratio corresponds to the ratio of the tie-line segments: CB/BD. Figures la and b show that on cooling or dilution of the initial protein solution A with water or its dialysis against water to the point B results in phase separation into two liquid phases, phase D and phase C. The protein concentration in the bottom phase can exceed 40%, while in the top phase it may be below 0.4%.

Since the phases strongly differ in protein concentration (and in refraction index), phase separation is accompanied by an increase in turbidity. Therefore, the cloud point technique can be used to determine the binodal points. Differences in concentration and density between the phases result in easy settling and formation of two liquid layers. Quantitation of the protein content in these layers gives the binodal and the tie-line positions. The phase separated turbid system C becomes transparent again after addition of a sufficient amount of salt (Fig. 1a) or by a sufficient increase in temperature (Fig. 1b) or in shear forces (Fig. 1c). The critical point shows either the maximum salt concentration (a) or the maximum temperature (b and c) at which phase separation occurs. At the critical point both coexisting phases have the same composition and volume (or mass, since their concentrations are low and densities are close to that of water). The dashed line coming vertically down from the critical points is the rectilinear diameter. It passes through the mid-tie-lines and represents the composition of systems demixing into phases of the same volume.

Systems containing native individual 11S and 7S seed globulins and their mixtures have upper critical points with the same critical protein concentration of 18%. The binodal curves are symmetrical about the rectilinear diameter. It has been assumed (Tolstoguzov, 1988c, 1991, 1992) that phase separation occurs between associated and nonassociated forms of the same protein. This means that associated (micelle-like) and nonassociated forms of the same proteins cannot recognize each other as being the same. In other words, the difference in concentration between the protein-enriched and the protein-impoverished phases (D and C), which are in osmotic equilibrium, reflects a relatively high degree of protein association. It has been also suggested that the main factors determining the association of protein molecules and phase separation are the high excluded volumes of oligomeric seed storage proteins and the dipole–dipole interaction between their molecules. The attraction of large sized micelle-like associated protein particles suspended in aqueous solution of incompatible biopolymer probably arises from depletion flocculation (Tolstoguzov, 1991, 1994a, 1997a). It should be noted that milk casein, that also occurs as large, like-charged protein associates, has a similar phase behavior in mixtures with polysaccharides. Phase separation of mixtures of 11S globulin molecules and their associates is likely similar to the phase separation of the denatured (aggregated) and native forms of the same protein (ovalbumin) (Tolstoguzov, 1988b, 1991). The crystallization of seed storage proteins in the highly concentrated phase (mesophase) is probably inhibited by the presence of isomeric protein forms. Formation of micelle-like associates of protein molecules, i.e., a process competing with protein crystallization, is of importance for formation of "protein bodies," which can be rapidly mobilized (hydrolyzed) during seed germination.

B. Phase Behavior of Biopolymer-1–Biopolymer-2–Water Systems

Conventionally, phase diagrams of three-component (ternary) systems are presented in triangular coordinates. Figure 2a shows a typical equilateral triangular diagram for a ternary system containing two biopolymers, A and B, and the solvent, water, C. Each corner of the equilateral triangle represents a pure component and its designation is marked at this corner. On the side opposite to this corner, the concentration of this component is zero. Each side of the triangle corresponds to a two-component system. The region inside the triangle represents mixtures of all three components. The system composition (e.g., X) is read along the axes, i.e., as a, b, and c coordinates whose sum equals unity.

However, an excess of solvent, water (compared to biopolymers), that is typical for biological systems, makes the use of phase diagrams in rectangular coordinates more practicable. This also simplifies plotting the effects of different variables. Figure 2b is a typical phase diagram for biopolymer-1–biopolymer-2–water systems in rectangular coordinates. Normally, the concentrations of biopolymers are plotted on the axis in weight percent; the rest is assumed to be water. Every point in a phase diagram corresponds to a system composition. The bold curve is a binodal. Biopolymers are fully miscible with each other in the concentration region under the binodal.

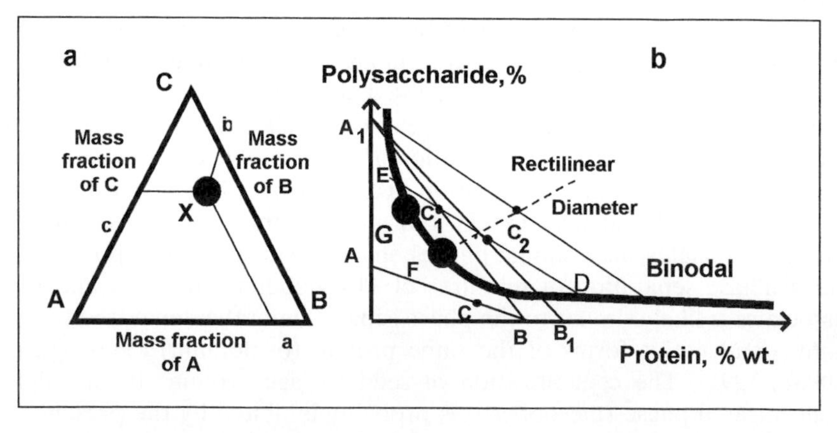

FIG. 2 Typical phase diagram for a ternary biopolymer 1 - biopolymer 2 - water system presented in the form of (a) triangular and (b) rectangular coordinates.

For instance, on mixing aqueous solutions of a polysaccharide solution A and a protein B, a single-phase stable mixture of composition C may be obtained. Mixture compositions lying above the binodal curve correspond to two-phase systems, where the phases are aqueous mixed solutions exhibiting limited co-solubility of biopolymers. For instance, on mixing a polysaccharide solution A_1 and protein solutions B or B_1, mixtures of composition C_1 and C_2 can be obtained. These mixed solutions spontaneously break down into two liquid phases, phase D and phase E. The phase D is rich in protein and another, E, is rich in the polysaccharide. On centrifugation, the protein-rich phase usually forms the lower liquid layer while the polysaccharide-rich phase forms the upper liquid layer. The composition of the more concentrated and denser bottom phase is usually plotted as the abscissa while the less dense top phase is presented as the ordinate.

The thin line DE is a tie-line. It connects the points representing the compositions of the coexisting equilibrium phases. Biopolymer mixed solutions whose compositions correspond to the same tie-line will break down into two phases of the same composition. For instance, all biopolymer mixtures of compositions DE will separate into the same phases D and E. The point of the initial biopolymer mixture (C_1) divides the tie-line in the two segments whose length ratio reflects the phase volume (weight) ratio. According to the inverse lever rule, the phase volume ratio corresponds to the ratio of the tie-line segments: EC_1/C_1D. The tie-lines can be nonparallel since an increase in concentration of biopolymers is usually accompanied by their self-association. The rectilinear diameter is the dashed line passing through the mid-tie-lines. It gives the system compositions splitting into phases of the same volume. In the vicinity of the rectilinear diameter, phase inversion can occur. For instance, this takes place when the system's composition changes from point C_1 to C_2. Lower values of the critical point F coordinates indicate lower co-solubility of the biopolymers and greater incompatibility. Table I gives critical point coordinates for several systems. The phase separation threshold G is the minimal total concentration of biopolymers required for phase separation to occur. The position of phase separation threshold geometrically corresponds to the point of contact between the binodal curve and the straight line cutting segments of the same length on the concentration axis of the biopolymers. The two-dimensional phase diagram corresponding to a certain constant temperature, pH, and ionic strength becomes three-dimensional when an additional variable, such as pH, salt concentration, or temperature is included (Tolstoguzov *et al.*, 1985).

C. Determination of Phase Diagrams

Normally, a preliminary qualitative study on phase behavior of a biopolymer mixture is very helpful. Techniques such as light microscopy, centrifuga-

TABLE I

Critical Point Coordinates for Biopolymer Mixtures

Biopolymer pair	Critical point coordinates Protein, %wt.: Pr; Polysaccharide, %wt.: Ps	Incompatibility conditions
Gelatin (mol.wt.170,000) + Pectin (330,000; DE = 67.2%)	Pr-1.0; Ps-2.6	pH 6.0; 40°C
	Pr-0.95; Ps-2.4	pH 8.0; 40°C
Gelatin (mol.wt.170,000) + alginate (400,000; MG = 50%)	Pr-0.25; Ps-1.65	pH 6.0; 40°C; 0.2 M NaCl
Gelatin (mol.wt.170,000) + alginate (400,000; MG = 20%)	Pr-0.30; Ps-1.25	pH 6.0; 40°C; 0.2 M NaCl
Gelatin (mol.wt.170,000) + methylcellulose (70,000)	Pr-0.5; Ps-1.2	pH 6.0; 40°C; 0.2 M NaCl
Gelatin (mol.wt.170,000) + dextran (2,000,000)	Pr-2.6; Ps-5.2	pH 6.0, 40°C; 0.2 M NaCl
Legumin (360,000) + dextran (270,000)	Pr-3.1; Ps-1.3	pH 7.8; 25°C; 0.1 M NaCl
Legumin (360,000) + k-Carrageenate-Na (disordered conformation)	Pr-2.10^{-2}; Ps-5.10^{-3}	pH 7.8; 0.1 M NaCl; 40°C
Legumin (360,000) + k-Carrageenate-Na (helical conformation)	Pr-5.10^{-3}; 1.10^{-3}	pH 7.8; 0.1 M NaCl; 10°C
Glycinin (360,000) + Pectinate-Na (DE = 58%)	Pr-9; Ps-0.28	PH 7.8; 0.3 M NaCl
Gelatin + *Vicia faba* globulins	$Pr_1 - 2.5$; $Pr_2 - 11.6$	pH 7.0; 40°C
Gelatin + 11S *Vicia faba* globulin	$Pr_1 - 2.1$; $Pr_2 - 12$	pH 7.0; 40°C
Gelatin + 11S *Vicia faba* globulin	$Pr_1 - 4.0$; $Pr_2 - 16.0$	pH 6.6; 40°C, NaCl 0.5 M°
Ovalbumin + *Vicia faba* globulins	Pr_1 <10; Pr_2 >15	pH 6.6; 20°C
Ovalbumin + thermodenatured ovalbumin	Pr_1 - 10.8; Pr_2 - 3.6	PH 6.7; 20°C

tion, turbidimetry, and viscometry can be used. Single-phase and biphasic systems can be identified by visual inspection (by turbidity and demixing into two layers) and by light microscopy. Spherical and nonspherical dispersed particles observed by an optical microscope correspond to liquid and solid phases, respectively. The cloud point method for construction of a phase diagram is especially useful when initial solutions of individual biopolymers have a low optical density. Normally, either cloud-point temperatures or cloud-point concentrations of an added biopolymer are determined. For qualitative determination of the binodal, a solution of one of the biopolymers is added to a solution of the second one until turbidity occurs. Then the solution of the second biopolymer is added until the mixture becomes clear. These operations are repeated using series of the individual biopolymer solutions of different concentrations to determine the binodal position (Grinberg and Tolstoguzov, 1972). A turbodimetric titrator was developed for determination of cloud points corresponding to a certain value of the turbidity change at fixed temperature, pH and salt concentration (Antonov *et al.,* 1975).

Viscosimetric titration can be used to construct phase diagrams of concentrated nontransparent systems (Tolstoguzov *et al.,* 1969). Phase separation is usually accompanied by a decrease in viscosity (Suchkov *et al.,* 1997), which can be used for finding the binodal points (e.g., in Fig. 1c, binodal II). Viscosimetry was also used to study the interactions of biopolymers in dilute mixed solutions (Varfolomeeva *et al.,* 1980).

Construction of phase diagrams usually starts with the preparation of series of mixed solutions sufficiently differing in bulk biopolymer concentration. Some of them can be single-phase solutions, others biphasic systems. A true equilibrium between the phases is experimentally obtained by mixing or shaking the water-in-water emulsions under different time–temperature conditions. Separation of the phases by centrifuge provides information about the state of matter and the volume ratio of the system phases. Centrifugation conditions vary for different water-in-water emulsions and can change greatly by phase inversion. The closer a system composition is to the critical point, the smaller the difference in density between the phases and the more difficult their separation by centrifugation.

The amount of each biopolymer in each phase separated by centrifugation can be quantified by various techniques. Estimation of protein concentration by UV absorbance at 280 nm (depending on the tyrosine, tryptophan, phenylalanine, and cysteine content) is widely used because of its simplicity and sensitivity. The disadvantage of the method is a strong contribution at the same wavelength from nucleic acids. An accurate method is binding of the anionic dye Coomassie blue to protein (to arginyl and lysyl side groups of the protein, but not to free amino acids) with an adsorbance maximum at 590 nm. The Biuret reaction, where the protein peptide bonds react with

Cu^{2+} under alkaline conditions, and the Lowry method, where the Folin reagent is additionally included, are not so widely applied because of strong interference from sugars, nucleic acids, and buffers. Kjeldahl's method is specially applicable for measuring the percentage of proteins. The concentration of the polysaccharide is usually determined at 490 nm by the phenol sulfuric method (Dubois *et al.,* 1956). Alternative methods have been recommended for analysis of the phases, e.g., optical activity, far-UV spectroscopy (190–220 nm), Fourier transform infrared spectroscopy, sedimentation, electrophoresis, HPLC, and other chromatographic and radiometric (radiolabelled biopolymer) techniques (Albertsson, 1972; Grinberg and Tolstoguzov, 1972; Polyakov *et al.,* 1980; Durrani *et al.,* 1993; Medin and Janson, 1993).

The phase volume ratio method was developed (Polyakov *et al.,* 1980; Antonov *et al.,* 1987) for determination of phase diagrams without chemical analysis. In this case, the volumes of the two separated phases and the phase volume ratio are determined for a large number of mixed solutions whose composition presented in the weight fraction of one of the biopolymers varies from zero to unity. Then to determine two points of the binodal curve the experimental dependence of the system composition-phase volume ratio is graphically extrapolated to zero and unity values of the phase volume ratio. The system composition with 0.5 phase volume ratio corresponds to the point of the rectilinear diameter.

In all cases, the experimental data are usually checked by the material balance of equilibrium phase separation. The tie-lines have to be straight lines, each connecting three points representing the compositions of the system (e.g., C) and of its phases (e.g., D and E). The phase volume ratio has to correspond to the inverse lever rule. The critical point can be determined by extrapolation of the rectilinear diameter. The phase separation threshold is graphically determined as the point of contact of the binodal and the line intercepting equal lengths on the coordinate axes.

III. Factors Affecting Phase Behavior of Biopolymer Mixtures

A. Effects of Biopolymer Composition and Environmental Medium (Salt and Sugar Concentration, pH, and Temperature)

Mixed biopolymer solutions can be quantitatively thermodynamically characterized by the interactions of nonidentical macromolecules with each other and with the solvent in terms of the second virial coefficient, experimentally obtained by light scattering. The cross second virial coefficient, reflects pair interactions of nonidentical biopolymers. Its positive value corresponds to a net repulsive biopolymer interaction and incompatibility. This approach has been used to study the interactions responsible for the incompatibility of different biopolymer pairs and the effect of additives, e.g., sucrose (Semenova *et al.,* 1990, 1991a,b; Tolstoguzov, 1991, 1992, 1994a,b; Tsapkina *et al.,* 1992). For instance, it was found that thermodynamic compatibility of many biopolymer pairs (e.g., legumin and dextran, legumin and ovalbumin, sodium caseinate and ovalbumin, sodium caseinate and sodium alginate) increases with sugar concentration. Incompatibility occurs under certain conditions (pH values, ionic strengths, and temperature) favoring association of identical macromolecules and inhibiting attraction between nonidentical macromolecules, i.e., formation of interbiopolymer complexes. Unfavorable interactions (repulsion) of chemically and structurally nonidentical macromolecules result in each macromolecule preferring to be surrounded by its own type. Compatibility usually arises from the formation of weak soluble complexes with an energy of attractive interbiopolymer interaction of about 2 kT or higher (Varfolomeeva *et al.,* 1980).

Most biopolymers are polyelectrolytes. Therefore, the major factors affecting interactions of biopolymers with each other and with the aqueous solvent are pH, salt concentration, the conformation and counterions of the biopolymers, their net charge, charge density, hydrophobicity, and H-bond forming ability. The counterions of biopolymers, as polyelectrolytes, contribute to an increase in mixing entropy and favor compatibility of biopolymers compared to classical polymers.

Normally, self-association of proteins increases in salt solutions when the pH approaches the protein's IEP. This corresponds to an enhancement of thermodynamic incompatibility between proteins and polysaccharides. When the salt concentration is below a certain critical value, the protein and the polysaccharide are completely compatible. The compatibility of the macromolecular components drops sharply when the salt concentration exceeds the critical point. The critical salt concentration required for phase separation is a function of the pH and salt composition. It increases in the following order for polysaccharides: carboxyl-containing$<$neutral$<$sulfated.

Generally, protein-neutral polysaccharide mixtures separate into two phases when the salt concentration exceeds 0.1 M. Proteins and carboxyl-containing polysaccharides have a limited compatibility when either the pH exceeds the protein's IEP (at any ionic strength) or the pH is equal to or less than the protein's IEP and the ionic strength exceeds 0.2 M. With sulfated polysaccharides, globular proteins are usually incompatible at ionic strengths above 0.3 M, irrespective of the pH (Antonov *et al.,* 1975, 1979,

1980; Grinberg and Tolstoguzov, 1972, 1997; Tolstoguzov *et al.*, 1985; Tolstoguzov, 1978, 1988b, 1997a,b).

B. Effect of Molecular Size. Excluded Volume Effects

When attraction between nonidentical macromolecules (e.g., between proteins and polysaccharides) is inhibited, the major factor determining phase separation threshold is the excluded volume that depends upon the size and shape of the macromolecules (Semenova *et al.*, 1990, 1991a,b; Tolstoguzov, 1978, 1991, 1992).

Because molecules are not penetrable by each other, the volume of a solution occupied by a macromolecule is not accessible to other macromolecules. Therefore, a minimal distance between two adjacent spherical molecules of a globular protein equals the sum of their radii, or the diameter of one of them. This signifies that around each protein molecule there is an excluded volume not accessible to other molecules. This excluded volume is more than eightfold greater than the protein molecule itself since it includes the hydration water attached to this protein molecule. It is significantly greater for nonspherical macromolecules and depends upon the flexibility of the macromolecular chain, and its configurational, rotational, and vibrational properties (Tanford, 1961). The effects of spatial limitations are enhanced by the transition from a dilute mixed solution, where individual macromolecules are independent of each other, to a semi-dilute biopolymer solution where molecules interact and compete for the same space. Phase separation usually takes place when the total concentration of a mixed biopolymer solution exceeds a certain critical value corresponding to the regime of semi-dilute solutions. The phase behavior of biopolymer mixtures can be predicted from the excluded volume of the macromolecules (Semenova *et al.*, 1990, 1991a,b; Tolstoguzov, 1991, 1992). Phase separation occurs at about 1–3% for mixtures of rigid, rod-like polysaccharides; about 2–4% for mixtures of linear polysaccharides with proteins of unfolded structure, such as gelatin or casein; about 4% or higher for globular protein-polysaccharide mixtures; and exceeds 10% for mixtures of globular proteins (Tolstoguzov, 1978, 1988c, 1990, 1991).

1. Protein Mixtures

From the viewpoint of structure, size, and molecular weight, globular proteins are macromolecular compounds, but are not typical polymers. Normally, mixed solutions of classical polymers tend to be completely separated into nearly pure phases containing individual polymers. This reflects a very low co-solubility of classical flexible chain polymers. Unlike mixtures of flexible chain polymers in a common solvent, significantly (at least tenfold) higher phase separation threshold values, a high phase diagram asymmetry and quite similar phase behavior of their mixtures are typical of mixed protein solutions. Table I shows critical point coordinates for several protein pairs. Figure 3 gives phase diagrams for mixed solutions of some proteins of different conformations. The remarkably higher co-solubility of proteins compared to that of protein-polysaccharide mixtures seems to be of importance for biological functions of proteins, especially enzymes and enzyme inhibitors. The compactness, rigidity, rounded shape, and limited number of accessible ionizable side groups of a protein molecule make the phase behavior of proteins greatly different from common polymers and polyelectrolytes. Besides a relatively low excluded volume effect and polyelectrolyte nature, a significant difference in size (molecular weight) is probably of importance to higher co-solubility of globular proteins, especially those belonging to the same class within the Osborne classification. For instance, the small size of molecules of low molecular weight proteins,

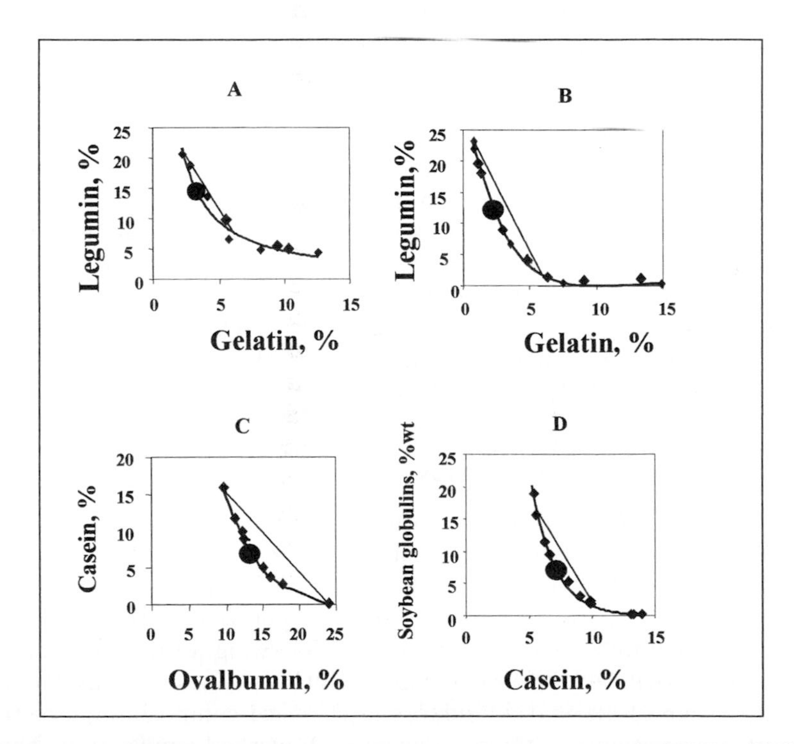

FIG. 3 Phase diagrams of some protein mixtures. A-Gelatin + legumin (pH 7.0, 40°C); B-Gelatin + legumin (pH 7.0, 0.5 M NaCl, 40°C); C-Ovalbumin + casein (pH 6.6, 20°C); D-Soybean globulins + casein (pH 6.9, 25°C). ●-critical point.

such as trypsin inhibitor, seems to be responsible for an increase in their compatibility with other proteins. This is probably due to an increase in mixing entropy of particles with similar shape and surface but differing in size. One more specific feature, which could contribute to high phase separation thresholds of globular proteins, is that their chemical information is mainly concealed in the hydrophobic interior of the protein globule. The tertiary structure of coiled and pleated polypeptides mimics on a molecular level that of proteins by hiding their chemical differences inside the globules. This molecular (or thermodynamic) mimicry minimizes chemical differences between proteins with respect to their interactions with both other macromolecules and solvent water. This results in better solubility and especially co-solubility of proteins in aqueous media. The same mechanism of thermodynamic mimicry is also probably used by polysaccharides forming helical structures and by polynucleotides coated by proteins to form viruses. The polyelectrolyte nature of biopolymers enhances co-solubility due to the contribution of low molecular weight counterions to an increase in mixing entropy. However, the effects of pH, salt concentration, and temperature on compatibility of proteins are less pronounced than for linear polyelectrolytes. Compatibility of proteins usually increases with ionic strength. Thermal denaturation of proteins and association or dissociation of oligomeric proteins greatly affects protein compatibility. Thermal denaturation can result in incompatibility of proteins of the same class according to the Osborne classification and even between the native and denatured forms of the same protein. The dissociation of protein complexes (such as casein arising from changes in pH, ionic strength, and binding of Ca^{2+} by complexons) and oligomeric proteins (e.g., seed storage globulins) results in an increase in compatibility. For instance, the compatibility of casein with soybean globulins and with pectin increases when the pH increases from 6.0 to 8.0 and sodium citrate is added (Andersson et al., 1985; Tolstoguzov, 1988c; Polyakov *et al.,* 1997).

2. Protein-Polysaccharide Mixtures

Generally, the chemical composition, structure, molecular weight and polydispersity of polysaccharides affect their co-solubility with each other and with proteins. Compatibility in protein-polysaccharide mixtures increases in the following order of polysaccharides: carboxyl-containing>neutral>sulfated; and in the following order of proteins: albumins>globulin>casein>prolamines>glutelins. These trends are observed in spite of marked differences in chemical composition, structure and molecular weight of the biopolymers. A marked similarity in phase behavior of protein mixtures with neutral and anionic polysaccharides reflects the formation of inter-biopolymer protein-neutral polysaccharide complexes at low ionic strengths and pH differing from the protein's IEP. Dissociation conditions correspond to those of phase separation. Generally, protein-polysaccharide mixtures show a lower critical phase separation temperature (Grinberg and Tolstoguzov, 1972, 1997; Tolstoguzov *et al.,* 1985; Varfolomeeva *et al.,* 1980).

An increase in excluded volume (molecular weight) of proteins and polysaccharides results in a decrease in compatibility. Biopolymer incompatibility increases under conditions favorable for enlargement of macromolecules, for instance, under a statistical coil-helix conformational transition. Linear polysaccharides (e.g., pectin and alginate) are less compatible with proteins than branched polysaccharides (e.g., gum arabic) of the same molecular weight. This obviously reflects the smaller excluded volume of branched macromolecules compared to linear molecules of the same molecular weight (Tolstoguzov, 1991).

Biopolymer incompatibility decreases with partial hydrolysis of macromolecules. Strong changes in compatibility resulting from a limited proteolysis (Danilenko *et al.,* 1992) could be of interest for modeling the enzymatic activation of inactive forms of many proteins, such as trypsinogen (secreted form) modified by an enzyme to yield trypsin.

3. Phase Diagram Asymmetry

The competition between macromolecules for space in a mixed solution determines both the critical conditions of phase separation and the water and biopolymer partition between the system phases, i.e., phase diagram asymmetry. Normally, phase diagrams of biopolymer mixture are markedly asymmetric concerning compositions of the coexisting phases. Phase diagram asymmetry can be evaluated by the ratio of the critical point coordinates, by the angle made by the tie-lines with the concentration axis of one of the biopolymers and by the length of the segment of a binodal curve between the critical point and the phase separation threshold (Tolstoguzov, 1986, 1988b, 1991). The role of the size of macromolecules is quite evident from Table I and Fig. 3. The concentration of a smaller size biopolymer (protein) is higher at the critical point. The critical point is usually shifted from the phase separation threshold toward the axis of the biopolymer with lower hydrophilicity. The binodal is always closer to the concentration axis of the biopolymer of lower excluded volume. The greater the difference in the biopolymer molecular masses, the greater the shift of the binodal towards the concentration axis of the lower molecular mass biopolymer or the greater the shift of the phase separation threshold from the coordinate angle bisector in the same direction. Self-association of macromolecules can change both the excluded volume effects and the affinity of supermolecular structural domains for the solvent water. As a result, the slope of tie-lines changes with an increase in the bulk biopolymer concentration.

4. Concentration and Fractionation of Biopolymers

The greater the phase diagram asymmetry the larger the difference in water content between the phases. The water content is higher in the phase of a more hydrophilic biopolymer with a larger excluded volume. Figure 2b shows that phase separation is accompanied by an increase in the concentration of each biopolymer in the coexisting phases. Protein concentration in phase D is higher than in the initial protein solution B and the mixed solution C_1. The initial polysaccharide solution (A_1) is diluted to point E. Phase separation can give rise to a strong increase in the concentration of one of the biopolymer phases and a corresponding dilution of the other phase. This phenomenon underlies a new method for concentrating biopolymer solutions called "membraneless osmosis." Membraneless osmosis occurs between immiscible solutions. Therefore, the semipermeable membrane used for conventional osmosis is replaced by an interfacial surface between the immiscible solutions. Membraneless osmosis was first employed to concentrate skimmed milk proteins using apple pectin. Figure 4a shows that a mixture of skimmed milk and 1% pectin solution breaks down into two liquid phases with the protein-rich phase containing about 20% of milk casein (Antonov *et al.,* 1982; Tolstoguzov *et al.,* 1985; Tolstoguzov, 1988a,c, 1995, 1996b, 1997b; Zhuravskaya *et al.,* 1986). Phase separation also enables biopolymers to be fractionated. Two approaches are used. The first is based on the phase separation threshold as determined by the less compatible of the biopolymer components. Partition of the other components corresponds to their relative affinities for the two phases and the interfacial layer. The second approach is based on the difference in the phase separation threshold of various proteins with the same polysaccharide. Accordingly, stepwise addition of a polysaccharide may induce separation and fractionation of proteins. For instance, the purification of baker's yeast proteins from nucleic acids and some lipids to be used for human consumption was achieved by gradual addition of pectin (Bogracheva *et al.,* 1983; Tolstoguzov, 1986).

5. Effect of Gelation

Phase equilibrium can only be achieved when liquid–liquid (Figs. 2, 3, and 4a) phase separation occurs (Tolstoguzov, 1988b, 1995). Figure 4b shows the competition between phase separation and lyotropic gelation, i.e., from separation of the solvent. The equilibrium between the phases is not achiev-

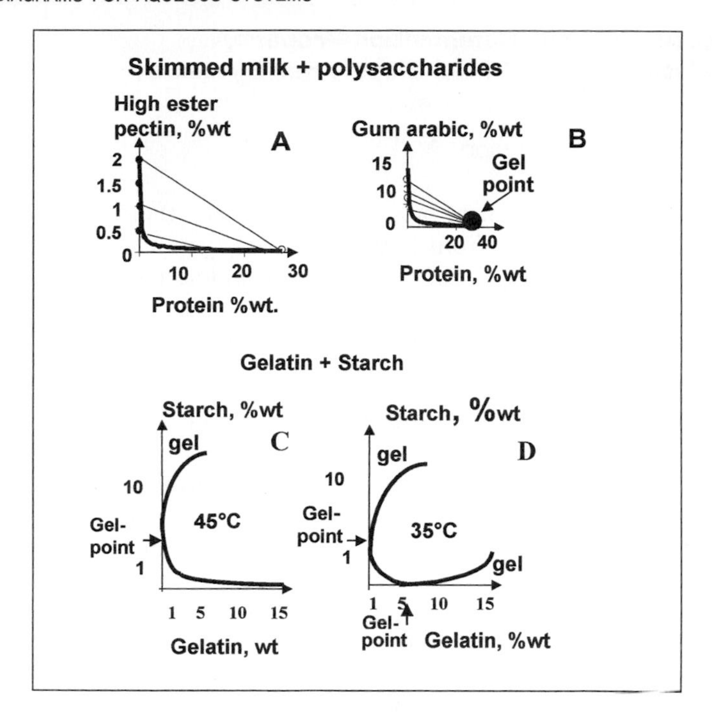

FIG. 4 Typical phase diagram for (A) an equilibrium protein-polysaccharide-water system (skimmed milk + high ester pectin) and systems with gelation of (B) the dispersed phase (skimmed milk + gum arabic), (C) the dispersion medium (gelatin + starch at 45°C) and (D) both phases (gelatin + starch at 35°C).

able when one or both phases of the system are gelled. Gelation stops the partitioning of the solvent and some solutes between the phases. Figures 4b, 4c and 4d illustrate phase diagram anomalies corresponding to gelation of the dispersed phase, the dispersion medium, and both phases, respectively. These anomalies can be observed, e.g., during phase separation of casein-polysaccharide (arabic gum and arabinogalactan) and gelatin-starch systems (Grinberg *et al.,* 1971; Antonov *et al.,* 1982). Gelation of the dispersed phase interrupts balancing of the osmotic pressure between coexisting phases. This is due to the back pressure of the gel stopping membraneless osmosis, water partition and establishing an osmotic equilibrium between the phases. The tie-lines converge to a certain point corresponding to the composition of the gelled dispersed phase. Gelation of both phases has the same consequences: an incomplete separation of the phases and nonequilibrium with respect to the biopolymer distribution. The dispersed particles act as filler in the matrix (continuous) gelled phase.

IV. Features of the Composition-Property Relationships in Mixed Biopolymer Systems

A. Low Interfacial Tension

The first feature of two-phase biopolymer systems is their very low interfacial tension. This results from both close compositions of the coexisting phases whose main component is the solvent, water, and a strong biopolymer co-solubility (Tolstoguzov *et al.*, 1974b; Tolstoguzov, 1994a). Low interfacial tension favors the stability of water-in-water emulsions.

B. Interfacial Layers in Biphasic Biopolymer Systems

The next feature of two-phase biopolymer systems is the presence of interfacial layers of low biopolymer concentration, low density, and viscosity, discussed as well in the chapter by Cabezas in this volume. The formation of interfacial layers results from unfavorable interactions of incompatible macromolecules. This reflects trends towards surroundings of the same type for each macromolecule. Nonidentical macromolecules are mutually depleted from the contact area between aqueous phases. The presence of an interfacial (or depletion) layer with a thickness of the order of magnitude of the size of the macromolecule and macromolecular aggregate has been demonstrated by both electron-microscopy of gels formed from a two-phase gelatin-dextran-water system (Tolstoguzov *et al.*, 1974a) and the measurement of steady-state viscosity in a two-phase liquid casein-sodium alginate-water system (Suchkov *et al.*, 1981). Electron microscopy revealed an interesting interface-induced structure formation in gelatin gels filled with droplets of dextran-rich phase. Within the interfacial layer between the droplets of dextran phase and an isotropic network of the bulk of the gelatin gel, all fibrous aggregates forming the gel network are oriented normally to the surface of the droplets (Tolstoguzov *et al.*, 1974a). This structure-forming effect of dispersed particles filling the gel, probably reflects the minimization of contacts between immiscible biopolymers.

The interfacial (or depletion) layers are responsible for depletion flocculation and coalescence of dispersed droplets of water-in-water emulsion as well as for an interfacial adsorption of macromolecules and colloidal particles between the aqueous phases (sections IV C and D).

C. Probable Surfactant for Water-in-Water Emulsions

Compounds such as glycoconjugates and synthetic protein-polysaccharide conjugates could be adsorbed at the water/water interfaces and play the role of surfactants for biphasic protein-polysaccharide systems. Macromolecules comprising two or several classes of incompatible biopolymers (e.g., polysaccharides or heterooligosaccharides covalently bound to the polypeptide chain) can have an affinity for both coexisting biopolymer phases. Adsorption of protein-polysaccharide conjugates (as surfactants for water-water emulsions) can increase adhesion between the aqueous protein-rich and polysaccharide-rich phases, improve thermodynamic stability of water-water emulsions and the co-solubility (compatibility) of biopolymers (Tolstoguzov, 1993a,c, 1994b).

The attraction and repulsion between the protein and polysaccharide parts of a hybrid macromolecule are demonstrated by its folding (collapse) and unfolding. This could greatly change the shape and size of protein-polysaccharide hybrids, their excluded volume, solubility, co-solubility, and the viscosity of their solutions and the interfacial layers. Attraction and repulsion between the protein and polysaccharide parts of hybrid macromolecules and control of their nonspecific inter- and intramolecular interactions could be used to model the folding–unfolding (denaturation–renaturation) of globular proteins and biological functions of glycoconjugates. Accordingly, surfactants for water-in-water emulsions in which both phases are protein-rich could be proteins with widely differing amino acid compositions or their aggregates.

D. Interfacial Adsorption of Lipids

Lipid dispersed particles added to water-in-water emulsions may be either encapsulated by the protein-rich phase (section E1) and/or concentrated at the interface between the two aqueous phases. Concentration and coalescence of lipid droplets at the interfacial layers can result in thin lipid layers between aqueous phases. Thus, interfacial adsorption of lipids in a water-in-water emulsion provides the physical basis for the formation of lipid/protein two-dimensional layers and three-dimensional honeycomb-like lipid structures. Formation of a continuous lipid phase is used for large scale production of low-fat butter replacers based on gelatin-polysaccharide mixtures (Tolstoguzov, 1994a, 1996b).

E. Composition-Property Relationships in Single- and Two-Phase Mixed Biopolymer Solutions

Owing to competition for space occupancy and excluded volume effects, different macromolecules mutually affect each other's behavior. This determines composition-property relationships in single-phase mixed solutions:

the main reaction upon increasing biopolymer concentration in a mixed biopolymer solution is reduction of excluded volume effects. According to Le Châtelier's principle, an equilibrium system subjected to a perturbation reacts in a way that tends to nullify the effect of this perturbation. A reduction of excluded volume effects is achievable by a decrease in the size, concentration and mobility of biopolymer space-filling particles due to, e.g., compactization, self-association of macromolecules, formation of interbiopolymer complexes, gelation, and crystallization of biopolymers. An increase in excluded volume effects in biopolymer mixtures can have the following manifestations: (i) increase in thermodynamic activity, (ii) enhancement of association of biopolymers, (iii) reduction of co-solubility of macromolecules and phase separation, (iv) enhancement of phase diagram asymmetry, (v) enhancement of protein adsorption at the oil/water and gas/water interfaces, (vi) increase in the rate of gelation and reduction in the critical concentration for gelation of mixed solutions compared to solutions of the individual biopolymers, and (vii) increase in the crystallization rate of polysaccharides. Mixing of biopolymers can greatly (synergistically or antagonistically) change physicochemical properties of liquid and gelled systems. This relates to both bulk and surface properties of biopolymer solutions with the compositions lying on both sides of the binodal curve (Tolstoguzov and Braudo, 1983; Morris, 1990; Tolstoguzov, 1978, 1990, 1994a,b, 1995; Zayspkin *et al.,* 1997).

1. Effect of Incompatibility on Protein Adsorption at the Oil/Water Interface

The relationship between solubility and surface activity differs between biopolymers and surfactants of low molecular weight. For instance, methanol and acetic acid are mixable with water in any proportion. But when the length of the hydrocarbon radical is increased, solubility in water decreases and affinity for an oil phase increases. Such a chemical modification leads to surfactants (fatty acids and alcohols) of low molecular weight. By contrast, the solubility and surface activity of a biopolymer can be controlled without chemical modification by addition of an incompatible biopolymer. The least soluble biopolymer component of a mixed aqueous solution is the most rapidly adsorbed at the interfaces. The competitive adsorption of biopolymers could include a stage of phase separation in the microvolume at the oil/water interface. Partition of a biopolymer between the bulk phase and the interfacial layer, including its adsorption at oil-water interfaces, can be managed by the addition of another biopolymer (Tolstoguzov, 1991). Additionally, complexing with lipids and/or anionic polysaccharides may reduce protein conformational stability, decrease solubility of the biopolymer (proteins and polysaccharides), increase biopolymer surface activity, and result in stronger emulsion stabilizing layers.

Adsorbed (on the oil/water interface) and dissolved (in the dispersion medium) molecules of the same biopolymer may differ in their conformational arrangement and charge distribution. It was, therefore, assumed that adsorbed and dissolved molecules of the same protein cannot recognize each other as being the same and can exibit incompatible behavior with each other. Accordingly, the monolayer of already adsorbed protein molecules can inhibit further adsorption of dissolved molecules of the same protein. This would cause an extended characteristic plateau covering a wide concentration region between critical concentrations for monolayer and multilayer adsorption in the adsorption isotherm of a globular protein. This is not, however, the case in casein, i.e., a protein complex (of thermodynamically compatible subunits), where there is a continuous increase in the amount of adsorbed protein with an increase in concentration of the casein solution (Tolstoguzov, 1991, 1992, 1994a).

Addition of a polysaccharide can affect the stability of oil-in-water emulsions stabilized by protein in various ways (Burova *et al.,* 1992; Pavlovskaya *et al.,* 1993; Tolstoguzov, 1991, 1992; Tsapkina *et al.,* 1992). An increase in protein adsorption on the oil/water interface can occur. This effect has been shown in mixtures of 11S broad bean globulins with dextran. Second is a decrease in the critical concentration required to produce multilayer protein adsorption with increasing amount and molecular weight of the polysaccharide. This results from phase separation of the continuous aqueous phase of the oil-in-water emulsion leading to a dispersed phase rich in protein-encapsulated lipid droplets. It has been shown that the stability of oil-in-water emulsions, stabilized by 11S broad bean globulin, is increased by addition of a polysaccharide that is incompatible with the protein. Third, depletion flocculation, i.e., reversible association of colloidal biopolymer particles within a nonwettable medium, i.e., a solution of incompatible biopolymer. Since colloidal dispersions are thermodynamically unstable, separation of mixed colloidal dispersions usually occurs at significantly lower concentrations than that of molecularly dispersed biopolymers (Tolstoguzov, 1988c, 1997a).

2. Effect of Incompatibility on Conformational Changes and Gelation of Biopolymers

Since thermal denaturation of a globular protein is not accompanied by any substantial change in its molecular volume, addition of a polysaccharide does not markedly affect the conformational stability of proteins, but can greatly accelerate both aggregation of denatured molecules and gelation (Tolstoguzov, 1991; Burova *et al.,* 1992). In mixtures, biopolymers behave as if they were in a solution of a higher concentration. For instance, gelation of a 5% gelatin solution is greatly accelerated (from 7 days to 2 hours) and the elastic modulus of the gelatin gel increases by 2- to 5-fold due to addition of 0.1–0.2% of dextran or another polysaccharide, such as methylcellulose or agarose (Tolstoguzov *et al.,* 1974a; Tolstoguzov, 1988a, 1990, 1995; Zasypkin *et al.,* 1997). Another example is amylopectin crystallization in gelatin gels. Doi (1965) showed that an increase in gelatin concentration from 16 to 46% reduces the time for complete crystallization of 0.5% amylopectin from 3 days to 20 hours at 4°C. Mechanical properties (e.g., the shear modulus and tensile strength) of filled gels are proportional to the volume fraction of the dispersed phase, i.e., they obey additivity laws (Gotlib *et al.,* 1988; Suchkov *et al.,* 1988; Tolstoguzov, 1978, 1988b).

Biopolymers can be more "co-soluble" in the bulk of the gel than in the liquid phase. Normally, biopolymer gelation results in a decrease in excluded volume effects due to a decrease in the amount and mobility of space-filling particles. This makes the dispersion medium of the gel a better solvent than the initial mixed solution. Therefore, the gel dispersed phase in Fig. 4a can be regarded as a compartment with a better solvent quality than the bulk of the continuous liquid phase. This decrease in the excluded volume of biopolymers in the gel state results in the formation of interpenetrated networks by biopolymers immiscible in the liquid solution (Tolstoguzov, 1990, 1995). Presumably, for this reason, the rate of enzymatic hydrolysis of a solid (gel-like) food usually seems to be proportional to its volume rather than to its surface area.

V. Concluding Remarks

Phase behavior of biopolymer mixed solutions has been considered to illuminate the possibility that phase separation could occur in cytoplasm. There is already a large variety of experimental evidence for the limited thermodynamic compatibility of biopolymers and for phase separation under conditions typical of cytoplasm. The concentration of biopolymers in the cytoplasm of cells and extracellular biological systems is probably comparable, or exceeds typical values of phase separation thresholds of biopolymers. Accordingly, two main aspects of incompatibility in biological systems are: (i) control of the thermodynamic activity of biopolymers and (ii) control of cytoplasm heterogeneity. Knowledge about composition-property relationships of aqueous biphasic systems may be of importance for a better understanding of the formation of cytoplasmic structures.

References

Ålbertsson, P.-A. (1958). Partition of proteins in liquid polymer-polymer two-phase systems. *Nature* (*London*) **177,** 709–711.

Ålbertsson, P.-A. (1972). "Partition of Cell Particles and Macromolecules." Wiley-Interscience, New York.

Ålbertsson, P.-A. (1995). Aqueous polymer phase systems: Properties and applications in bioseparation. *In* "Biopolymer Mixtures" (S. E. Harding, S. E. Hill, and J. R. Mitchell, eds.), pp. 1–12. Nottingham University Press, Nottingham.

Andersson, O., Schmandke, H., Polyakov, V. I., Grinberg, V. Ya., Bikbov, T. M., Danilenko, A. N., Leontjev, A. L., and Tolstoguzov, V. B. (1985). Compatibility of gelatin with broad bean globulin in an aqueous medium. *J. Food Sci.* **50,** 1133–1135.

Antonov, Yu. A., Grinberg, V. Ya., and Tolstoguzov, V. B. (1975). Phasengleichgewichte in Wasser/Eiweiss/Polysaccharid Systemen. *Staerke* **27,** 424–431.

Antonov, Yu. A., Grinberg, V. Ya., and Tolstoguzov, V. B. (1979). Thermodynamische Aspekte der Vertraeglichkeit von Eiweissen und Polysacchariden in waessrigen Medien. *Nahrung* **23,** 207–214, 597–610.

Antonov, Yu., Grinberg, V., Zhuravskaya, N., and Tolstoguzov, V. (1980). Liquid two-phase water-protein-polysaccharide systems and their processing into texturated protein products. *J. Text. Stud.* **11,** 199–215.

Antonov, Yu. A., Grinberg, V. Ya., Zhuravskaya, N. A., and Tolstoguzov, V. B. (1982). Concentration of protein skimmed milk by the method of membraneless isobaric osmosis. *Carbohydr. Polym.* **2,** 81–90.

Antonov, Yu. A., Pletenko, M. G., and Tolstoguzov, V. B. (1987). Thermodynamic compatibility of polysaccharides in aqueous media. *Vysokomol. Soedin., Ser. A* **29,** 2477–2481, 2482–2486.

Asherie, N., Lomakin, A. and Benedek, G. (1996). Phase diagram of colloidal solutions. *Phys. Rev. Lett.* **77,** 4832–4835.

Beijerinck, M. W. (1896). Ueber eine eigentümlichkeit der löslichen stärke. *Centralb. Bakteriol. Parasitenkd. Infektionskr.* **2,** 697–699.

Beijerinck, M. W. (1910). Ueber Emulsionsbildung bei der Vermischung wässeriger lösungen gewisser gelatinierender kolloide. *Kolloid-Z.* **7,** 16–20.

Bogracheva, T. Ya., Grinberg, V. Ya., and Tolstoguzov, V. B. (1983). Ueber die thermodynamische Vertraeglichkeit von Gummiarabicum und makromolecularen Komponenten der Globulinfraktion der Baeckerhefe. *Nahrung* **27,** 735–740.

Broide, M. L., Berland, C. R., Pande, J., Ogun, O., and Benedek, G. B. (1991). Binary-liquid phase separation of lens protein solution. *Proc. Natl. Acad. Sci. U.S.A.* **88,** 5660–5664.

Bungenberg de Jong, H. G. (1936). "La coaceruation, les coacervates et leur importance en biologie" Hermann, Paris.

Burova, T. V., Grinberg, N. V., Grinberg, V. Ya., Leontiev, A. L., and Tolstoguzov, V. B. (1992). Effects of polysaccharides upon the functional properties of 11S globulin from broad beans. *Carbohydr. Polym.* **18,** 101–108.

Clewlow, A. C., Rowe, A. J., and Tombs, M. P. (1995). Pectin-gelatin phase separation: The influence of polydispersity. *In* "Biopolymer Mixtures" (S. E. Harding, S. E. Hill, and J. R. Mitchell, eds.), pp. 173–191. Nottingham University Press, Nottingham.

Danilenko, A. N., Vetrov, V. Yu., Dmitrochenko, A. P., Leontev, A. L., Braudo, E. E., and Tolstoguzov, V. B. (1992). Restricted proteolysis of legumin of broad beans: Effect on thermodynamic properties of aqueous solutions and interaction with ficoll. *Nahrung* **36,** 105–111.

Dobry, A., and Boyer-Kawenoki, F. (1948). Sur l'incompatibilité des macromolécules en solution aqueuse, *Bull. Soc. Chim. Belg.* **57,** 280–285.
Doi, K. (1965). Formation of amylopectin granules in gelatin gel as a model of starch precipitation in plant plastids. *Biochim. Biophys. Acta* **94,** 557–565.
Doi, K., and Nikuni, Z. (1962). Formation of starch granules from an aqueous gelatinous solution. *Staerke 14,* 461–465.
Dubois, M., Gilles, K. A., Hamilton, J. K., Rebers, P. A., and Smith, F. (1956). Colorimetric method for determination of sugars and related substances. *Anal. Chem.* **28,** 350–356.
Durrani, C. M., Prystupa, D. A., Donald, A. M., and Clark, A. H. (1993). Phase diagram of mixtures of polymers in aqueous solution using Fourier transform infrared stectroscopy. *Macromolecules* **26,** 981–987.
Edmond, E., and Ogston, A. G. (1968). An approach to the study of phase separation in ternary aqueous systems. *Biochem. J.* **109,** 569–576.
Flory, P. J. (1953). "Principles of Polymer Chemistry." Cornell University Press, Ithaca, NY.
Gotlieb, A. M., Plashchina, I. G., Braudo, E. E., Titova, E. F., Belavtseva, E. M., and Tolstoguzov, V. B. (1988). Investigation of mixed agarose-gelatin gels. *Nahrung,* **32,** 927–937.
Grinberg, V. Ya., Schwenke, K. D., and Tolstoguzov, V. B. (1970). Phase state of the gelatin-dextran-water system. *Izv. Akad. Nauk SSSR, Ser. Khim.,* pp. 1430–1432.
Grinberg, V. Ya., Tolstoguzov, V. B., and Slonimskii, G. L. (1971). Thermodynamic compatibility of gelatin and certain poly-D-glucanes in aqueous medium. *Kolloid. Zh.* **33,** 666–669.
Grinberg, V. Ya., and Tolstoguzov, V. B. (1972). Thermodynamic compatibility of gelatin and some D-glucans in aqueous media. *Carbohydr. Res.* **25,** 313–320.
Grinberg, V. Ya., and Tolstoguzov, V. B. (1997). Thermodynamic incompatibility of proteins and polysaccharides in solutions. *Food Hydrocolloids* **11,** 145–158.
Ismond, M. A. H., Murray, E. D., and Arntfield, S. D. (1988). The role of noncovalent forces in micelle formation by vicilin from *Vicia faba.* III. The effect of urea, guanidine hydrochloride and sucrose on protein interactions. *Food Chem.* **29,** 189–198.
Ismond, M. A. H., Georgiou, C., Arntfield, S. D., and Murray, E. D. (1990). Role of noncovalent forces in micellization using legumin from *Vicia faba* as a study system. *J. Food Sci.* **55,** 1638–1642.
Kalichevsky, M. T., and Ring, S. G. (1987). Incompatibility of amylose and amylopectin in aqueous solution. *Carbohydr. Res.* **162,** 323–328.
Kalichevsky, M. T., Orford, P. D., and Ring, S. G. (1986). The incompatibility of concentrated aqueous solutions of dextran and amylose and its effect on amylose gelation. *Carbohydr. Polym.* **6,** 145–154.
Kasapis, S., Morris, E. R., Norton, I. T., and Gidley, M. J. (1993). Phase equilibria and gelation in gelatin/maltodextrin systems-2. *Carbohydr. Polym.* **21,** 249–259.
Khokhlov, A. R., and Nyrkova, I. A. (1992). Compatibility enhancement and microdomain structuring in weakly charged polyelectrolyte mixtures. *Macromolecules* **25,** 1493–1502.
Krause, S. (1978). Polymer-polymer compatibility. *In* "Polymer Blends" (D. R. Paul and S. Newman, eds.), Vol. 1, pp. 16–113. Academic Press, New York.
Kwei, T. K., and Wang, T. T. (1978). Phase separation behavior of polymer-polymer mixtures. *In* "Polymer Blends" (D. R. Paul and S. Newman, eds.), Vol. 1, pp. 141–184. Academic Press, New York.
Laurent, T. C., and Ogston, A. G. (1963). The interaction between polysaccharides and other macromolecules. The osmotic pressure of mixtures of serum albumin and hyaluronic acid. *Biochem. J.* **89,** 249–253.
Ledward, D. A. (1993). Creating textures from mixed biopolymer systems. *Trends Food Sci. Technol.* **4,** 402–405.
Liu, C., Lomakin, A., Thurston, G. M., Hayden, D., Pande, A., Pande, J., Ogun, O., Asherie, N., and Benedek, G. B. (1995). Phase separation in multicomponent aqueous-protein solutions. *J. Phys. Chem.* **99,** 454–461.

Lomakin, A., Asherie, N., and Benedek, G. B. (1996). Monte Carlo study of phase separation in aqueous protein solutions. *J. Chem. Phys.* **104,** 1646–1656.

Medin, A. S., and Janson, J.-C. (1993). Studies on aqueous polymer two-phase systems containing agarose. *Carbohydr. Polym.* **22,** 127–136.

Morris, E. R. (1990). Mixed polymer gels. *In* "Food Gels" (P. Harris, ed.), pp. 291–359. Elsevier, London.

Ogston, A. G. (1937). Some observations on mixtures of serum albumin and globulin. *Biochem. J.* **31,** 1952–1957.

Ostwald, W., and Hertel, R. H. (1929). Kolloidchemische reaktionen zwischen solen von eiweisskörpern und polymeren kohlehydraten, *Kolloid-Z.* **47,** 258, 357.

Pavlovskaya, G. E., Semenova, M. G, Thzapkina, E. N., and Tolstoguzov, V. B. (1993). The influence of dextran on the interfacial pressure of adsorbing layers of 11S globulin *Vicia faba* at the planar n-decane/aqueous solution interface. *Food Hydrocolloids* **7,** 1–10.

Polyakov, V. I., Grinberg, V. Ya., Antonov, Yu. A., and Tolstoguzov, V. B. (1979). Limited thermodynamic compatibility of proteins in aqueous solutions. *Polym. Bull.* **1,** 593–597.

Polyakov, V. I., Grinberg, V. Y., and Tolstoguzov, V. B. (1980). Application of phase-volume-ratio method for determining the phase diagram of water-casein-soybean globulins system. *Polym. Bull.* **2,** 757–760.

Polyakov, V. I., Kireyeva, O. K., Grinberg, V. Ya., and Tolstoguzov, V. B. (1985a). Thermodynamic compatibility of proteins in aqueous media.-1. *Nahrung* **29,** 153–160.

Polyakov, V. I., Popello, I. A., Grinberg, V. Ya., and Tolstoguzov, V. B. (1985b). Thermodynamic compatibility of proteins in aqueous media.-2. *Nahrung* **29,** 323–333.

Polyakov, V. I., Grinberg, V. Ya., Popello, I. A., and Tolstoguzov, V. B. (1986a). Thermodynamic compatibility of proteins in aqueous media.-3. *Nahrung* **30,** 81–88.

Polyakov, V. I., Grinberg, V. Ya., Popello, I. A., and Tolstoguzov, V. B. (1986b). Thermodynamic compatibility of proteins in aqueous medium. *Nahrung* **30,** 365–368.

Polyakov, V. I., Grinberg, V. Ya., and Tolstoguzov, V. B. (1997). Thermodynamic incompatibility of proteins. *Food Hydrocolloids* **11,** 171–180.

Poole, S., West, S. I., and Walters, C. L. (1984). Protein-protein interactions: Their importance in the foaming of heterogeneous protein systems. *J. Sci. Food Agric.* **35,** 701–711.

Popello, I. A., Suchkov, V. V., Grinberg, V. Ya., and Tolstoguzov, V. B. (1990). Liquid/liquid phase equilibrium in globulin/salt/water systems. Legumin. *J. Sci. Food Agric.* **51,** 345–353.

Popello, I. A., Suchkov, V. V., Grinberg, V. Ya., and Tolstoguzov, V. B. (1991). Liquid/liquid phase equilibrium in globulin/salt/water systems. Vicilin. *J. Sci. Food Agric.* **54,** 239–244.

Popello, I. A., Suchkov, V. V., Grinberg, V. Ya., and Tolstoguzov, V. B. (1992). Effects of pH upon the liquid-liquid phase equilibria in solutions of legumins and vicilins from broad beans and peas. *Food Hydrocolloids* **6,** 147–152.

Samant, S. K., Singhal, R. S., Kulkarni, P. R., and Rege, D. V. (1993). Protein-polysaccharide interactions: A new approach in food formulations. *Int. J. Food Sci. Technol.* **28,** 547–562.

Semenova, M. G., Bolotina, V. S., Grinberg, V. Ya., and Tolstoguzov, V. B. (1990). Thermodynamic incompatibility of the 11S fraction of soybean globulin and pectinate in aqueous medium. *Food Hydrocolloids* **3,** 447–456.

Semenova, M. G., Bolotina, V. S., Dmitrochenko, A. P., Leontiev, A. L., Polyakov, V. I., Braudo, E. E., and Tolstoguzov, V. B. (1991a). The factors affecting the compatibility of serum-albumin and pectinate in aqueous medium. *Carbohydr. Polym.* **15,** 367–385.

Semenova, M. G., Pavlovskaya, G. E., and Tolstoguzov, V. B. (1991b). Light scattering and thermodynamic phase behaviour of the system 11S globulin-k-carrageenan-water. *Food Hydrocolloids* **4,** 469–479.

Suchkov, V. V., Grinberg, V. Ya., and Tolstoguzov, V. B. (1981). Steady-state viscosity of the liquid two-phase disperse system water-casein-sodium alginate. *Carbohydr. Polym.* **1,** 39–53.

Suchkov, V. V., Grinberg, V. Ya., Muschiolik, G., Schmandke, H., and Tolstoguzov, V. B. (1988). Mechanical and functional properties of anisotropic geleous fibres obtained from the two-phase system of water-casein-sodium alginate. *Nahrung* **32,** 661–668.

Suchkov, V. V., Popello, I. A., Grinberg, V. Ya., and Tolstoguzov, V. B. (1990). Isolation and purification of 7S globulin and 11S globulin from broad beans and peas. *J. Agric. Food Chem.* **38,** 92–95.

Suchkov, V. V., Popello, I. A., Grinberg, V. Ya., and Tolstoguzov, V. B. (1997). Shear effects on phase behaviour of the legumin-salt-water system. Modelling protein recovery. *Food Hydrocolloids* **11,** 135–144.

Tanford, C. (1961). "Physical Chemistry of Macromolecules," pp. 192–202. Wiley, New York.

Tolstoguzov, V. B. (1978). "Artificial Foodstuffs." Nauka, Moscow.

Tolstoguzov, V. B. (1986). Functional properties of protein-polysaccharides mixtures. *In* "Functional Properties of Food Macromolecules" (J. R. Mitchell and D. A. Ledward, eds.), pp. 385–415. Elsevier Appl. Sci., London.

Tolstoguzov, V. B. (1988a). Creation of fibrous structure by spinneretless spinning. *In* "Food Structure—Its Creation and Evaluation" (J. M. V. Blanshard and J. R. Mitchell, eds.), pp. 181–196. Butterworth, London.

Tolstoguzov, V. B. (1988b). Concentration and purification of proteins by means of two-phase systems. *Food Hydrocolloids* **2,** 195–207.

Tolstoguzov, V. B. (1988c). Some physico-chemical aspects of protein processing into foodstuffs. *Food Hydrocolloids* **2,** 339–370.

Tolstoguzov, V. B. (1990). Interactions of gelatin with polysaccharides. *In* "Gums and Stabilisers for the Food Industry" (G. O. Phillips, P. A. Williams, and D. J. Wedlock, eds.), Vol. 5, pp. 157–175. IRL Press, Oxford.

Tolstoguzov, V. (1991). Functional properties of food proteins and role of protein-polysaccharide interaction. *Food Hydrocolloids* **4,** 429–468.

Tolstoguzov, V. (1992). The functional properties of food proteins. *In* "Gums and Stabilisers for the Food Industry" (G. O. Phillips, P. A. Williams, and D. J. Wedlock, eds.), Vol. 6, pp. 241–266. IRL Press, Oxford.

Tolstoguzov, V. (1993a). Functional properties of food proteins. Role of interactions in protein systems. *In* "Food Proteins. Structure and Functionality" (K. D. Schwenke and R. Mothes, eds.), pp. 203–209. VCH, Weinheim.

Tolstoguzov, V. (1993b). Thermoplastic extrusion—the mechanism of the formation of extrudate structure and properties. *J. Am. Oil Chem. Soc.* **70,** 417–424.

Tolstoguzov, V. (1993c). Thermodynamic incompatibility of food macromolecules. *In* "Food Colloids and Polymers: Stability and Mechanical Properties" (E. Dickinson and P. Walstra, eds.), Spec. Publ. No. 113, pp. 94–102. Royal Society of Chemistry, Cambrige, UK.

Tolstoguzov, V. (1994a). Thermodynamic aspects of food protein functionality. *In* "Food Hydrocolloids: Structure, Properties and Functions" (K. Nashinari and E. Doi, eds.), pp. 327–340. Plenum, New York.

Tolstoguzov, V. (1994b). Some physico-chemical aspects of protein processing into foods. *In* "Gums and Stabilisers for the Food Industry" (G. O. Phillips, P. A. Williams, and D. J. Wedlock, eds.), Vol. 7, pp. 115–124. IRL Press, Oxford.

Tolstoguzov, V. (1995). Some physico-chemical aspects of protein processing in foods. Multicomponent gels. *Food Hydrocolloids* **9,** 317–332.

Tolstoguzov, V. B. (1996a). Structure-property relationships in foods. *In* "Macromolecular Interactions in Food Technology" (N. Parris, A. Kato, L. K. Creamer, and J. Pearce, eds.), pp. 2–14. Am. Chem. Soc. Washington, DC.

Tolstoguzov, V. (1996b). Applications of phase separated biopolymer systems. *In* "Gums and Stabilisers for the Food Industry" (G. O. Phillips, P. A. Williams, and D. J. Wedlock, eds.), Vol. 8, pp. 151–160. IRL Press, Oxford.

Tolstoguzov, V. B. (1997a). Protein-polysaccharide Interactions. *In* "Food Proteins and Their Applications in Foods" (S. Damodaran and A. Paraf, eds.), pp. 171–198. Dekker, New York.
Tolstoguzov, V. (1997b). Thermodynamic aspects of dough formation and functionality. *Food Hydrocolloids* **11,** 181–193.
Tolstoguzov, V. B., and Braudo, E. E. (1983). Fabricated foodstuffs as multicomponent gels. *J. Tex. Stud.* **14,** 183–212.
Tolstoguzov, V. B., Grinberg, V. Ya., and Fedotova, L. I. (1969). Thermodynamic compatibility of gelatin and human serum albumin with some poly-D-glucans in aqueous medium. *Izv. Akad. Nauk SSSR, Ser. Khim.,* pp. 2839–2841.
Tolstoguzov, V. B., Belkina, V. P., Gulov, V. Ya., Titova, E. F., Belavtseva, E. M., and Grinberg, V. Ya. (1974a). Phasen Zustand, Struktur und mechanische Eigenschaften des gelartigen Systems Wasser-Gelatin-Dextran. *Staerke* **26,** 130–137.
Tolstoguzov, V. B., Mzel'sky, A. I., and Gulov, V. Ya. (1974b). Deformation of emulsion droplets in flow. *Colloid Polym. Sci.* **252,** 124–132.
Tolstoguzov, V. B., Grinberg, V. Ya., and Gurov, A. N. (1985). Some physicochemical approaches to the problem of protein texturization. *J. Agric. Food Chem.* **33,** 151–159.
Tombs, M. P. (1970a). Alterations to proteins during processing and the formation of structures. *In* "Protein as Human Food" (R. A. Lawrie, ed.), pp. 126–138. Butterworth, London.
Tombs, M. P. (1970b). Protein products. Ger. Pat. 1,940,561.
Tombs, M. P. (1972). Protein products. Br. Pat. 1,265,661.
Tombs, M. P. (1975). Aqueous protein composition. U.S. Pat. 3,870,801.
Tombs, M. P. (1985). Phase separation in protein—water systems and the formation of structure. *In* "Properties of Water in Foods" (D. Simatos and J. L. Multon, eds.), pp. 25–36. Martinus Nijhoff Publ., Dordrecht, The Netherlands.
Tombs, M. P., Newsom, B. G., and Wilding, P. (1974). Protein solubility: Phase separation in arachin-salt-water systems. *Int. J. Pept. Protein Res.* **6,** 253–277.
Tsapkina, E. N., Semenova, M. G., Pavlovskaya, G. E., Leontiev, A. L., and Tolstoguzov, V. B. (1992). The influence of incompatibility on the formation of adsorbing layers and dispersion of n-decane emulsion droplets in aqueous solution containing a mixture of 11S globulin from *Vicia faba* and dextran. *Food Hydrocolloids* **6,** 237–251.
Varfolomeeva, E. P., Grinberg, V. Ya., and Tolstoguzov, V. B. (1980). On the possibility of estimating weak interactions of macromolecules in solution from the experimental viscous flow activation energies data. *Polym. Bull.* **2,** 613–618.
Walter, H., and Brooks, D. (1995). Phase separation in cytoplasm, due to macromolecular crowding, is the basis for microcompartmentation. *FEBS Lett.* **361,** 135–139.
Walter, H., Brooks, D. E., and Fisher, D. (1985). "Partitioning in Aqueous Two-phase Systems: Theory, Methods, Uses and Applications to Biotechnology." Academic Press, Orlando, FL.
Walter, H., Johansson, G., and Brooks, D. E. (1991). Partitioning in aqueous two-phase systems: Recent results. Review. *Anal. Biochem.* **197,** 1–18.
Zasypkin, D. V., Braudo, E. E., and Tolstoguzov, V. B. (1997). Multicomponent biopolymer gels. *Food Hydrocolloids* **11,** 159–170.
Zhuravskaya, N. A., Kiknadze, E. V., Antonov, Y. A., and Tolstoguzov, V. B. (1986). Concentration of proteins as a result of the phase-separation of water-protein-polysaccharide systems, *Nahrung* **30,** 591–599, 601–613.

Partitioning and Concentrating Biomaterials in Aqueous Phase Systems

Göte Johansson and Harry Walter
Department of Biochemistry, University of Lund, S-22100, Lund, Sweden and Aqueous Phase Systems, Washington D.C. 20008

Aqueous phase separation is a general phenomenon which occurs when structurally distinct water-soluble macromolecules are dissolved, above certain concentrations, in water. The number of aqueous phases obtained depends on the number of such distinct macromolecular species used. Aqueous two-phase systems, primarily those containing poly(ethylene glycol) and dextran, have been widely used for the separation of biomaterials (macromolecules, membranes, organelles, cells) by partitioning. The polymer and salt compositions and concentrations chosen greatly affect the physical properties of the phases. These, in turn, interact with the physical properties of biomaterials included in the phases and affect their partitioning. Specific extractions of biomaterials can be effected by including affinity ligands in the systems. The phase systems can also be used to obtain information on the surface properties of materials partitioned in them; to study interactions between biomaterials; and to concentrate such materials.

KEY WORDS: Affinity partitioning, Aqueous phase systems, Cells, Concentration of biomaterials, Membranes, Nucleic acids, Partitioning, Proteins, Purification, Separation.

I. Introduction

A. Immiscible Phases

Aqueous phase separation generally occurs when structurally distinct water-soluble macromolecules are dissolved, above certain concentrations, in water. First described by Beijerinck about one hundred years ago (1896), two-polymer aqueous-aqueous two-phase systems were applied by Albertsson, starting in the 1950s, to partitioning biomaterials (Albertsson, 1986; Walter *et al.,* 1985; Walter and Johansson, 1994; Zaslavsky, 1995). Thus, differential partitioning of components of a mixture between or among immiscible liquid phases, one of the classical separatory methods, became available for the separation and fractionation of labile biomaterials: macromolecules, membranes, organelles, and even cells.

B. Partitioning between Liquid Phases and Its Use in the Separation of Mixtures

The separation of components of a mixture can be effected when they have different solubilities in the top and bottom phases of an immiscible phase system to which the mixture is added. The separation of two soluble components, for example, can be induced by mixing the phases, letting the phases settle and then physically separating them. Each phase, top or bottom, will be enriched with respect to the component which has a greater solubility in it. When the solubility of each component differs greatly in the two phases (i.e., one component is essentially soluble only in one phase and the other component only in the other) a virtually complete separation is obtained in a single, or bulk, extraction step. Smaller differences in the solubility of components in the two phases still permit their separation but require that multiple extraction steps be carried out [as, for example, in countercurrent distribution (CCD) (Åkerlund and Albertsson, 1994)]. In this procedure each of the phases, top and bottom, is reextracted, respectively, with bottom and top phases. The number of sequential extraction steps required to effect a separation is determined by the relative solubilities of the components in the two phases. The distribution of each material between the phases is quantitatively described by a partition coefficient, K, which is defined as the concentration of the material in the top phase/concentration of material in the bottom phase.

C. Aqueous Two-Phase and Multiphase Systems

Aqueous two-phase systems are formed, as indicated above, when two structurally distinct polymers are dissolved in water above certain concentrations. Immiscible aqueous multiphase systems can be obtained by increasing the number of different polymers used.

1. Partitioning of Biomaterials

The most widely used and studied aqueous two-phase systems contain dextran and poly(ethylene glycol) (PEG). These have been found to be mild and nondeleterious to biological materials and often even exhibit protective effects on materials partitioned in them. They have very low interfacial tensions; moderately low viscosities; densities close to that of water; and can be buffered and rendered isotonic, if necessary (Albertsson, 1986; Walter *et al.,* 1985; Walter and Johansson, 1994).

Both soluble and particulate materials can be partitioned in these systems. While soluble materials partition according to their relative solubilities in the top and bottom phases, the partitioning of particulates depends on their relative affinity for the bulk phases and the interface (for discussion, see Walter *et al.,* 1992). The partitioning of macromolecules and small particulates (e.g., viruses) is influenced by both surface properties and surface area. The partitioning of particulates with surface areas larger than about 0.2 μm^2 appears to depend predominantly on surface properties (Walter *et al.,* 1990). While soluble materials and small particulates partition between the bulk phases, particulates tend to partition, with increasing size, among the two bulk phases and the interface and, finally, between one bulk phase and the interface.

The partitioning of larger particulates is time-dependent and is thus measured at the shortest time after mixing deemed adequate for virtually complete phase settling. The time required for phase separation depends, among other things, on phase composition (e.g., polymers and salt used and their concentrations), on the volume ratio (top to bottom phase), and on the presence of the particulates themselves. The partitioning is usually described as the quantity of particulate in a bulk phase as a percentage of total particulate added, *P*.

The physical properties of the phases can be manipulated (as described below) so as to involve non-charge related, charge-associated, or biospecific groups as determinants of the partitioning behavior of the added biomaterial.

2. Concentration by Partitioning between Phases

Any material which, due to a change in the properties of the system, changes its partitioning from being distributed in both phases to being primarily in one phase (by altering partitioning conditions) will cause it to be concentrated. Concentration also results when a partitioning material is forced from a larger into a smaller phase. If the volume ratio is extreme and the collecting phase is small, concentration of several hundredfold can be obtained.

II. Aqueous Phase Systems

The separation of a solvent such as water into two liquid phases by addition of two soluble polymers is a very general phenomenon. The polymers must, however, distinctively differ in their chemical structure. The concentrations of polymers necessary decrease with increasing molecular weights. In the dextran-PEG systems, the concentration of water is normally 92–98% in the top phase and 80–90% in the bottom phase. With some polymer pairs, two-phase systems with up to 98% water in each phase have been obtained (Albertsson, 1986; Tjerneld and Johansson, 1990).

A. Macromolecular Compositions and Aqueous Phase Separation

The formation of two phases in water has been observed for a great number of polymer pairs (Tjerneld and Johansson, 1990). Two synthetic polymers like PEG and polyvinylalcohol (PVA), which both consist of linear molecules, generate two phases at relatively high concentrations. The most used systems have been composed of one synthetic polymer, typically PEG, and one polysaccharide, e.g., dextran. The popularity of these systems is mainly due to phase formation at moderate concentration of polymers and the relatively rapid settling of the phases. Several systems based on two polysaccharides are known. Dextran and modified dextran, e.g., hydroxypropyl dextran, benzoyl dextran, or valeryl dextran, phase separate. Also the same derivative of dextran, e.g., benzoyl dextran, but with differing degrees of substitution may, when mixed, yield phase separation (Lu *et al.,* 1994).

Proteins are known to form liquid–liquid two-phase systems (at certain concentrations) with a number of polymers. Synthetic polymers which have been used for protein precipitation, e.g., PEG, PVA, and poly(vinylpyrrolidone) (PVP), generate, at moderate concentrations, liquid systems with many proteins and protein mixtures (Johansson, 1996). Polysaccharides may also form two-phase systems with proteins (see chapter by Tolstoguzov, this volume). Certain proteins give rise to protein–protein aqueous two-phase systems (see chapter by Tolstoguzov, this volume) while others form aqueous two-phase systems composed of a protein-rich and a protein-poor phase (see chapter by Clark and Clark, this volume).

A third polymer added to a two-phase system will give rise to a third phase if the concentrations of polymers are high enough. When a third polymer is introduced at low concentration into a PEG-dextran solution that is sufficiently dilute to form a single phase, it may cause the phases to separate. At low concentration, the third polymer will distribute between the two phases but it may influence their properties such as density and viscosity. The introduction of surfaces in the form of particulates may cause phase separation to occur below the concentrations expected from the phase diagram (G. Johansson, unpublished results).

The solubility of some synthetic polymers, e.g., copolymers of ethylene oxide and propylene oxide, in water depends strongly on temperature. By heating a solution of such a polymer phase separation can occur. The liquid–liquid system consists of two phases: one polymer-rich, the other polymer-poor. By use of a polymer of appropriate structure and by including moderate concentrations of salt, such a phase transition can be made to occur at room temperature (Alred *et al.,* 1992).

B. Some Typical Properties of Aqueous Phases Exemplified by Dextran-Poly(ethylene Glycol) Systems

The composition and concentration of phase-forming polymers, salts, and other additives, as well as the temperature, determine the physical properties of a phase system. Manipulation of these parameters provides the means for selecting phase system properties.

1. Polymer Composition and Concentration

For the separation of biomaterials, phase-forming polymers can be chosen on the basis of structure, molecular weight, and ionic groups. The concentration of polymers required for phase separation to occur increases with decreasing polymer molecular weight. Phase diagrams are useful in characterizing phase systems and Fig. 1 presents such a diagram for solutions containing the two polymers dextran T500 (MW 500,000) and poly(ethylene glycol) (PEG) 8000 (MW 8000) at 20°C. Below certain concentrations (i.e., those at or below the curved binodal line), only a single phase exists. Concentrations above the binodal line give rise to phase separation. The straight lines (Fig. 1) are called tie lines. The points on these lines give the polymer concentrations of the entire system and, where they meet the binodal curve, indicate, at one end, the concentration of each of the phase-separating polymers in the top, PEG-rich and, on the other, that in the bottom, dextran-rich phase. The two aqueous phases obtained thus differ, at equilibrium, in polymer composition. The top phase behaves as if it were more hydrophobic than the bottom since PEG is more hydrophobic than dextran (Albertsson, 1986). Increases in polymer concentrations (increased tie line length) result in larger differences between the properties of the two phases (e.g., density, viscosity) as well as an increase in the interfacial tension. Phase-forming polymer distribution between the phases, and thus the phase diagrams, is influenced by the incorporation of certain salts and small additives into the phase system. Phase diagrams also show characteristic shifts with changes in temperature. With dextran-PEG phases, reduced temperature results in phase separation at lower polymer concentrations.

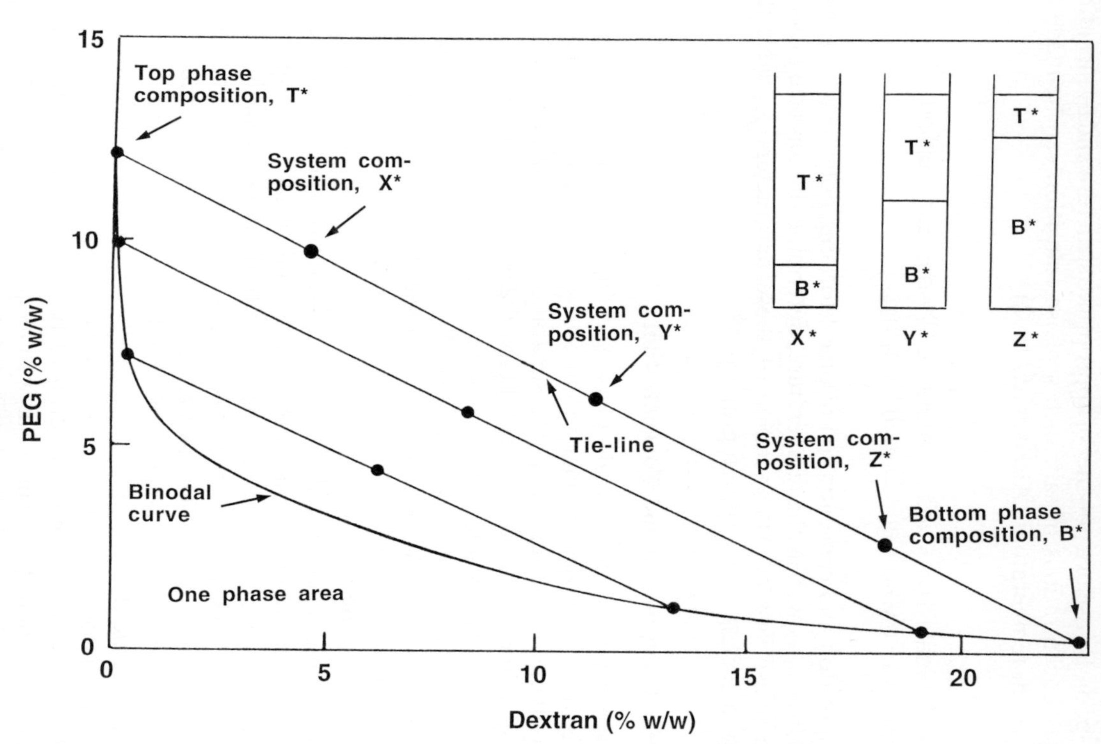

FIG. 1 Phase diagram for the system dextran 500—PEG 8000 at 20°C. Adapted from Albertsson (1986) with permission.

2. Salt Composition and Concentration

Salt, both composition and concentration, included in dextran-PEG aqueous phase systems has a major effect on the partitioning behavior of charged biomaterials added to the system (Albertsson, 1986; Walter *et al.,* 1985; Walter and Johansson, 1994). Thus, alkali phosphates and halides have opposite effects on the partitioning of negatively charged materials with the former causing higher and the latter lower K values.

Depending on their net charge, materials increase or decrease their K or P in a sequence similar to the Hofmeister series. For example, the K values of negatively charged proteins increase with $ClO_4^- < SCN^- < I^- < Br^- < Cl^- < CH_3CO_2^- < F^- < SO_4^{2-} < HPO_4^{2-}$ and $K^+ < Na^+ < NH_4^+ < Li^+$.

Among the advantages of using aqueous phase systems are that they can readily be buffered and that their tonicity can be adjusted. Since the net charge of many materials is pH-dependent, pH adjustment, together with salt selection, can often be used to obtain desirable K values (e.g., of proteins). Selection and maintenance of phase system pH is important in pH-dependent studies or when partitioning pH-sensitive materials (e.g., some molecular interactions; partitioning mammalian cells).

The tonicity of two-polymer aqueous phase systems is relatively low because of the large molecular weights of the phase-forming polymers (Bamberger *et al.,* 1985). Incorporation of salt or other small molecules (either charged or uncharged) can be used to bring the phase system to a tonicity optimal for a given experiment (e.g., to isotonicity when partitioning mammalian cells or organelles).

3. Salt Partitioning: Donnan Potential Difference between the Phases

Many salts added at moderate concentrations (up to 0.1 M) give fairly similar concentrations in the two phases. No separation of the ions of the salt can occur since each phase has to be electroneutral. This, however, does not require that cation and anion of a salt have the same relative affinity for the two phases. This relative affinity may differ and can be expressed in terms of hypothetical K values, K^+ and K^- (see Fig. 2). The measured partition coefficient of the salt, K_{salt} (= K for both cation and anion) will be the weighted geometrical mean of the K^+ and K^- values. For a 1–1 electrolyte like NaCl, the partition coefficient of the salt is $K_{NaCl} = \sqrt{K^+_{Na} K^-_{Cl}}$. A difference in K^+ and K^- gives rise to an electrical potential difference between the two phases (a Donnan potential difference) proportional to K^-/K^+ (Albertsson, 1986; Johansson, 1974a). For further discussion see below.

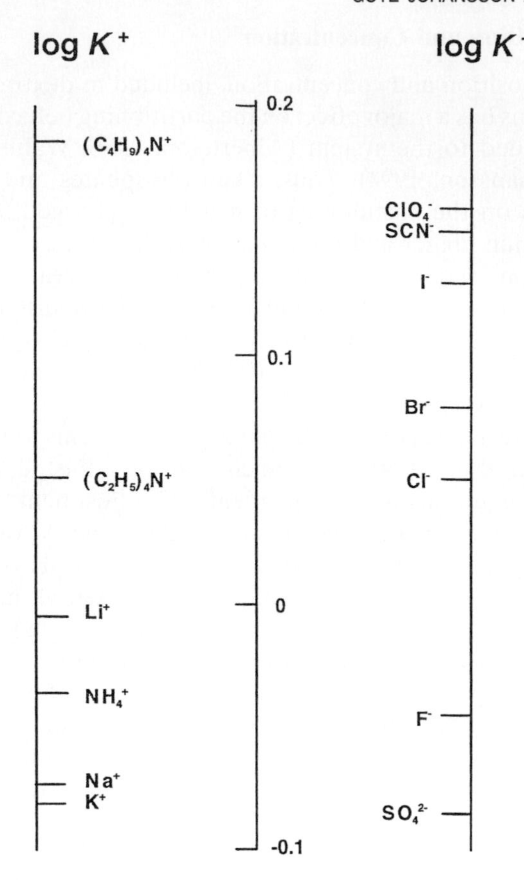

FIG. 2 The relative affinity of some ions for the top phase, expressed as the logarithm of the hypothetical K^+ (for cations) and K^- (for anions) in a system containing 8% (w/w) dextran 500, 8% (w/w) PEG 3400, and 25 mM salt. Adapted from Johansson (1974b).

The partition coefficient of halides (with a common cation) in dextran-PEG systems increases in the order $F^- < Cl^- < Br^- < I^-$. The alkali metal ions increase their partitioning in the order $K^+ < Na^+ < Li^+$. Salts of (weak) di- and tribasic acids like carbonates, malates, citrates, etc. change ionic composition and therefore also the partition coefficient with pH. As a rule, more negatively charged forms of anions have lower *K* values (higher affinity for the dextran-rich phase). Extremely high partition coefficients have been reported for salts containing ions such as TcO_4^- and HgI_4^{2-} (Rogers *et al.,* 1995; Rogers and Zhang, 1996; Griffin and Rogers, 1998). Such salts give rise to very high Donnan potential differences.

A salt (e.g., Ca^{2+}) that gives rise, in two-phase systems, to differing concentrations of that ion in the two-phases can, in a multiphase system, have different concentrations in each phase. If the affinities of the ion for the phases parallel the phase densities, a gradient of the ion is formed.

4. Mixing–Demixing: Effects of Interfacial Tension, Density and Viscosity

The time needed to reach partition equilibrium is on the scale of a few seconds if the systems are mixed in an effective way. This is due to the low interfacial tension and the small difference in the densities between the phases which result in very small phase domains for modest mixing energies. For these same reasons, the settling time of the phases is often slow in the system as a whole. The settling time is a consequence of the coalescence of phase droplets which depends on the viscosity of the phases and the interfacial tension. The latter increases strongly with the concentration of polymers in the system and speeds the formation of bulk phases (see Brooks *et al.,* 1985, for discussion). A PEG-dextran system, not too close to the critical point, with a height of 5 cm must be left undisturbed for 5–15 min to get well separated bulk phases. The settling time depends on the volume ratio and the presence of partitioned material, especially particles, in the system.

5. Affinity and Hydrophobic Groups Restricted by Partitioning to One Phase

The fact that the phase-forming polymers are concentrated in opposite phases makes them useful carriers for chemical groups with affinity for a given protein. Higher concentrations of the phase-forming polymers will give rise to phases, each of which contains less of the other polymer. An attached group will then be increasingly concentrated in one phase. With a hydrophobic ligand bound to PEG, the upper phase will attract proteins with hydrophobic pockets; when bound to dextran, the same proteins will be extracted into the bottom phase. By using specific affinity ligands, a protein which binds the ligand can be specifically and effectively extracted into the ligand-rich phase (see chapter by Kopperschläger, this volume).

III. Partitioning of Biomaterials

A. In Dextran/Poly(ethylene Glycol) Systems

The dextran-PEG two-phase system is the most studied. The materials partitioned cover a wide range of particle sizes: from small ions and organic molecules to viruses and whole cells.

1. Partitioning of Small Molecules

Small molecules like sugars, carboxylic acids, and amino acids partition fairly equally between the phases of a dextran-PEG system and they normally have *K* values between 0.6 and 1.5. As mentioned above, the same holds for most simple salts. Substances which interact with one of the phase-forming polymers may display a more extreme preference for one of the phases, e.g., large aromatic molecules, like dyes, often partition preferentially into the PEG-rich upper phase.

2. Partitioning of Proteins and Nucleic Acids

Soluble biopolymers are usually recovered after partitioning only in the two phases and not, as is the case with particulate materials, at the interface. Some exceptions to this have been reported. High molecular weight nucleic acids and proteins after prolonged incubations in dextran-PEG systems are also partially found at the interface (Kimura and Kobayashi, 1996; Grimonprez and Johansson, 1996).

a. Relative Solubility in Two Phases (including Protein Molecular Weight Effects) Since PEG at high concentrations, 8–30%, can be used to precipitate soluble proteins, it is not surprising that protein solubility decreases in the top phase with increasing PEG concentration (i.e., with increasing tie-line length) (Johansson, 1978). Dextran is much less of a precipitating agent and the relative solubilities of proteins in the two phases are such that the *K* value is less than one in PEG-dextran systems for most proteins, i.e., proteins tend to accumulate in the bottom, dextran-rich phase. The *K* values for a protein, human serum albumin (HSA), as a function of polymer concentrations, are shown in Fig. 3. The affinity of HSA for the top phase is increased by including PEG-palmitate in the system. The *K* values of proteins with low molecular weight are, in general, less dependent on polymer concentration (Albertsson *et al.,* 1987). The relative solubilizations are best studied in systems with a stable zero potential difference (i.e., in the absence of electrostatic interactions).

The partitioning of nucleic acids is strongly dependent on their molecular weights. Low molecular weight RNA (e.g., t-RNA) does not partition extremely. High molecular weight nucleic acids, like DNA from the cell nucleus, may attain very high or very low *K* values (Albertsson, 1986).

b. Influence of Electrolytes A PEG-dextran two-phase system containing one protein is a four-component system. Incorporation of a neutral salt adds two more constituents, the cation and the anion. As stated earlier, the main phase constituent is water followed by phase polymer. Proteins,

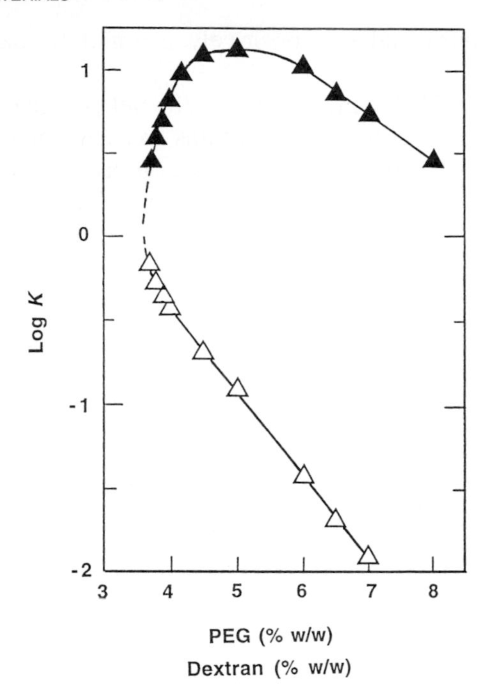

FIG. 3 The logarithmic partition coefficient, log K, of human serum albumin as a function of the polymer concentrations; concentration of PEG 8000 = concentration of dextran 500. Proteins: ▲, In the presence of PEG-palmitate; △, in the absence of PEG-palmitate. The systems contained 100 mM KCl and 5 mM Na phosphate buffer, pH 7.2. Temperature, 23°C. From Johansson and Shanbhag (1984) with permission.

by contrast, are usually present at low concentrations, 0.1–2%, giving a molar concentration in the range of 0.01–1 mM. The salt concentration used to steer the partitioning of proteins is typically 10–100 mM. In the systems interactions may, in principle, take place between each kind of ion and all the other components. These interactions will minimize the free energy (G) of the system and therefore determine the partitioning of the proteins. Each ion has its own relative affinity for the phases depending on more or less favorable interactions between ion and polymer, between ion and "bulk" water, or between ion and solvation water of the polymer. The partition coefficient of a substance can be shifted (either increased or decreased) by changing salt(s). However, at a pH value equal to the isoelectric point of the protein, i.e., when its net charge is zero, its K value (K_0) is generally independent of salt used. A salt which gives high K values for a negative protein will give a low K values for a positive protein and vice versa. The effect of some neutral salts on the partitioning of one negatively charged (ovalbumin) and one positively charged (lysozyme) protein is shown in Fig. 4.

By shifting the pH value in the system, the net charge of the protein can be adjusted. When the system also contains an excess of a neutral salt, e.g., sodium chloride, it has been found for several proteins that the change in logarithmic partition coefficient, log K, is proportional to the change in net charge, Z, of the protein induced by the pH shift. In other words

$$\log K = \gamma Z + \log K_0$$

where K_0 is the K value at the isoelectric point. The factor γ depends on the salt used. Since it is possible to use salts which have a γ value very close to zero, the K values then become practically independent of pH and equal to K_0.

The effect of salts on the partitioning has been correlated with an interfacial or Donnan potential difference. The relation between K of a protein and the Donnan potential difference, $\Delta\varphi$, is given by the following expression

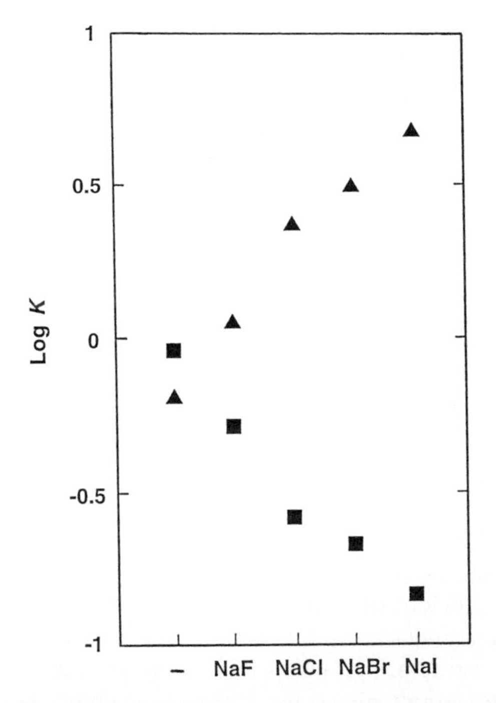

FIG. 4 Partitioning of lysozyme, 1 g/liter (▲) and ovalbumin, 2 g/liter (■) in 8% dextran 500 and 8% PEG 3400 systems containing salt (25 mM) and 0.5 mM sodium phosphate buffer, pH 6.9. Temperature, 23°C. From Johansson (1971) with permission.

(Albertsson, 1986), which assumes the protein behaves like a simple electrolyte:

$$\log K = ZF\Delta\varphi/RT + \log K_0$$

where F is the Faraday constant, R the gas constant, T the absolute temperature, and Z the exposed net charge which may be less than predicted by the titration curve of the protein (Johansson, 1985). In this case, the salt has to be in large excess over the protein and it is the salt ions which determine the interfacial potential difference.

The K values for nucleic acids, when of high molecular weights, are strongly dependent on the salt content and the pH of the system (see Fig. 5). This is a consequence of their high negative charge. Double and single stranded DNA differ in partition coefficient and can easily be separated by partitioning.

c. Effects of Conformational Changes Conformational changes in the protein molecule induced by pH variation, temperature, effectors, etc., may be detected as shifts in the partition coefficient of the protein. Thus, the affinity of the protein for the two phases is altered by the restructuring of the molecule. The change in affinity may be due to a change in the relative

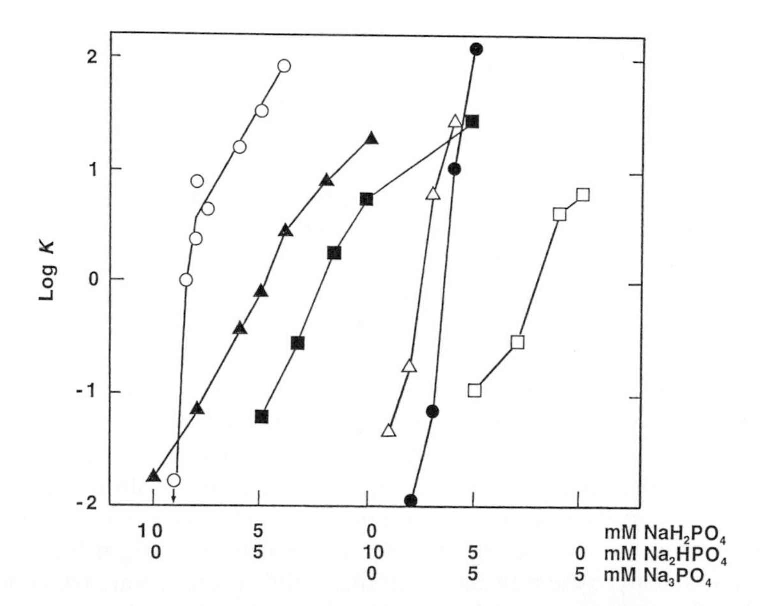

FIG. 5 Partitioning of nucleic acids in systems containing various mixtures of phosphates. ○, DNA; ▲, denatured DNA; ■, polyU; △, polyC; ●, polyA; and □, polyH. System composition: 5% (w/w) dextran 500-4% (w/w) PEG 8000 at 4°C. From Albertsson (1965) with permission.

exposure of hydrophobic sites or other chemical groups on the surface of the protein molecule. The association of proteins to form dimers or larger aggregates or their dissociation into subunits may also be detected as changes in the *K* value. Both conformational changes and association/dissociation phenomena are best studied by partitioning in systems insensitive to protein charge (Johansson, 1974b). Unfolding a protein prior to partitioning (followed by refolding), e.g., insulin-like growth factor-I from *E. coli,* has been used as part of the purification process (Lester, 1997). Lebreton *et al.* (1998) have shown that serum albumin undergoes conformational changes when transferred from one phase to another in a PEG-salt system.

d. Effects of Groups Attached to Proteins Chemical groups which are attached to (or detached from) proteins in biological systems, e.g., phosphate (phosphorylation) or sugar (glycosylation), may change the partitioning of the protein. Whether charged groups (phosphate) have been added or removed can be observed by partitioning in systems with salts that make the *K* value charge sensitive (e.g., $KClO_4$ or lithium phosphate buffer) or by using systems containing charged PEG (e.g., trimethylamino-PEG). Noncharged sugar groups may change K_0 values in favor of the dextran-rich phase.

e. Interactions with Other Proteins and Cell Components In addition to association of molecules of a protein (forming dimers, trimers, etc.), different proteins may also interact. When the protein–protein interaction is specific, it may reflect an interaction of biological significance. An example of specific enzyme–enzyme interaction is the weak complex formed between malate dehydrogenase and aspartate aminotransferase detected by phase partitioning (Backman and Johansson, 1976). Only when the two enzymes come from the same cell compartment, cytosol or mitochondria, is such an interaction observed. When the protein–protein interaction is strong, a change in the *K* value of one protein is observed when the second protein is added to the system. When very weak, the interaction can still be demonstrated by using multistep procedures like countercurrent distribution (Backman *et al.,* 1977). The two proteins will then each give a distribution pattern displaced along the extraction train when compared to those obtained when each protein is distributed in the absence of other protein. Another possibility is to arrange the CCD in such way that a zone of one protein is moving through systems which all contain the second protein and analyze the retardation of the mobile protein zone (see Fig. 6).

Interactions between nucleic acids or interaction of proteins with nucleic acids or cell particles can be detected in the same way. Examples are interactions of RNA with DNA (Mak *et al.,* 1976), histones with DNA

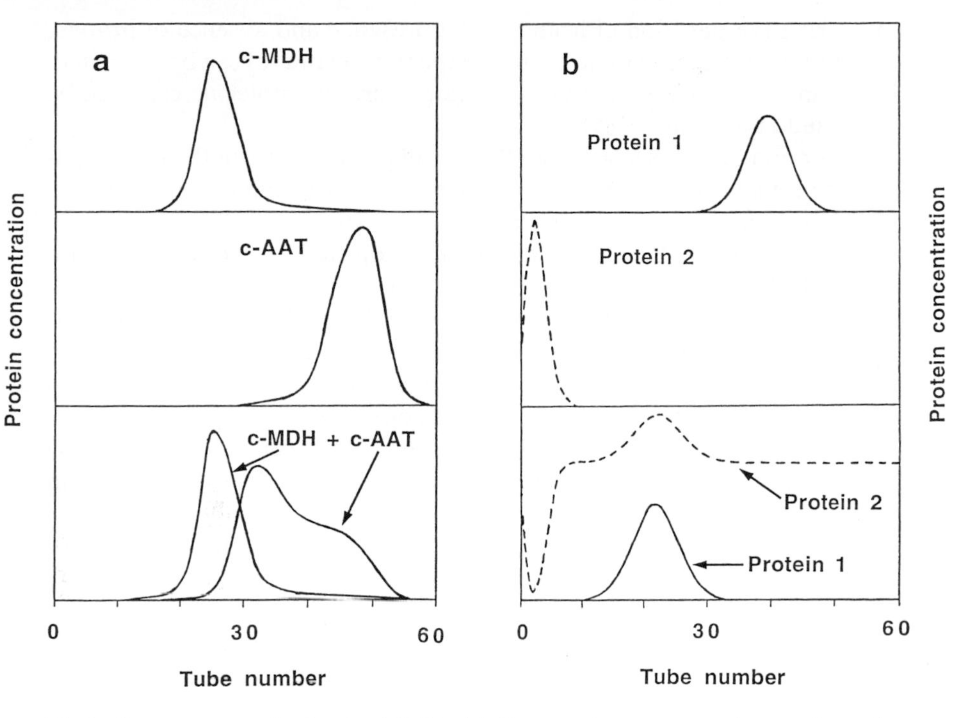

FIG. 6 Countercurrent distribution (CCD). (a) Cytoplasmic aspartate aminotransferase (c-AAT) and malate dehydrogenase (c-MDH) separate or in mixture. From Backman and Johansson (1976) with permission. (b) Example of how interaction between two proteins (1 and 2) can be studied by running a zone of protein 1 through systems containing a constant concentration of protein 2. The figure shows also the profiles of each protein, originally applied in tube 0, after CCD.

(Gineitis *et al.,* 1984), and serum albumin and fibrinogen with bacteria (Miörner *et al.,* 1980).

The chiral specific interaction between a protein and one of the enantiomers of a racemic mixture has been used to resolve optically active components. Serum albumin, in aqueous phase systems in which it strongly partitions into one phase, has been used to resolve D- and L-kynurenine (Sellergren *et al.,* 1988).

f. Interactions with Biospecific Ligands Interaction between a protein and a molecule with specific affinity for a site on it has been studied by two-phase partitioning in a number of ways. Proteins may be strongly directed toward the bottom phase while low molecular weight compounds (like many protein ligands) have a partition coefficient close to one. By comparing the partition of ligand in the presence and absence of protein, the amount of protein-bound ligand (in the bottom phase) can be estimated. The number of ligand molecules bound per protein molecule can then be calculated (Backman, 1985).

PEG-bound ligands which, like PEG, will predominate in the top (PEG-rich) phase strongly increase the *K* of proteins with binding sites for the ligand (see chapter by Kopperschläger, this volume).

The partitioning of nucleic acids has been adjusted to favor the top phase by use of PEG-bound affinity ligands (Müller, 1985). The ligands used have been dyes which intercalate certain base pairs or base sequences. This results in a very selective extraction and has been applied as a chromatographic method using adsorbed bottom phase as the stationary phase and the ligand-containing top phase as the mobile one.

The relative binding strengths of different ligands can be determined by combining affinity partitioning and the addition of free ligand. The protein is given a high *K* value by being partitioned with a PEG-bound ligand and the decrease in the log *K* value as a function of the concentration of added free ligand is determined. Interactions of different nucleotides with the enzyme phosphofructokinase from baker's yeast have been studied in this manner by using PEG-bound triazine dye as ligand (Kopperschläger and Lorenz, 1985).

3. Partitioning of Particulates

While soluble materials partition according to their solubilities between the two bulk phases, particulates, partition, as their size increases, among the two bulk phases and the interface and, finally, between one bulk phase and the interface. Since interface formation requires energy proportional to its area and the presence of particulate materials reduces this area, the resting position of such materials is in the interface. Only an interaction between the surface properties of the particulate with the chemical or physical properties of the phase system will cause it to be "pulled" out of the interface and into one or both bulk phases and thus to partition (Walter and Larsson, 1994). Most membranes and organelles partition, in dextran-PEG systems, between the top (PEG-rich) phase and the interface. Partitioning of some materials is also to the bottom phase.

Phase variables that affect (and, hence, can be used to manipulate) partitioning include polymer composition and concentration, salt composition and concentration, and the incorporation of biospecific (or nonspecific) ligands into the system. Thus, hydrophobic–hydrophilic, charge and ligand-binding interactions between surface and phase system properties demonstrably affect membrane/organelle partitioning behavior.

Membranes and organelles which, in a given phase system, may initially be at the interface often partition as the polymer (dextran, PEG) concentration, and thus the interfacial tension, is reduced. A lowered interfacial tension causes the biomaterial to be held less tightly at the interface and allows the hydrophobic–hydrophilic interactions between membrane or organelle surface properties and the physical properties of the phases to be adequate for partitioning to take place (Walter and Larsson, 1994). Partitioning is also influenced by selection of salt. As indicated above certain salts (e.g., alkali phosphate buffers) give rise to a measurable Donnan potential between the phases (top phase positive for the salts indicated). Since all membranes and organelles are negatively charged at neutral pH, they tend to partition toward or into the relatively more positively charged top phase. The partitioning behavior of membranes and organelles is, however, complex with the hydrophobic–hydrophilic properties sometimes outweighing and sometimes subservient to surface charge as partitioning determinant. Thus, it is often not possible to specify which surface properties dominate a particular material's partitioning behavior (Walter and Larsson, 1994).

The attachment of a biospecific (or nonspecific) ligand to one of the phase-forming polymers which binds specifically (or nonspecifically) to a receptor or site on the surface of a membrane or organelle of interest can be used to extract it into the phase in which the polymer-ligand predominates (e.g., Johansson, 1994a; Persson and Jergil, 1994). The phase system into which the polymer-ligand is incorporated is selected (polymer concentrations, salt composition, and concentration) such that no partitioning of the biomaterial occurs in the absence of the polymer-ligand.

The partitioning behavior of larger particulates has been shown to depend on the kinetics and extent of their adsorption to the phase droplets that occur on mixing immiscible phases and on the rate at which the particulate is delivered on the droplets to the bulk interface as the phases settle (for a discussion, see Brooks *et al.,* 1985; Fisher *et al.,* 1991). As a consequence of this partitioning mechanism for larger particulates, the rapidity of phase settling has an effect on the efficiency of their separation (Walter *et al.,* 1991). The more rapid the phase separation, the poorer the efficiency. Phase column height is one of the factors influencing the speed of phase separation, and hence the efficiency of particulate segregation, with higher phase columns requiring longer settling times.

While bulk extraction (i.e., single vessel partitioning) can yield a desired membrane or organelle preparation, the heterogeneity of these biomaterials often requires that a multiple extraction procedure (e.g., countercurrent distribution) be carried out (see Åkerlund and Albertsson, 1994) to obtain adequate purifications.

a. Membrane Partitioning: Examples Not only can different membranes be separated by partitioning in aqueous phase systems but even fragments from a given membrane have characteristic partition ratios reflecting differences in the fragments' domain markers (surface properties). A method has been developed, based on these findings, which permits the determination of domain relative proximity in a given membrane (Albertsson, 1988, 1994; Gierow, 1994).

Examples of membrane separations and/or purifications by partitioning include plasma membranes from L cells in a dextran-PEG system containing phosphate buffer (Brunette and Till, 1971); synaptic membranes from different regions of calf brain (Cebrián-Pérez *et al.,* 1991), and plasma and internal membranes from cultured mammalian cells separated with the plasma membranes partitioning to the top and the internal membranes to the bottom phase of an appropriately selected dextran-PEG system (Morré *et al.,* 1994). Rightside-out and inside-out membranes of both mammalian (Walter and Krob, 1976a) and plant origin (Larsson *et al.,* 1994) have different partition ratios and can thus be segregated.

Affinity partitioning studies have been carried out on nicotinic cholinergic receptor membranes from *Torpedo californica* using trimethylamino-PEG (Flanagan, 1985) or bis(triethylaminoethoxy)aniline linked to PEG via azelaic acid (Johansson *et al.,* 1981); synaptic membranes rich in opiate-binding receptors have been extracted using PEG-naloxone (Johansson, 1994a); rat liver plasma membranes were separated from other membranes using the lectin wheat germ agglutinin bound covalently to dextran (Persson *et al.,* 1991); and rightside-out and inside-out erythrocyte membranes could be segregated by use of a phase system containing PEG-palmitate which extracts membranes by interacting with hydrophobic surface areas (Walter and Krob, 1976b).

b. Organelle Partitioning: Examples Examples of organelle separations and/or purifications by partitioning include the partial purification of liver mitochondria using a dextran-PEG system (Ericson, 1974); the separation of mitochondria and synaptosomes from mammalian brain in a system in which the synaptosomes partition to the top and the mitochondria to the bottom phase (López-Pérez, 1994); fractionation of smooth, light rough and heavy rough microsomal membranes from rat liver (Ohlsson *et al.,* 1978); and a study of the heterogeneity of smooth endoplasmic reticulum (Gierow and Jergil, 1989) and resolution by partitioning of vesicles derived from the Golgi apparatus, which are highly heterogeneous, into several fractions bearing distinct marker enzymes (Hino *et al.,* 1978a,b).

Highly purified chloroplasts can be recovered from the bottom phase and interface of an appropriately selected dextran-PEG system (Larsson *et al.,* 1994). Actually, a scheme which includes partitioning has been devised for the isolation of all major organelles and membranous cell components (nuclei, chloroplasts, mitochondria, plasma membranes, endoplasmic reticulum and Golgi apparatus, tonoplasts) from a single homogenate of green leaves (Morré and Andersson, 1994). Furthermore, quite generally in plants, membrane and organelle partitioning into the top phase increases in the following order: inside-out thylakoids $<$ intact chloroplasts $<$ mitochondria $\sim$ peroxisomes $<$ endoplasmic reticulum $<$ thylakoids $<$ Golgi $\ll$ multiorganelle complexes = protoplasts = plasma membranes (Larsson *et al.*, 1985).

c. ***Other Particulate Partitioning: Example*** Liposomes of known composition have been used to model aspects of the partitioning behavior of membranes and organelles. Eriksson and Albertsson (1978) concluded that the most important parameter determining partitioning of liposomes composed of pure phospholipids (or mixtures of two lipids) was the polar group with lipid unsaturation or cholesterol (when included) of lesser importance. Incorporation of glycolipids caused a shift in liposome distribution (Sharpe and Warren, 1984).

B. Systems Containing Charged Polymers

Charged (ionic) groups on one of the polymers strongly reduces its ability to form two-phase systems with another polymer. By addition of salt to the system, electrostatic interaction can be reduced or eliminated and phase separation then occurs at nearly the same concentrations as for uncharged polymers (Johansson, 1970). By using only a fraction of a phase-forming polymer as carrier for charged groups or by using only one or a few charged groups per polymer molecule, the electrostatic forces can be used to affect the partitioning of proteins and nucleic acids (Johansson and Hartman, 1974).

1. Systems with Charged Poly(ethylene Glycol) or Dextran

The partitioning of a protein, e.g., hemoglobin (Fig. 7), between the phases of a dextran-PEG system is strongly affected by introducing a single positively charged group on PEG molecules (Johansson *et al.*, 1973). The protein has strong affinity for the bottom phase at pH values at which hemoglobin is positively charged. Negatively charged hemoglobin (i.e., at higher pHs) partitions mainly to the upper phase. The positively charged group, trimethylamino, and the negatively charged group, sulfonate, have opposite effects on protein partitioning and the partitioning is, therefore, strongly pH-dependent. The most pronounced effect is obtained at low

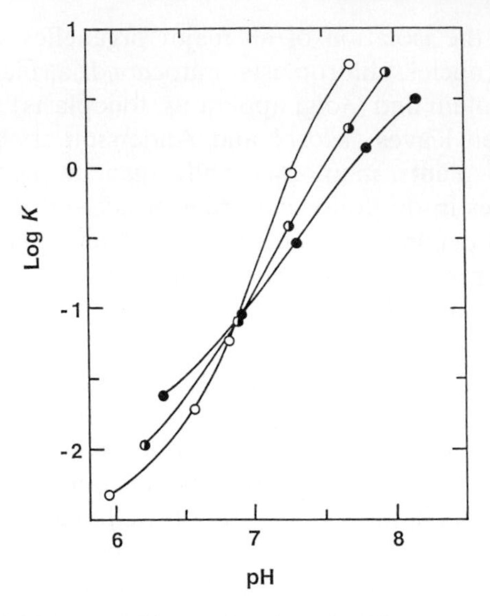

FIG. 7 Partitioning of CO-hemoglobin as a function of pH in a system containing positively charged trimethylamino-PEG (8% w/w), dextran 500 (8% w/w) and K-phosphate buffer: ○, 2 mM; ◑, 5 mM; and ●, 10 mM. From Johansson *et al.* (1973) with permission.

salt (buffer) concentration in the system, e.g., 1–2 mM. At higher salt concentrations, the effect of the charged PEG becomes markedly reduced (Johansson and Hartman, 1974).

Strongly charged dextrans, such as diethylaminoethyl dextran (DEAE-dextran) or carboxymethyldextran (CM-dextran) when added in small quantities to a two-phase (PEG-dextran) system often show a strong effect on the partitioning of biopolymers as well as bioparticles. The partitioning of these polyelectrolytes can furthermore be varied widely by the choice of salt added to the system.

2. Systems in Which One or Both Polymers are Polyelectrolytes

A polyelectrolyte polymer phase separates in water with an uncharged polymer only when high salt concentrations are present in the system. When both polymers have the same charge (e.g., anionic) the phase separation occurs also in salt-free systems but addition of salt causes a slight shift in the position of the binodal curve (Piculell *et al.,* 1991), as demonstrated with systems composed of dextran sulfate and polystyrene sulfonate. Two polyelectrolytes of opposite charge (i.e., one cationic and one anionic), when mixed in water, can form a liquid phase, often of small volume, which contains the majority of both species. This phenomenon is observed with proteins of opposite charge or with nucleic acids in combination with positively charged proteins (Veis, 1970). The phenomenon is called complex coacervation. Addition of salt reduces the interaction between the two species so that at very high salt concentrations a single phase results.

3. Polymer-Bovine Serum Albumin System

PEG and some other polymers are used as agents for protein precipitation. However, at certain polymer and protein concentrations, also dependent on pH and salt present, protein excluded from the solution does not form a precipitate but forms instead a liquid phase containing 65–80% (w/w) of water. The protein, thus, behaves like any other polymer which, with a second polymer, generates a liquid–liquid two-phase system. Two-phase formation is enhanced by salts such as KSCN. The PEG-bovine serum albumin system has been studied (Johansson, 1996). Enzymes partitioned in these systems have a strong affinity for the lower, protein-rich phase.

C. Aqueous Multiphase Systems

The partitioning of proteins and particles has been studied in some three- and four-phase systems (Albertsson and Birkenmeier, 1988; Lu *et al.*, 1991). The partitioning of a protein between the three phases (top, middle, and bottom) of a three-phase system can be described by the two partition coefficients $K_{t/m}$ and $K_{m/b}$. The third partition coefficient, $K_{t/b}$, is the product of the first two K values. The partitioning of proteins in three-phase systems made up of dextran, Ficoll, and PEG is affected both by salts and charged PEG (Hartman *et al.*, 1974; Hartman, 1976). The enzyme enolase from yeast consists of several isoenzymes and these can be partially resolved by partitioning in such a three-phase system, Fig. 8 (Hartman *et al.*, 1974). While the three polymers are concentrated in different phases, PEG in the top, Ficoll in the middle, and dextran in the bottom, each one can be used as a carrier for a ligand which will then be concentrated in the phase in which the particular polymer predominates. Such systems have been used for the directed partitioning of blood plasma proteins (Albertsson and Birkenmeier, 1988).

Dextran esterified with benzoic acid (benzoyl dextran) phase separates both with dextran, PEG and with benzoyl dextran with another degree of substitution. When all these four components are dissolved in water, a four-phase system is obtained (Lu *et al.*, 1991). This shows that even slight modifications of a polymer are sufficient to cause it to lose identity and to

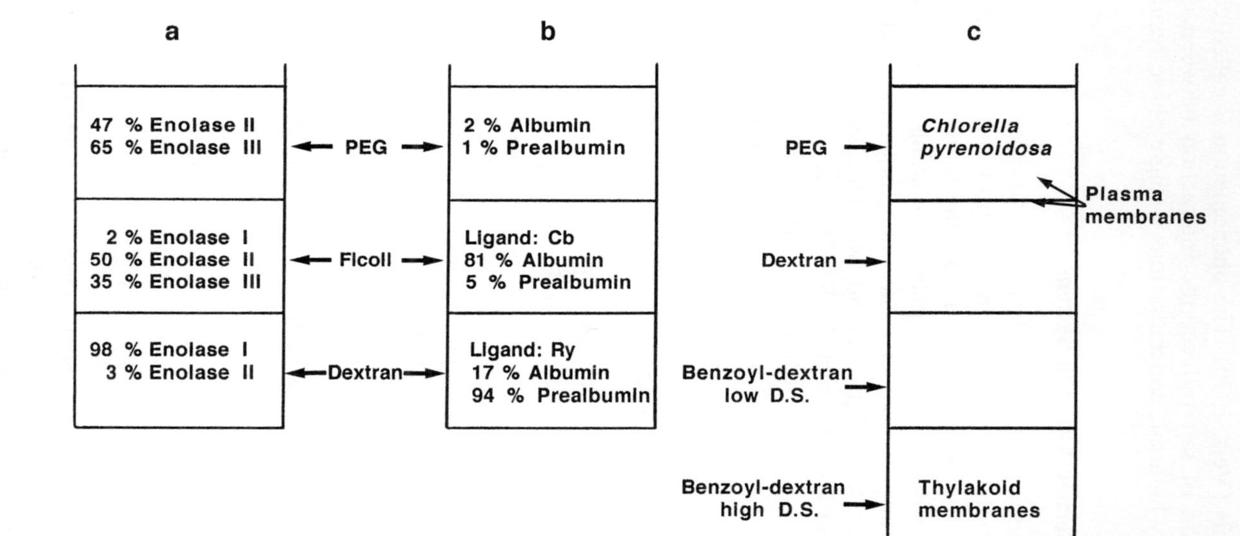

FIG. 8 (a) Partitioning of three isoenzymes of enolase among a three-phase system. From Hartman (1976) with permission. (b) Partitioning of blood plasma proteins in a three-phase system containing two affinity ligands: Cibacron blue F3G-A (Cb) is Ficoll bound ligand (middle phase) and Remazol yellow GGL (Ry) is dextran bound ligand (bottom phase). From Albertsson and Birkenmeier (1988) with permission. (c) Partitioning of some biomaterials in a four-phase system. From Lu *et al.* (1991) with permission.

phase separate with "itself." The benzoyl group gives a hydrophobic character to phases.

D. Effect of System/Polymers on Biological Activities

The presence of polymer (or the reduced water activity which ensues) has been shown to tend to stabilize biological activities, such as those of enzymes. PEG can be used as an additive for storage of enzymes (Berry, 1970). Microorganisms are viable in solutions of the polymers used for the formation of two-phase systems. It is possible to grow both mammalian cells (Zijlstra *et al.,* 1996) and microorganisms in the two-phase systems (Hahn-Hägerdal and Gruvegård, 1987).

In some cases, enzyme activity has been found to be strongly but reversibly reduced when the enzyme is incubated in the systems in the cold. Examples of such enzymes are yeast phosphofructokinase (G. Kopperschläger, personal communication) and mitochondrial aspartate aminotransferase from swine heart (L. Backman, personal communication).

The stabilizing effect of the polymers makes the aqueous phase systems especially useful for extraction of sensitive biological materials. By homogenizing cells in the systems, instead of in a buffer solution, the exposure to destabilizing solvent surroundings can be avoided and separation can be carried out directly by partitioning.

IV. Concentration of Proteins

Two-phase systems in which a protein partitions extremely into one of the phases can be used not only for purification but also for protein concentration. The extent of concentration depends on the volume of the original protein solution and that of the receiving phase. Good recovery is dependent on the K value. The smaller the volume of the receiving phase, the more extreme the partition coefficient must be.

A. Concentration by Manipulation of Partition Coefficient (K)

1. Extreme K

Proteins in PEG-dextran systems can be forced into the lower, dextran-rich phase (low K value) by using systems containing PEG with high molecular weight, dextran with low molecular weight and systems with a long tie-line. The *K* value can be made more extreme by appropriate choice of salt, e.g., KSCN or $KClO_4$ at high pH values, 8–10.

2. Extreme *K* and Volume Ratio

To concentrate proteins using a very small bottom phase requires that the system contain a high concentration of PEG, 16–22%, and little dextran (or any other phase-forming polysaccharide), 0.1–0.5%. In such systems, the top to bottom phase ratio of 500 or more can easily be reached (Johansson and Reczey, 1998). By partitioning in this manner, proteins can be concentrated several hundred-fold. The enzyme β-glucosidase has been concentrated by this method and it was simultaneously purified in relation to both other proteins and pigments (Johansson and Reczey, 1998).

B. Binding to Another Component with an Extreme *K*

Another way to get an extreme partitioning of one (or all) proteins in a mixture is to use a polymeric substance which both partitions extremely to one phase and also interacts with the target proteins. Some of the specific interactions which can be used are found in the chapter by Kopperschläger (discussion of affinity partitioning). For nonspecific (electrostatic) interactions, dextran charged with positive groups (e.g., diethylaminoethyl, DEAE) can be used at high pH where the proteins are negatively charged (Johansson, 1994b).

V. Concluding Remarks

Aqueous phase separation is a physical phenomenon that occurs, quite generally, when two or more structurally distinct macromolecules are dissolved, above certain concentrations, in water. The number of aqueous phases obtained can be the same as the number of different macromolecules used. Biomaterials including proteins, membranes, and organelles, can be distributed between or among such phases (i.e., separated and/or compartmentalized). The systems can also be used to study the interactions between biomolecules. The partitioning of biomaterials is dependent on the phase-forming polymers used, their concentration, and their molecular weights as well as the temperature, salts, and pH employed. Biospecific extraction of biomaterials can be greatly enhanced through use of affinity ligands.

References

Åkerlund, H.-E., and Albertsson, P.-Å. (1994). Thin-layer countercurrent distribution and centrifugal countercurrent distribution apparatus. *In* "Methods in Enzymology" (H. Walter and G. Johansson, eds.), Vol. 228, pp. 87–99. Academic Press, San Diego, CA.

Albertsson, P.-Å. (1965). Partition studies on nucleic acids. I. Influences of electrolytes, polymer concentration and nucleic acid conformation on the partition in dextran-polyethylene glycol. *Biochim. Biophys. Acta* **103,** 1–12.

Albertsson, P.-Å. (1986). "Partition of Cell Particles and Macromolecules," 3rd ed. Wiley-Interscience, New York.

Albertsson, P.-Å. (1988). Analysis of the domain structure of membranes by fragmentation and separation in aqueous polymer two-phase systems. *Q. Rev. Biophys.* **21,** 61–98.

Albertsson, P.-Å. (1994). Domain structure of biological membranes obtained by fragmentation and separation analysis. *In* "Methods in Enzymology" (H. Walter and G. Johansson, eds.), Vol. 228, pp. 503–511. Academic Press, San Diego, CA.

Albertsson, P.-Å., and Birkenmeirer, G. (1988). Affinity separation of proteins in aqueous three-phase systems. *Anal. Biochem.* **175,** 154–161.

Albertsson, P.-Å., Cajarville, A., Brooks, D. E., and Tjerneld, F. (1987). Partition of proteins in aqueous polymer two-phase systems and the effect of molecular weight of the polymer. *Biochim. Biophys. Acta* **926,** 87–93.

Alred, P., Tjerneld, F., Kozlowski, A., and Harris, J. M. (1992). Synthesis of dye conjugates of ethylene oxide-propylene oxide copolymers and application in temperature-induced phase partitioning. *BioSeparation* **2,** 363–373.

Backman, L. (1985). Interacting systems and binding studied by partitioning. *In* "Partitioning in Aqueous Two-Phase Systems: Theory, Methods, Uses, and Applications to Biotechnology" (H. Walter, D. E. Brooks, and D. Fisher, eds.), pp. 267–314. Academic Press, Orlando, FL.

Backman, L., and Johansson, G. (1976). Enzyme-enzyme complexes between aspartate aminotransferase and malate dehydrogenase from pig heart muscle. *FEBS Lett.* **65,** 39–43.

Backman, L., Shanbhag, V., and Johansson, G. (1977). A method to detect protein-protein interactions. *Biochem. Soc. Trans.* **5,** 748–750.

Bamberger, S. B., Brooks, D. E., Sharp, K. A., Van Alstine, J. M., and Webber, T. J. (1985). Preparation of phase systems and measurement of their physicochemical properties. *In* "Partitioning in Aqueous Two-Phase Systems: Theory, Methods, Uses, and Applications to Biotechnology" (H. Walter, D. E. Brooks, and D. Fisher, eds.), pp. 85–130. Academic Press, Orlando, FL.

Beijerinck, M. W. (1896). Ueber eine Eigentuemlichkeit der loeslichen Staerke. *Centralbl. Bakteriol. Parasiten. Infektionskr.* **2,** 697–699.

Berry, J. S. (1970). Stabilisierte wässrige Enzymmischung. BRD Pat. 1,964,088.

Brooks, D. E., Sharp, K. A., and Fisher, D. (1985). Theoretical aspects of partitioning. *In* "Partitioning in Aqueous Two-Phase Systems: Theory, Methods, Uses, and Applications to Biotechnology" (H. Walter, D. E. Brooks, and D. Fisher, eds.), pp. 11–84. Academic Press, Orlando, FL.

Brunette, D. M., and Till, J. E. (1971). A rapid method for the isolation of L-cell surface membranes using an aqueous two-phase polymer system. *J. Membr. Biol.* **5,** 215–224.

Cebrián-Pérez, J. A., Muiño-Blanco, M. T., and Johansson, G. (1991). Heterogeneity of synaptosomal membrane preparations from different regions of calf brain studied by partitioning and counter-current distribution. *Int. J. Biochem.* **23,** 1491–1495.

Ericson, I. (1974). Determination of the isoelectric point of rat liver mitochondria by cross-partition. *Biochim. Biophys. Acta* **356,** 100–107.

Eriksson, E., and Albertsson, P.-Å. (1978). The effect of lipid composition on the partition of liposomes in aqueous two-phase systems. *Biochim. Biophys. Acta* **507,** 425–432.

Fisher, D., Raymond, F. D., and Walter, H. (1991). Factors in cell separation by partitioning in two-polymer aqueous phase systems. *ACS Symp. Ser.* **464,** 175–188.

Flanagan, S. D. (1985). Partitioning of animal membranes and organelles. *In* "Partitioning in Aqueous Two-Phase Systems: Theory, Methods, Uses, and Applications to Biotechnology" (H. Walter, D. E. Brooks, and D. Fisher, eds.), pp. 453–495. Academic Press, Orlando, FL.

Gierow, J. P. (1994). Relative proximity of domains in plasma membrane and smooth endoplasmic reticulum from rat liver. *In* "Methods in Enzymology" (H. Walter and G. Johansson, eds.), Vol. 228, pp. 512–522. Academic Press, San Diego, CA.

Gierow, J. P., and Jergil, B. (1989). Heterogeneity of smooth endoplasmic reticulum from rat liver studied by two-phase partitioning. *Biochem. J.* **262,** 55–61.

Gineitis, A. A., Suciliene, S. P., and Shanbhag, V. P. (1984). Dissociation and isolation of chromatin proteins in salt solutions by an aqueous two-phase system. *Anal. Biochem.* **139,** 400–403.

Grimonprez, B., and Johansson, G. (1996). Liquid-liquid partitioning of some enzymes, especially phosphofructokinase, from *Saccharomyces cerevisiae* at sub-zero temperature. *J. Chromatogr.* **B680,** 55–63.

Hahn-Hägerdal, B., and Gruvegård, G., eds. (1987). "Microorganisms in Aqueous Two-Phase Systems," Lect. Workshop, 1986, IVA-Rep. 330, Royal Swed. Acad. Sci., Stockholm.

Hartman, A. (1976). Partition of cell particles in three-phase systems. *Acta Chem. Scand., Ser. B* **B30,** 585–594.

Hartman, A., Johansson, G., and Albertsson, P.-Å. (1974). Partition of proteins in a three-phase system. *Eur. J. Biochem.* **46,** 75–81.

Hino, Y., Asano, A., and Sato, R. (1978a). Biochemical studies on rat liver Golgi apparatus. II. Further characterization of isolated Golgi fractions. *J. Biochem.* (*Tokyo*) **83,** 925–934.

Hino, Y., Asano, A., and Sato, R. (1978b). Biochemical studies on rat liver Golgi apparatus. III. Subfractionation of fragmented Golgi apparatus by counter-current distribution. *J. Biochem.* (*Tokyo*) **83,** 935–942.

Johansson, G. (1970). Studies on aqueous dextran-poly(ethylene glycol) two-phase systems containing charged poly(ethylene glycol). *Biochim. Biophys. Acta* **222,** 381–389.

Johansson, G. (1971). Effects of different ions on the partition of proteins in a dextran-poly(ethylene glycol) two-phase system. *In* "Proceedings of the International Solvent Extraction Conference" (J. G. Gregory, B. Evans., and P. C. Weston, eds.), Vol. 2, pp 928–935. Soc. Chem. Ind., London.

Johansson, G. (1974a). Partition of proteins and micro-organisms in aqueous biphasic systems. *Mol. Cell. Biochem.* **4,** 169–180.

Johansson, G. (1974b). Effects of salts on the partition of proteins in aqueous polymeric biphasic systems. *Acta Chem. Scand., Ser. B* **B28,** 873–882.

Johansson, G. (1978). Comparison of two aqueous biphasic systems used for the partition of biological material. *J. Chromatogr.* **150,** 63–71.

Johansson, G. (1985). Determination of ionic charge by liquid-liquid partition. *J. Chromatogr.* **322,** 425–432.

Johansson, G. (1994a). Synaptic membranes. *In* "Methods in Enzymology" (H. Walter and G. Johansson, eds.), Vol. 228, pp. 496–503. Academic Press, San Diego, CA.

Johansson, G. (1994b). Recovery of proteins and phase-forming chemicals. *In* "Methods in Enzymology" (H. Walter and G. Johansson, eds.), Vol. 228, pp. 569–573. Academic Press, San Diego, CA.

Johansson, G. (1996). Aqueous two-phase systems with a liquid protein (bovine serum albumin) phase for partitioning of enzymes. *J. Chromatogr.* **B680,** 123–130.

Johansson, G., and Hartman, A. (1974). Partition of proteins in aqueous biphasic systems. *In* "Proceedings of the International Solvent Extraction Conference" (T. C. J. Gribnau, J. Visser, and R. J. F. Nivard, eds.), Vol. 1, pp. 927–942. Soc. Chem. Ind., London.

Johansson, G., and Reczey, K. (1998). Concentration and purification of β-glucosidase from *Aspergillus niger* by using aqueous two-phase partitioning. *J. Chromatogr.* **B711,** 161–172.

Johansson, G., and Shanbhag, V. P. (1984). Affinity partitioning of proteins in aqueous two-phase systems containing polymer-bound fatty acids. 1. Effect of polythylene glycol palmitate on the partition of human serum albumin and α-lactalbumin. *J. Chromatogr.* **284,** 63–72.

Johansson, G., Gysin, R., and Flanagan, S. D. (1981). Affinity partitioning of membranes: Evidence for discrete membrane domains containing cholinergic receptor. *J. Biol. Chem.* **256,** 9126–9135.

Johansson, G., Hartman, A., and Albertsson, P.-Å. (1973). Partition of proteins in two-phase systems containing charged poly(ethylene glycol). *Eur. J. Biochem.* **33,** 379–386.

Kimura, K., and Kobayashi, H. (1996). RNA partitioning accompanied by adsorption: High-molecular-mass RNA adsorbed at the interface like a particle. *J. Chromatogr.* **B680,** 213–219.

Kopperschläger, G., and Lorenz, G. (1985). Interaction of yeast glucose 6-phosphate dehydrogenase with diverse triazine dyes: A study by means of affinity partitioning. *Biomed. Biochim. Acta* **44,** 517–525.

Larsson, C., Andersson, B., and Åkerlund, H.-E. (1985). Partitioning of plant cells, cell walls, membranes, and organelles. *In* "Partitioning in Aqueous Two-Phase Systems: Theory, Methods, Uses, and Applications to Biotechnology" (H. Walter, D. E. Brooks, and D. Fisher, eds.), pp. 497–527. Academic Press, Orlando, FL.

Larsson, C., Sommarin, M., and Widell, S. (1994). Isolation of highly purified plant plasma membranes and the separation of inside-out and right-side-out vesicles. *In* "Methods in Enzymology" (H. Walter and G. Johansson, eds.), Vol. 228, pp. 451–469. Academic Press, San Diego, CA.

Lebreton, B., Huddleston, J., and Lydiatt, A. (1998). Polymer-protein interactions in aqueous two-phase systems: fluorescent studies of the partition behaviour of human serum albumin. *J. Chromatogr.* **B711,** 69–79.

Lester, P. M. (1997). Scale-up of an aqueous two-phase separation for the isolation of insulin-like growth factor-I from *E. coli.* Abstracts, *10th Int. Conf. Partitioning Aqueous Two-Phase Syst.,* Reading, England.

López-Pérez, M. J. (1994). Preparation of synaptosomes and mitochondria from mammalian brain. *In* "Methods in Enzymology" (H. Walter and G. Johansson, eds.), vol. 228, pp. 403–411. Academic Press, San Diego, CA.

Lu, M., Tjerneld, F., Johansson, G., and Albertsson, P.-Å. (1991). Preparation of benzoyl dextran and its use in aqueous two-phase systems. *BioSeparation* **2,** 247–255.

Lu, M., Albertsson, P.-Å., Johansson, G., and Tjerneld, F. (1994). Partitioning of proteins and thylakoid membrane vesicles in aqueous two-phase systems with hydrophobically modified dextran. *J. Chromatogr.* **A668,** 215–228.

Mak, S., Öberg, B., Johansson, K., and Philipson, L. (1976). Purification of adenovirus messenger ribonucleic acid by an aqueous polymer two-phase system. *Biochemistry* **15,** 5754–5761.

Miörner, H., Myhre, E., Björk, L., and Kronvall, G. (1980). Effect of specific binding of human albumin, fibrinogen and immunoglobulin G on surface characteristics of bacterial strains as revealed by partition experiments in polymer phase systems. *Infect. Immun.* **29,** 879–885.

Morré, D. J., and Andersson, B. (1994). Isolation of all major organelles and membranous cell components from a single homogenate of green leaves. *In* "Methods in Enzymology" (H. Walter and G. Johansson, eds.), Vol. 228, pp. 412–419. Academic Press, San Diego, CA.

Morré, D. J., Reust, T., and Morré, D. M. (1994). Plasma and internal membranes from cultured mammalian cells. *In* "Methods in Enzymology" (H. Walter and G. Johansson, eds.), Vol. 228, pp. 448–450. Academic Press, San Diego, CA.

Müller, W. (1985). Partitioning of nucleic acids. *In* "Partitioning in Aqueous Two-Phase Systems: Theory, Methods, Uses, and Applications to Biotechnology" (H. Walter, D. E. Brooks, and D. Fisher, eds.), pp. 227–266. Academic Press, Orlando, FL.

Ohlsson, R., Jergil, B., and Walter, H. (1978). Fractionation of microsomal membranes on the basis of their surface properties. *Biochem. J.* **172,** 189–192.

Persson, A., and Jergil, B. (1994). Rat liver plasma membranes. *In* "Methods in Enzymology" (H. Walter and G. Johansson, eds.), Vol. 228, pp. 489–496. Academic Press, San Diego, CA.

Persson, A., Johansson, B., Olsson, H., and Jergil, B. (1991). Purification of rat liver plasma membranes by wheat-germ-agglutinin affinity partitioning. *Biochem. J.* **273,** 176–177.

Piculell, L., Nilsson, S., Falck, L., and Tjerneld, F. (1991). Phase separation in aqueous mixtures of similarly charged polyelectrolytes. *Polym. Commun.* **32,** 158–160.

Rogers, R. D., and Griffin, S. T. (1998). Partitioning of mercury in aqueous biphasic systems and on ABEC^{TM} resins. *J. Chromatogr.* **B711,** 277–283.

Rogers, R. D., and Zhang, J. H. (1996). Effects of increasing polymer hydrophobicity on distribution ratios of $TcO4^-$ in polyethylene/poly(propylene glycol)-based aqueous biphasic systems. *J. Chromatogr.* **B680,** 231–236.

Rogers, R. D., Zhang, J. H., Bond, A. H., Bauer, C. B., Jezl, M. L., and Roden, D. M. (1995). Selective and quantitative partitioning of pertechnate in polyethylene-glycol based aqueous biphasic systems. *Solvent Extr. Ion Exch.* **13,** 665–688.

Sellergren, B., Ekberg, B., Albertsson, P.-Å., and Mosbach, K. (1988). Preparative chiral separation in an aqueous two-phase system by a few counter-current extractions. *J. Chromatogr.* **450,** 277–280.

Sharpe, P. T., and Warren, G. S. (1984). The incorporation of glycolipids with defined carbohydrate sequence into liposomes and the effects on partition in aqueous two-phase systems. *Biochim. Biophys. Acta* **772,** 176–182.

Tjerneld, F., and Johansson, G. (1990). Aqueous two-phase systems for biotechnical use. *BioSeparation* **1,** 255–263.

Veis, A. (1970). Phase equilibria in systems of interacting polyelectrolytes. *Biol. Macromol.* **3,** 211–273.

Walter, H., and Johansson, G., eds. (1994). "Methods in Enzymology," Vol. 228. Academic Press, San Diego, CA.

Walter, H. and Krob, E. J. (1976a). Partition in two-polymer aqueous phases reflects differences between surface properties of erythrocytes, ghosts and membrane vesicles. *Biochim. Biophys. Acta* **455,** 8–23.

Walter, H., and Krob, E. J. (1976b). Hydrophobic affinity partition in aqueous two-phase systems containing poly(ethylene glycol) -palmitate of rightside-out and inside-out vesicles from human erythrocyte membranes. *FEBS Lett.* **61,** 290–293.

Walter, H., and Larsson, C. (1994). Partitioning procedures and techniques: Cells, organelles, and membranes. *In* "Methods in Enzymology" (H. Walter and G. Johansson, eds.), Vol. 228, pp. 42–63. Academic Press, San Diego, CA.

Walter, H., Brooks, D. E., and Fisher, D., eds. (1985). "Partitioning in Aqueous Two-Phase Systems: Theory, Methods, Uses, and Applications to Biotechnology." Academic Press, Orlando, FL.

Walter, H., Fisher, D., and Tilcock, C. (1990). On the relation between surface area and partitioning of particulates in two-polymer aqueous phase systems. *FEBS Lett.* **270,** 1–3.

Walter, H., Krob, E. J., and Wollenberger, L. (1991). Partitioning of cells in dextran-poly (ethylene glycol) aqueous phase systems: Study of settling time, vessel geometry and sedimentation effects on the efficiency of separation. *J. Chromatogr.* **542,** 397–411.

Walter, H., Raymond, F. D., and Fisher, D. (1992). Erythrocyte partitioning in dextran-poly(ethylene glycol) aqueous phase systems: Events in phase and cell separation. *J. Chromatogr.* **609,** 219–227.

Zaslavsky, B. Y. (1995). "Aqueous Two-Phase Partitioning: Physical Chemistry and Bioanalytical Applications." Dekker, New York.

Zijlstra, G. M., Michielsen, M. J. F., De Gooijer, C. D., Van der Pol, L. A., and Tramper, J. (1996). Hybridoma and CHO cell partitioning in aqueous two-phase systems. *Biotechnol. Prog.* **12,** 363–370.

Effects of Specific Binding Reactions on the Partitioning Behavior of Biomaterials

Gerhard Kopperschläger
Institute of Biochemistry, Medical School, University of Leipzig, Germany

Affinity partitioning is a special branch of biomaterials separations using aqueous two-phase systems. It combines the capability of diverse biomolecules to partition in aqueous two-phase systems using the principle of biorecognition. As a result, the macromolecule exhibiting affinity for a certain ligand is transferred to that phase where the ligand is present. This chapter describes the present status of the theoretical background of this approach and the properties of various natural and artificial compounds which act as affinity ligands in liquid–liquid systems. The affinity partitioning of proteins (enzymes and plasma proteins), cell membranes, cells, and nucleic acids are described as typical examples. The results are discussed in terms of theoretical understanding and practical application.

KEY WORDS: Affinity partitioning, Proteins, Nucleic acids, Cell membrances, Natural ligands, Artificial ligands, Pseudo-biospecific dyes.

I. Introduction*

A. Interaction of Biomaterials with Certain Ligands

The term biorecognition expresses a phenomenon which is found throughout the entire field of biology. It characterizes the interactions of natural or artificial ligands with their biological counterparts, which may be various

* Abbreviations: PEG, polyethylene glycol; IDA-PEG, imidodiacetate-PEG; EDTA, ethylenediamine tetraacetic acid; WGA, wheat-germ agglutinin; PFK, phosphofructokinase; LDH, lactate dehydrogenase; APDE, alkaline phosphodiesterase: 5′-N, 5′-nucleotidase.

International Review of Cytology, Vol. 192
0074-7696/00 $30.00

biomolecules such as proteins, nucleic acids, membranes, or cells. The principle of biorecognition is accomplished, for example, by binding substrates to their enzymes, by docking hormones to their corresponding receptors, or by antigen–antibody interaction. The strength of interaction can be expressed by an association constant which is based on the law of mass action and ranges widely from about 10^4 to 10^{15} M^{-1}. The specificity of biorecognition is achieved by certain bonds formed between the ligand and the corresponding biomaterial if structural requirements on both partners exist.

Biospecific ligands are chemically diverse. There are natural ligands of enzymes, e.g., substrates, effectors or inhibitors of low molecular weight, but ligands of higher-molecular weight like polypeptides, proteins, oligosaccharides, or lipids might also be mentioned. The complex formation is achieved by noncovalent binding forces and is found, as a rule, to be reversible.

In addition to natural ligands, another class of ligands, called "pseudobiospecific" ligands, is known. This class comprises chemically synthesized compounds not existing in biological systems per se, but which exhibit surprising bioaffinity for certain macromolecules. Reactive dyes which are commonly used in textile staining and metal chelates belong to this class. In some cases, the interaction of a ligand with its counterpart was found to be absolutely specific. This is realized, for example, by antibody–antigen or hormone–receptor interaction, or if a substrate has an absolute specificity for its enzyme due to the existence of distinct structural requirements for interacting. However, many ligands are characterized as group-specific or "general ligands." That means structurally similar groups of diverse substances bind to a certain class of proteins which share a common motif on their surface, e.g., the nucleotide-binding domain.

Biorecognition is an elementary principle of metabolism and its regulation existing on the level of a single cell and between cells and organs in more complex systems. Deeper insight into the chemical mechanism of this phenomenon gives rise to a comprehensive understanding of biological processes and allows, moreover, numerous applications of this principle in the field of biotechnology.

B. Principle of Affinity Partitioning

Partitioning of biological macromolecules in aqueous two-phase systems was found to be a simple method for their separation and fractionation. The success of partitioning depends on numerous intrinsic properties, such as size, electrochemical properties, hydrophobic and hydrophilic surface properties, conformation properties, and other factors. Extrinsic parameters influencing partitioning of macromolecule are: kind and molecular weight of the polymers used, concentration of the polymers, kind of buffer components, ionic strength, pH of the buffer, temperature, and others.

If a biospecific ligand is bound to one of the phase-forming polymers or is partitioned in favor of one phase, the partition of its corresponding biomolecule is preferentially into the phase containing the ligand. This principle is called affinity partitioning and combines the property of any biomaterial to partition in aqueous two-phase systems with that of biorecognition.

Biospecific affinity partition in aqueous two-phase systems was introduced by Shanbhag and Johansson (1974), who demonstrated that human serum albumin could be selectively extracted from plasma by palmitoyl-PEG. At the same time, Takerkart *et al.* (1974) showed that partition of trypsin is changed if a competitive trypsin inhibitor was coupled to the PEG polymer. Later, Flanagan and Barondes (1975) applied dinitrophenol as ligand for affinity partitioning of S-23 myeloma protein. The introduction of pseudo-biospecific dyes as ligands in affinity partition was first reported by Kroner *et al.* (1982) and Kopperschläger and Johansson (1982). A fundamental advantage of an approach such as this is its predictive property because the ligand of choice can be chemically designed to achieve adequate affinity to the target biomaterial. The construction of such a class of "biomimetic dyes" has been successfully demonstrated by Lowe *et al.* (1992).

II. Theoretical Considerations of Affinity Partitioning

The partition coefficient, K, defined as the ratio of the macromolecule concentration in the top to that in the bottom phase, is a characteristic property of the macromolecule and can be viewed as the sum of individual terms,

$$\ln K = \ln K^{o} + \ln K_{el} + \ln K_{hphob} + \ln K_{hphil} + \ln K_{conf} + \ln K_{size} + \ln K_{lig} \quad (1)$$

where the subscripts refer to effects on partitioning owing to electrostatic effects (el), to hydrophobic (hphob) or hydrophilic (hphil) surface properties, to conformational states (conf), to size (size) of the macromolecule, to ligand macromolecule interaction (lig). K^{o} summarizes other factors. When the strength of ligand binding is high, the term $\ln K_{lig}$ becomes dominant, and partition of the macromolecule is governed mainly by the formation of the ligand–macromolecule complex. At saturating ligand concentrations and optimal conditions, the target macromolecule can be transferred completely from one phase into the other.

The effect of affinity partitioning is quantified by measuring the difference of the partition coefficient, K, of a macromolecule obtained in the presence and in the absence of the ligand, mostly expressed on a logarithmic scale (i.e., $\Delta\log K$). When $\Delta\log K$ is plotted against increasing concentration of the ligand-polymer, as shown in Fig. 1, saturation is observed for many proteins, from which the value $\Delta\log K_{max}$ (maximum extraction power) and the half-saturation point $0.5 \times \Delta \log K_{max}$ can be calculated. While $\Delta\log K_{max}$ is somehow related to the number of binding sites (see below), the parameter of half-saturation is assumed to be a measure of the relative affinity of the ligand for the target material (Kirchberger *et al.,* 1991; Kopperschläger *et al.,* 1983; Kopperschläger and Johansson, 1985; Kopperschäger and Lorenz, 1985). In order to quantify affinity partitioning of proteins, a first attempt at a mathematical model was worked out by Flanagan and Barondes (1975).

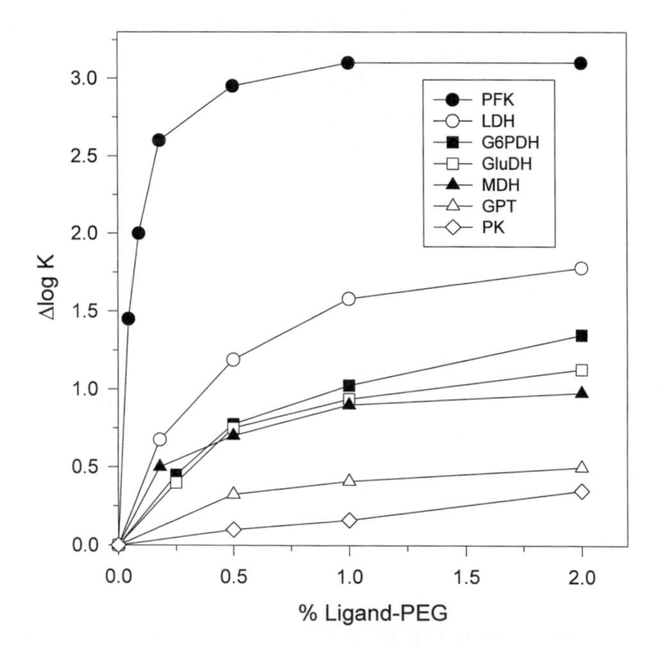

FIG. 1 Affinity partitioning of enzymes as a function of increasing concentration of Cibacron Blue F3G-A-poly(ethylene glycol). The systems are composed of 5% PEG 6000, 7% dextran T500, and 50 mM sodium phosphate buffer, pH 7.0. PFK, phosphofructokinase; LDH, lactate dehydrogenase; G6PDH, glucose-6-phosphate dehydrogenase; GluDH, glutamate dehydrogenase; MDH, malate dehydrogenase; GPT, glutamate pyruvate transaminase; PK, pyruvate kinase. From Kopperschläger and Birkenmeier (1993) with permission.

Based on multiple equilibria existing in the aqueous two-phase systems (Fig. 2), the change in the standard state Gibbs' free energy, ΔG^o_5, required to transfer one mole of the ligand-polymer-protein complex from the top into the bottom phase may be calculated from the sum of the individual free energies of the respective species.

$$\Delta G^o_5 = \Delta G^o_1 + \Delta G^o_2 + \Delta G^o_3 + \Delta G^o_4 \tag{2}$$

ΔG^o_1 and ΔG^o_4 are related to the dissociation constants of the ligand-polymer-protein complex in the top phase $(K_d)_t$ and in the bottom phase $(K_d)_b$ according to Gibbs' equation $\Delta G^o = -\,RT \ln k$.

ΔG^o_2, ΔG^o_3, and ΔG^o_5 are related to the partition coefficients of the free ligand-polymer (K_L), the free protein (K_P), and the ligand-polymer-protein complex (K_c), according to Albertsson (1971), by $\Delta G^o_2 = RT \ln K_L$; $\Delta G^o_3 = RT \ln K_P$; $\Delta G^o_5 = RT \ln K_c$.
The following equation results by substitution of these parameters in Eq. (2):

$K_C = [(k_d)_b\, K_L /(k_d)_t]\, K_P$, or for n-binding sites of equal dissociation constants $K_c = [(k_d)_1\, K_L /(k_2)_t]^n\, K_P$ (3)

If the binding constants for the ligand are independent and equal in both phases, Eq. (3) reduces to

$$K_C = (K_L)^n\, K_P \tag{4}$$

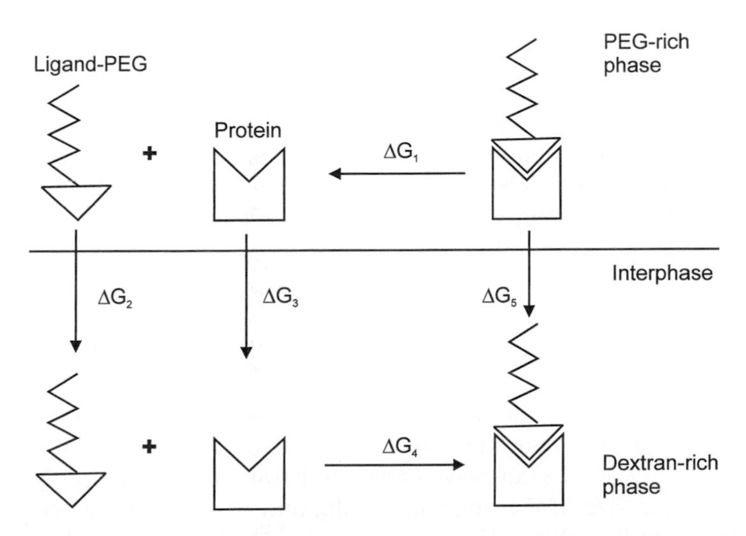

FIG. 2 Diagram depicting partition of monovalent protein and an interacting monovalent polymer-ligand in aqueous two-phase system. From Flanagan and Barondes (1975) with permission.

or in the logarithmic form under the condition of ligand saturation to

$$\Delta \log K_{max} = \log (K_C/K_P) = n \log K_L. \quad (5)$$

A more general expression describing partition of macromolecules in aqueous two-phase systems at any ligand concentration has been derived by Brooks *et al.* (1985),

$$K_C = K_P [1 + C_{Lt} (1_d)_t]^n/[1 + C_{Lb}(2_d)_b]^n \quad (6)$$

where C_{Lt}, C_{Lb} are the concentration of the ligand in the top or bottom phase, respectively. This approach also assumes the same binding constant for each binding site. At saturating ligand concentration with $(1_d)_t = (2_d)_b$, Eq. (6) reduces to Eq. (5).

The assumptions for Eq. 5 are not valid for calculating the number of binding sites for the ligand-protein complex in most cases. There are several reasons for this. First, the dissociation constant of the complex is not identical in the top and bottom phases. In general, the ligand-PEG-protein complex is less stable in the PEG phase than in the dextran-containing phase. This has been confirmed by affinity partitioning of formate dehydrogenase in systems containing PEG and dextran and in the presence of Cibacron Blue F3G-A-PEG, Procion Red H-E3B-PEG, and NADH-PEG (Cordes *et al.,* 1987). Identical results were obtained by fluorescence titration of lactate dehydrogenase with Procion Yellow H-E3G in presence of dextran and PEG (Alred *et al.,* 1992).

Second, different classes of ligand binding sites of the macromolecule exist in most cases. This may be caused by intrinsic properties of the biomaterial or could be achieved by steric hindrance of further complex formation caused by the large polymer tail after binding the first ligand-polymer. A third influence may arise from self-aggregation of the ligand-polymer due to hydrophobic interactions of the ligand, which reduces the affinity partitioning effect. Based on Eq. (6), a similar model has been developed to describe protein partitioning in a Cu^{2+}-chelate-PEG/dextran system (Suh and Arnold, 1990). This working model has been extended to account for metal chelate affinity partitioning of proteins. The model was successfully tested at low ligand concentration with various proteins whose number of binding sites were known using protons as inhibitors.

Baskir *et al.* (1989) derived a lattice model for examining the interaction of the ligand-polymer tail of a bound ligand-polymer with the macromolecule and the surrounding solution. The model allows the calculation of the binding energy of a ligand-polymer as a function of the ligand-polymer tail chemistry, the size of the macromolecule, and the polymer concentration of the surrounding phase. The model calculation also indicates that binding the ligand-polymer to the protein should be weaker in the phase which is rich in the given polymer than in the other phase. In addition, the strength of ligand-polymer binding in the phase rich in the given polymer is expected to decrease with increasing concentration of the phase-forming polymer. These results agree with the findings of Cordes *et al.* (1987). The theoretical treatment of the effects of affinity ligands on partitioning of cell fragments or whole cells is much more complicated than for proteins. Unlike soluble proteins, the substantially larger cell particles exhibit a large number of binding sites for the ligand and are often adsorbed into the interphase region. Adsorption of particles into the interphase is driven by interfacial surface tension forces which are derived from the decrease in the interfacial surface area upon adsorption of a particle into the interface. Analyzing the role of interfacial forces is rather complicated because one cannot assume that equilibrium is established in the distribution of particles between the bulk phase and the interphase. The free energy difference $\Delta\gamma$ between particles immersed completely in the top phase and in the bottom phase and those adsorbed in the interphase (where they are in contact with both phases) have been estimated by measuring the interfacial tension of the liquid–liquid boundary and the contact angle formed between the interphase and the cell surface. To introduce the affinity partitioning effect into this approach, it is necessary to determine how the binding of the ligand-polymer to cell surfaces changes the $\Delta\mu$ value. A theoretical treatment has been derived and successfully tested by Brooks *et al.* (1985).

III. Selection of Ligands and Mode of Coupling

A. Natural Low Molecular Weight Ligands

There is a large variety of low molecular weight compounds acting as substrates, cofactors, or inhibitors of enzymes or of other proteins but only a few of them have found application in affinity partitioning. One reason might be the more or less laborious chemical coupling procedure, e.g., the attachment of a natural ligand to the polymer in a functional active state. Another reason is due to the relatively high costs of these substances, particularly when affinity partitioning is used on a technical scale.

In principle, all phase-forming polymers are capable of carrying covalently linked affinity ligands. But PEG has been preferentially used because of its availability in a large variety of molecular weights, its low price, and its stoichiometry of binding maximally two ligands per molecule. For a single derivatization, monoalkylated PEG may be used. The coupling procedure should be performed under mild chemical conditions to avoid breakdown of the polymer and to protect the structure of the ligand. In order to increase the reactivity of PEG, the terminal hydroxyl groups have been substituted by reactive electrophiles, such as halides (Johansson, 1970; Bückmann *et al.,* 1981).

In some cases a spacer was introduced between the polymer and the ligand. As an example, N^6-(2-aminoethyl)-NADH-PEG has been prepared by the reaction of carboxylated PEG with *N*(1)-(2-aminoethyl)-NAD^+ in the presence of carbodiimide and subsequent reduction of the product (Bückmann *et al.,* 1987). Another biospecific ligand-polymer, ATP-PEG, was synthesized by activation of the respective polymer to the adiponyldihydrazo derivative which easily can react with periodate-oxidized ATP (Persson and Olde, 1988).

Hubert *et al.* (1976) described the synthesis of PEG-bound estradiol, by coupling estradiol-7α-butyric acid with an amino group modified PEG by a substitution reaction.

B. Pseudo-Biospecific Ligands

1. Reactive Dyes

The ability of sulfonated polyaromatic reactive dyes to act as "general ligands" in many fields of affinity separation technology is well established (Kopperschläger *et al.,* 1982; Clonis *et al.,* 1987). This class of dyes reveals several advantages over their natural counterparts. It includes the moderate to high selectivity for enzymes and other proteins, the availability as low-cost chemicals exhibiting a large diversity of structures, the high reactivity to permit substitution to many polymers, and the chemical and biological stability. Reactive dyes are composed of a chromophoric and a reactive part which are linked by auxochromic groups in a variety of chemical combinations (Fig. 3). Although a large number of compounds have been synthesized, the Cibacron and Procion dyes have found the greatest range of application. Reactive dyes of the mono- and dichlorotriazinyl type and of the vinylsulfonic acid type can be attached under mild alkaline conditions directly to PEG or to other polymers (Johansson *et al.,* 1983; Johansson and Joelsson, 1985a; Patricia *et al.,* 1992) or by using amino-PEG (Cordes and Kula, 1986).

2. Metal Chelates

Metal chelates have turned out to be another group of "general ligands" capable of binding various proteins, cell fragments and cells (Poráth *et al.,* 1975). Chelated metals such as Cu^{2+}, Zn^{2+} or Ni^{2+} form coordination compounds with electron-donating N, S, and O atoms in peptides and proteins. The chelating PEG is synthesized by reacting chloromonome-

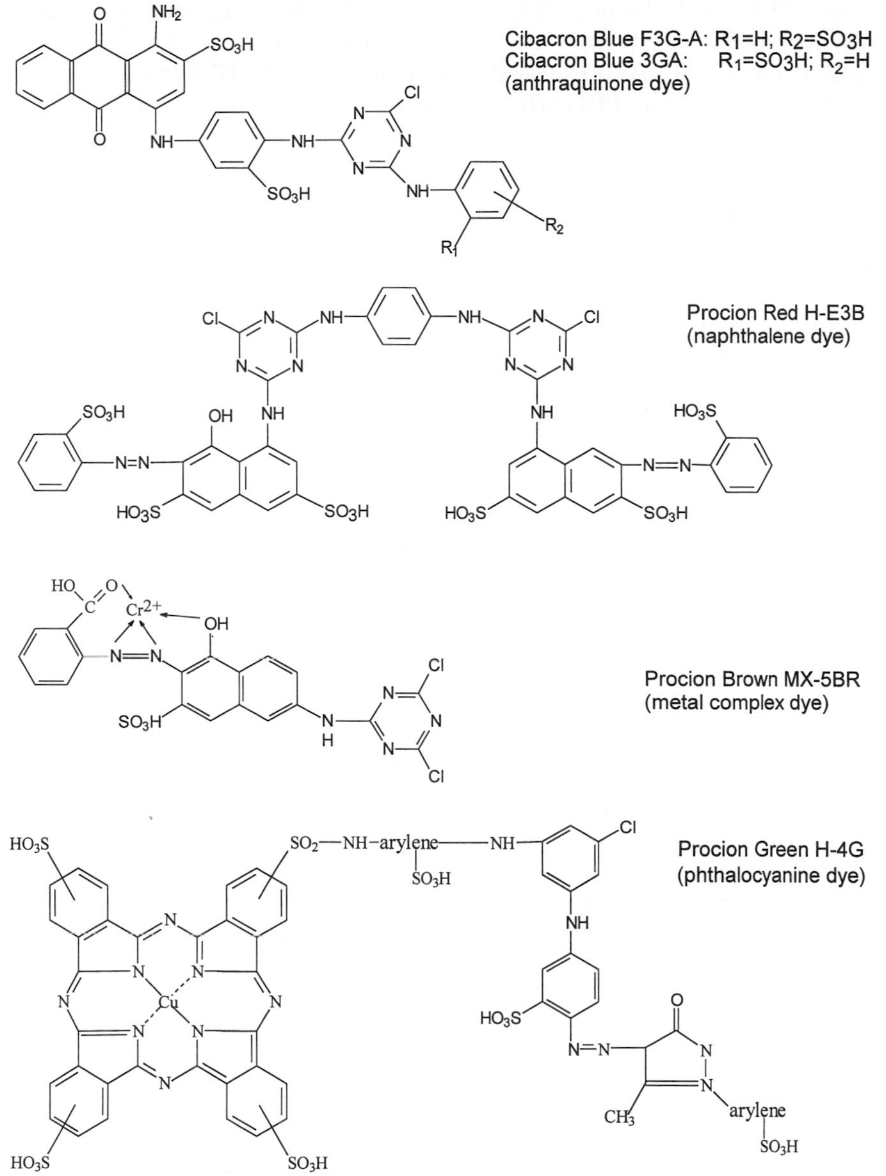

Cibacron Blue F3G-A: R_1=H; R_2=SO_3H
Cibacron Blue 3GA: R_1=SO_3H; R_2=H
(anthraquinone dye)

Procion Red H-E3B
(naphthalene dye)

Procion Brown MX-5BR
(metal complex dye)

Procion Green H-4G
(phthalocyanine dye)

FIG. 3 Structure of dye-ligands commonly used in affinity techniques.

thoxy-PEG with iminodiacetic acid (IDA) under alkaline conditions (Wuenschell *et al.,* 1990). Alternatively, IDA-PEG can be prepared from amino monomethoxy-PEG with bromoacetic acid (Birkenmeier *et al.,* 1991). The chelating PEG when saturated with Cu^{2+} forms a 1:1 complex with the metal ion.

3. Ligands for Synaptic Receptors

Charged bis-quaternary amines like $Et_3N(CH_2)_nNEt_2$ have been found to bind with synaptic domains containing specific receptors (Johansson *et al.,* 1981). A general scheme for the synthesis of quaternary amine-PEG is the reaction of PEG-halides with trialkylamines (Johansson *et al.,* 1973). For the synthesis of a special opiate receptor-binding polymer, 6-aminonaloxone was allowed to react with tresyl-activated PEG (Olde and Johansson, 1985).

C. Antibodies, Proteins, Peptides

For the employment of peptides and proteins as affinity ligands, PEG is preferentially used as carrier-polymer. In order to protect the functional integrity of these ligands, the coupling procedure has to be carried out under mild chemical reaction conditions. As a rule, monomethoxy-PEG is activated with the cyanuric chloride method (Abuchowski *et al.,* 1977), as described by Sharp *et al.* (1986) and Karr *et al.* (1986). Alternatively, PEG or dextran can be activated by tresyl chloride (Delgado *et al.,* 1991, 1992; Persson and Olde, 1988) or by synthesizing PEG-oxirame (Elling and Kula, 1991) followed by protein binding. Another way to couple an antibody to the PEG-phase is the use of PEG-modified protein A which reacts with the antibody to form a stable complex (Karr *et al.,* 1988). Similarly, a monoclonal antibody-PEG was used as a bridge to bind specifically IgG antibodies raised in rabbits (Stocks and Brooks, 1988).

D. Hydrophobic Ligands

Hydrophobic ligands are also characterized as general ligands because they do not interact very specifically with hydrophobic domains of cells, cell fragments, or proteins. Several standard procedures for coupling fatty acids to PEG have been elaborated (Shanbhag and Johansson, 1974; Pillai *et al.,* 1980; Müller *et al.,* 1981). Alternatively, dextran has been modified with benzoylchloride under mild alkaline conditions to form a polymer with a higher degree of hydrophobicity (Lu *et al.,* 1991). A new type of aqueous two-phase systems containing benzoyldextran as bottom phase polymer and the random copolymer of ethylene oxide and propylene oxide (Ucon) as top phase polymer has been composed. The system was successfully applied for the purification of 3-phosphoglcycerate kinase from baker's yeast (Lu *et al.,* 1996).

IV. Affinity Partitioning of Proteins

A. Dye Ligands and Natural Ligands

Affinity partitioning of proteins has proven to be of interest for both purification of proteins by extractive processes and for studying ligand–protein interactions. When the extraction power (Δlog K) exceeds a certain value (approximately 1, that is a 10-fold change in the partition of the target protein), single or multistep procedures are able to enrich the protein sufficiently. Among the large variety of ligands, which are listed in Table I, biomimetic dyes have found practical application. This class of pseudobiospecific ligands is capable of interacting with diverse nucleotide-dependent enzymes such as kinases and dehydrogenases by mimicking morc or less precisely the structure of the natural nucleotide. When in the absence of the dye-ligand the protein bulk is concentrated mainly in the bottom phase, a partial replacement of PEG by dye-PEG results in a significant transfer of those proteins from the bottom phase into the top phase, which exhibits affinity for the ligand. Based on this behavior, preparative extraction steps for the purification of formate dehydrogenase (Cordes and Kula, 1986; Walsdorf *et al.,* 1990), phosphofructokinase (Kopper-

TABLE I
Affinity Partitioning of Proteins

Affinity ligand	Coupling polymer	Target protein
Biomimetic dyes	PEG, dextran, UCON	Nucleotide dependent enzymes
NAD^+ NADH	PEG	Dehydrogenases
ATP	PEG	Nucleotide dependent enzymes
Fatty acids	PEG	Plasma proteins
Metal chelates	PEG	rHis-tag proteins
Protein A	PEG	Antibody labelled cells
Antibodies	PEG	Surface antigens of diverse cells
Receptor-ligands	PEG	Opiate receptors
Estradiol	PEG	3-Oxosteroid isomerase
Dinitrophenol	PEG	S-32 myeloma proteins
p-Aminobenzamidine	PEG	Trypsin

schläger and Johansson, 1982; Tejedor *et al.,* 1992), glucose-6-phosphate dehydrogenase (Johansson and Joelsson, 1985b; Delgado *et al.,* 1990), lactate dehydrogenase (Johansson and Joelsson, 1986), alcohol dehydrogenase (Pesliakas *et al.,* 1988), restriction endonucleases (Vlatakis and Bouriotis, 1991) have been elaborated. In addition, affinity partitioning is also suited for studying small changes in ligand–protein interaction or for the selection of ligands for other affinity separation techniques. For example, distinct partition of isoenzymes of human alkaline phosphatase in aqueous two-phase systems containing certain triazine dyes has been observed (Kirchberger *et al.,* 1992). The differences in the affinity towards the dye ligands are caused by either differences in the carbohydrate content of the isoenzymes, in particular by the terminal sialic acid, or by specific binding properties.

Another example of isoenzyme recognition in aqueous two-phase systems was found with lactate dehydrogenase LDH1(H_4) and LDH5(M_1). The triazine dye Procion Blue H-5G is able to distinguish between these isoenzymes, thus leading to different changes of their partition coefficients when the respective dye-PEG is added to the aqueous two-phase systems (Kirchberger *et al.,* 1991).

In contrast to the widespread application of dye ligands, there are only a few reports dealing with affinity partitioning of enzymes with natural ligands. Using NADH-PEG as affinity polymer, lactate dehydrogenase, alcohol dehydrogenase, aldehyde dehydrogenase, and formate dehydrogenase can be extracted into the top phase (Bückmann *et al.,* 1987). The application of ATP-PEG for the extraction of phosphoglycerate kinase from spinach chloroplasts was reported by Persson and Olde (1988).

Several pseudo-biospecific dyes are capable of interacting with certain plasma proteins. This property is surprising because these proteins do not exhibit specific nucleotide binding domains. Human albumin was found to interact rather strongly with Cibacron Blue F3G-A, while another dye, Remazol Yellow GGL, was found to bind specifically to prealbumin (Birkenmeier *et al.,* 1984a, 1986b; Birkenmeier and Kopperschläger, 1987). In Fig. 4, the effect of different dye-PEG derivatives on partitioning of prealbumin and albumin is demonstrated. The result indicates that only these two dyes are capable of distinguishing between the two plasma proteins sufficiently.

Although the physicochemical properties of human albumin and α-fetoprotein are similar, their behavior in aqueous two-phase systems containing Cibacron Blue F3G-A-PEG is quite different (Birkenmeier *et al.,* 1984b). As shown in Table II, most of the dye-PEG's listed interact with albumin but do not react with α-fetoprotein sufficiently, with the exception of Procion Navy H-ER. Thus, this particular property has been successfully

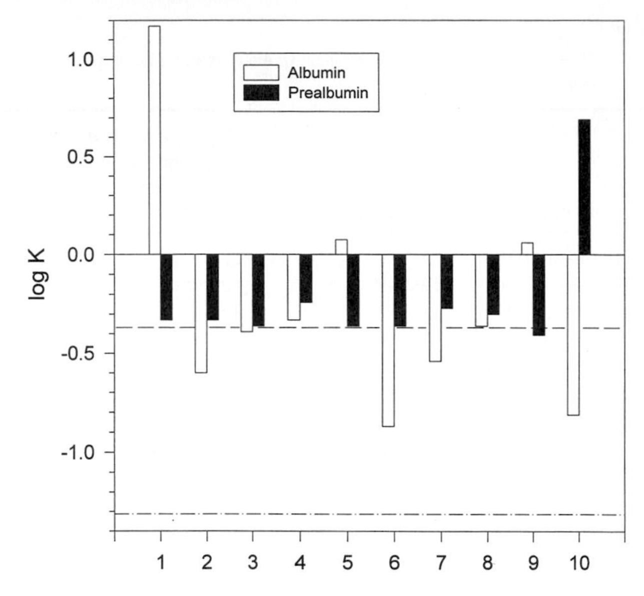

FIG. 4 Effect of different dye PEG derivatives on the partitioning of prealbumin (black bars) and albumin (white bars). Phase systems were composed of 10% dextran T500, 7.5% PEG 8000 including 1.6% of the respective dye PEG derivative, 10 mM sodium phosphate, pH 7.0, and protein. Temperature 0°C. Partitioning of prealbumin (---------) and albumin (-·-·-·-·) in absence of dye PEG is shown for comparison. From Birkenmeier *et al.* (1986a) with permission.

1 = Cibacron Blue F3G-A	6 = Procion Yellow MX-4G
2 = Procion Red H-3B	7 = Procion Scarlet MX-G
3 = Procion Orange MX-G	8 = Procion Yellow MX-GR
4 = Procion Yellow MX-R	9 = Procion Brown MX-5BR
5 = Procion Scarlet HR-N	10 = Remazol Yellow GGL

employed for separation of the large amount of albumin from traces of α-fetoprotein in blood from the umbilical cord (Huse *et al.,* 1983).

Other plasma proteins which show interaction with reactive dyes as recognized by affinity partitioning are thyroxine-binding protein (Birkenmeier *et al.,* 1886b), β-lipoprotein (Birkenmeier *et al.,* 1986a) and α_2-macroglobulin (Birkenmeier *et al.,* 1987).

One requirement for performing affinity partitioning is the asymmetric distribution of the respective ligand which is achieved, as a rule, by coupling the ligand of choice to one of the phase-forming polymers. However, in aqueous two-phase systems composed of maltodextrin (bottom phase) and polyvinylpyrrolidone (top phase), free triazine dyes also partition asymmetrically, favoring the less dense polyvinylpyrrolidone-rich phase. This system has been successfully applied to affinity partitioning of lysozyme (Giuliano, 1991).

TABLE II

Effect of Dye-poly(ethylene glycol) on Partition of Albumin and α-Fetoprotein Expressed in Terms of Δlog K[a]

Dye-poly(ethylene glycol) derivatives	Δlog K at 0.5% dye-poly (ethylene glycol)	
	Albumin	α-Fetoprotein
Cibacron Blue F3G-A	2.91	0.73
Cibacron Brilliant Blue FBR-P	2.59	0.47
Procion Blue MX-3G	1.93	0.54
Procion Navy H-ER	1.70	1.12
Procion Blue MX-R	1.70	0.49
Procion Blue H-E7B	1.36	0.65
Procion Red H-E3B	1.33	0.84
Procion Red H-3B	0.67	0

[a] Systems contain 7.5% (w/w) PEG 6000, 10% (w/w) dextran T500, 10 mM sodium phosphate, pH 7.1, and 20 mM sodium chloride. The Δlog K was calculated from the difference of the logK-values in presence and absence of the dye. From Birkenmeier *et al.* (1984b) with permission.

B. Metal Chelate Ligands

The first data exploiting metal chelate–protein interaction in aqueous two-phase systems were reported by Wuenschell *et al.* (1991) and Suh and Arnold (1990). Since then, various metal chelates coupled to PEG have been tested for affinity partitioning of proteins and cells (see below). The desired selectivity in binding proteins arises from the type of the chelating group, the transition metal, and the availability of the target amino acid residues on the protein surface. Moreover, the microenvironment of the electron donor group on the protein surface may result in a variable intensity of protein recognition by a metal ligand. In particular, the number and sometimes the distribution of histidyl groups on the protein surface dictate almost exclusively the strength of interaction (Arnold, 1991).

Immobilized metal ion affinity partitioning has been applied for the purification of D-hydroxyisocaproate dehydrogenase (Schustolla *et al.*, 1992) and for the separation of isoenzymes of lactate dehydrogenase (LDH) (Otto and Birkenmeier, 1993). Using Cu^{2+}-iminodiacetate-PEG (Cu^{2+}-IDA-PEG) the surface of LDH isoenzymes from different sources was probed for the presence of metal binding sites. It could be demonstrated that homotetramers (LDH-1)(H4) from rabbit bovine and pig display weak binding to chelated copper while the LDH-5 (M4) isoenzyme binds strongly to the ligand. In contrast, the human isoenzymes are bound to chelated copper in a reversed sequence.

Affinity partitioning of yeast alcohol dehydrogenase, muscle malate dehydrogenase and lactate dehydrogenase in the presence of Cu^{2+}-IDA-PEG decrease in the order listed. It has been shown that the variations in the affinity of the enzymes to the metal chelate are related to the content of histidine residues of the enzymes accessible to the solvent (Pesliakas *et al.,* 1994).

The interaction of various plasma proteins with immobilized metal chelates has been investigated comprehensively (Sulkowski, 1985). In order to study this class of proteins in metal chelate affinity partitioning, α_2-macroglobulin, tissue plasminogen activator, superoxide dismutase, and a monoclonal antibody (IgG) have been used as model proteins (Birkenmeier *et al.,* 1991). As shown in Table III, all four proteins interact with Cu^{2+}-IDA-PEG most strongly and a reduction of the affinity partition effect is observed when Ni^{2+} or Zn^{2+} is used instead of Cu^{2+}. α_2-Macroglobulin reveals the highest affinity partitioning effect with a 1000-fold increase of the partition coefficient.

Fe^{3+}-IDA complexes were found to have high affinity for phosphoryl groups and may be used as a complexing metal to target phosphorylated proteins in affinity partitioning experiments. However, proteins, rich in

TABLE III

Effect of Metal-IDA-PEG on Partitioning of Different Proteins in PEG/Dextran and PEG/Salt Systems[a]

	Δlog K							
	PEG/dextran: metal-IDA-PEG				PEG/salt: metal-IDA-PEG[b]			
Protein	Cu^{2+}	Zn^{2+}	Ni^{2+}	None	Cu^{2+}	Zn^{2+}	Ni^{2+}	None
α_2-Macroglobulin	3.00	1.50	1.10	0.17	2.40	0.06	0.18	0.02
Tissue plasminogen activator	1.59	0.78	1.20	0.14	1.23	0.46	0.31	0.03
Superoxide dismutase	1.30	0.05	0.18	0.08	0.83	0.10	0.20	0.05
Monoclonal antibody (IgG_{1K})	1.15	0.12	0.05	0.10	n.d.	n.d.	n.d.	n.d.

[a] PEG/dextran system: 5% PEG 6000, 7.5% dextran T70, 0.1 M Na_2SO_4, 0.01 M sodium phosphate buffer (pH 7.0). For the calculation of ΔlogK, 20% of the total PEG was replaced with uncharged IDA-PEG or metal-IDA-PEG. PEG-salt system: 11% PEG 1540, 2% PEG 6000, 13.5% K_2HPO_4-KH_2PO_4(pH 7.0). For the calculation of Δlog K, 50% of the total PEG 6000 was replaced with uncharged IDA-PEG or metal-IDA-PEG. From Birkenmeier *et al.* (1991).

[b] n.d. = not determined

carboxyl groups and tyrosine, are also bound to Fe^{2+}-IDA-PEG complexes (Chung and Arnold, 1991).

C. Hydrophobic Ligands

Among this group of general ligands, aliphatic fatty acids have preferentially found application in affinity partitioning of various plasma proteins like albumin (Shanbhag and Johansson, 1974; Johansson and Shanbhag, 1984; Birkenmeier *et al.,* 1986c), lactalbumin (Johansson and Shanbhag, 1984; Shanbhag *et al.,* 1991), α-fetoprotein (Birkenmeier *et al.,* 1986c), α_2-macroglobulin (Birkenmeier *et al.,* 1989; Jensen *et al.,* 1993, 1994), pregnancy zone protein (Birkenmeier *et al.,* 1989; Jensen *et al.,* 1993), cell membranes, and cells (see below).

For example, human albumin and α-lactalbumin were compared in a PEG/dextran system containing palmitoyl-PEG (Johansson and Shanbhag, 1984). With increasing concentration of dextran, the $\Delta\log K_{max}$ value for albumin rises from 1 (4% PEG/4% dextran) to 3 (4% PEG/8% dextran) while under the same conditions α-lactalbumin shows a value of $\Delta\log K_{max} = 0.5$ which is independent of the dextran concentration. This result reflects the different binding behavior of the fatty acid to both proteins.

Similarities in the interaction of fatty acids with albumin and α-fetoprotein are indicated by affinity partitioning of both proteins in systems with increasing chain length of the acyl residue (Birkenmeier *et al.,* 1986c). The high log K for both proteins as well as the tendency that these values increase with prolonging chain length of the PEG-bound acyl groups corroborates the structure–function relation of both proteins.

D. Other Ligands

There are a variety of other ligands capable of exhibiting affinity partitioning effects. Those that have been used for the partition of cells or membranes are described in Section VII. Chicken avidin-PEG was reported to be a general ligand for affinity partitioning of biotinylated proteins. The modified avidin is preferentially partitioned into the PEG-rich phase of a PEG/dextran system. Using avidin-PEG, the immune complex formed between biotinylated anti-mouse IgG and its antigen is significantly transferred into the PEG phase (Nishimura *et al.,* 1995).

Coupling the D-Ala-D-Ala peptide to methoxy-PEG leads to affinity partitioning of the antibiotic vancomycin in PEG/dextran aqueous two-phase systems (Singh and Clark, 1994).

p-Aminobenzamidine, a strong inhibitor of trypsin, may be used as a ligand for affinity partitioning of trypsin in an aqueous two-phase system containing a copolymer of N-isopropylacrylamide and glycidyl acrylate and either dextran or pullulan (Nguyen and Luong, 1990).

Affinity partitioning of Δ^{3-4}-3-oxosteroid isomerase from *Pseudomonas testosteroni* in PEG/dextran systems with estradiol-PEG has been studied, leading to a substantial enrichment of the enzyme in the top phase if negatively or positively charged dextran is used (Hubert *et al.,* 1976; Chaabouni and Dellacherie, 1979).

Enzyme-enzyme interaction between aspartate aminotransferase and malate dehydrogenase have been followed in a PEG/dextran system containing trimethylamino-PEG or carboxymethyl-PEG and applying the technique of counter-current distribution (Backman and Johansson, 1976).

V. Partitioning of Genetically Engineered Proteins

Increasing the affinity partitioning effect by changing the surface structure of the target protein by genetic engineering is a fascinating approach in aqueous two-phase systems. A first attempt was reported by fusion of β-galactosidase to immunoglobulin G binding domains of staphylococcal protein A (Köhler *et al.,* 1991a,b). Due to the extreme partition of β-galactosidase into the PEG-rich phase, the fusion products change their partition significantly in comparison to the control protein (Fig. 5). This result led to the development of a specific branch of affinity partitioning of proteins modified with a peptide tail, based on the hypothesis of specific interactions between PEG and tryptophan (Radzicka and Wolfenden, 1988; Köhler *et al.,* 1991b;). Beyond the classical systems composed of PEG/dextran or PEG/salt, other systems containing PEG/benzoyl dextran, PEG/valeryl dextran (Lu and Tjerneld, 1997), and Ucon/dextran have proved useful for partitioning of proteins fused with tryptophan-rich peptides (Carlsson *et al.,* 1996). In order to increase the metal chelate affinity partitioning effect of hirudin, a number of variants have been constructed by genetic engineering to contain additional accessible histidines (Chung *et al.,* 1994). Those hirudin variants with two or three surface-accessible histidines are more selectively partitioned into the PEG-rich phase (applying Cu^{2+}-IDA-PEG) than the control without a fusion peptide exhibiting only one single surface-accessible histidine.

VI. Factors Influencing Affinity Partitioning of Proteins

A. Competing Effectors

The interaction of proteins with their respective ligands is mostly reversible. Therefore, the strength of binding can be disturbed by a number of factors.

Plasmid	Protein	kDa	% non-ßGal	*K*
	[ß-galactosidase]	465	-	17 (C)
pFdK216-Lys/Arg	[Fdx N][ ] ↦ a.a.5 [ß-galactosidase]	492	6	17 (C)
pRIT1	[Sp A][ ] ↦ a.a.25 [ß-galactosidase]	592	21	3.4 (B)
pRIT1B	[Sp A][ ] ↦ a.a.7 [ß-galactosidase]	584	21	3.5 (B)
pRIT2	[Sp A][ ]	31	-	0.7 (E)
pRIT36	[Sp A][ ][C2C3] ↦ a.a.7 [ß-galactosidase]	656	28	2.8 (C)
pRIT35	[Sp A][ ][C2C3]	46	-	0.7 (D)

FIG. 5 Schematic description of the β-galactosidase fusion proteins and their partitioning in PEG 4000/potassium phosphate systems. The capital letters given with the partition coefficients (K values) indicate the following phase system compositions, in % (w/w) PEG 4000/% (w/w) potassium phosphate: B, 6.8/10.5; C, 6.9/10.6; D, 7.0/10.8; E, 7.0/10.9. FdxN, *Rhizobium meliloti* ferredoxin-like protein; Sp A, staphylococcal protein A; C2C3, IgG-binding regions from streptococcal protein G. From Köhler *et al.* (1991b) with permission.

One external parameter was found to be the ionic strength. For example, many dye PEG-protein complexes dissociate after adding salt to the top phase. As a result, a new aqueous two-phase system is formed consisting of PEG (and free ligand-PEG) as top phase and salt as bottom phase. The latter contains more than 90% of the protein formerly bound to the ligand (Johansson, 1984). In addition, there are findings that the effect of salt on affinity partitioning is also a matter of the type of the ion (Johansson *et al.,* 1983; Birkenmeier *et al.,* 1986a). As demonstrated in Table IV, affinity partitioning of prealbumin to Remazol Yellow GGL depends more on the chaotropic property of the respective anion than on its concentration. Higher selectivity in disturbing the affinity partitioning effect is achieved by adding interfering ligands (natural or artificial) to the two-phase system. Consequently, the Δlog K value is diminished or is even completely abolished if a proper amount of the competing effector is added. The strength of competition can be quantified by determining the concentration of the free ligand which has diminished the effect of affinity partition completely (Kirchberger *et al.,* 1989).

TABLE IV

Effect of Salts and Divalent Cations on Affinity Partitioning of Prealbumin[a]

Buffer composition	Δlog K (relative)
10 mM Na phosphate, pH 7.0	100
Additives:	
50 mM NaCl	125
150 mM NaCl	103
50 mM NaJ	31
150 mM NaJ	3
50 mM NaSCN	12
150 mM NaSCN	0
10 mM Tris-HCl, pH 7.0	107
Additives:	
5 mM Ca^{2+}	131
5 mM Mg^{2+}	126

[a] The phase systems were composed of 10% dextran T500, 7.5% PEG 8000 (with or without 1.6% Remazol Yellow GGL-PEG), protein, and buffer of different composition. The affinity partitioning effect is given, as the Δlog K value relative to the corresponding value of a system containing 10 mM sodium phosphate, pH 7.0 (= 100%). From Birkenmeier *et al.* (1986a) with permission.

Many examples may be mentioned that illustrate the competitive effect of natural ligands in affinity partitioning of enzymes or other proteins containing dye ligands, e.g., ATP and kinases (phosphofructokinase, pyruvate kinase), $NADH/NAD^+$ or $NADPH/NADP^+$ and dehydrogenases (glucose-6-phosphate dehydrogenase, lactate dehydrogenase, alcohol dehydrogenase, malate dehydrogenase, formate dehydrogenase), alkaline phosphatase and inorganic phosphate, ADP or AMP, prealbumin and L-thyroxine thyroxine binding protein and L-thyroxine. For example, the competitive effect of various natural ligands on affinity partitioning of alkaline phosphatase is shown in Fig. 6 (Kopperschläger *et al.,* 1988). Metal chelate-protein interactions are effectively disturbed by addition of EDTA or other metal complexing substances causing a decrease of Δlog K.

Competitive studies with natural ligands make it possible to distinguish between specific or nonspecific interactions of the ligand–polymer with the protein, which is often helpful for understanding the mechanism of binding an artificial ligand. As a result, the best ligand may be selected or synthesized which fits well the binding site of the natural counterpart of an enzyme or which makes it possible to distinguish between multiple enzyme forms (Kirchberger *et al.,* 1991).

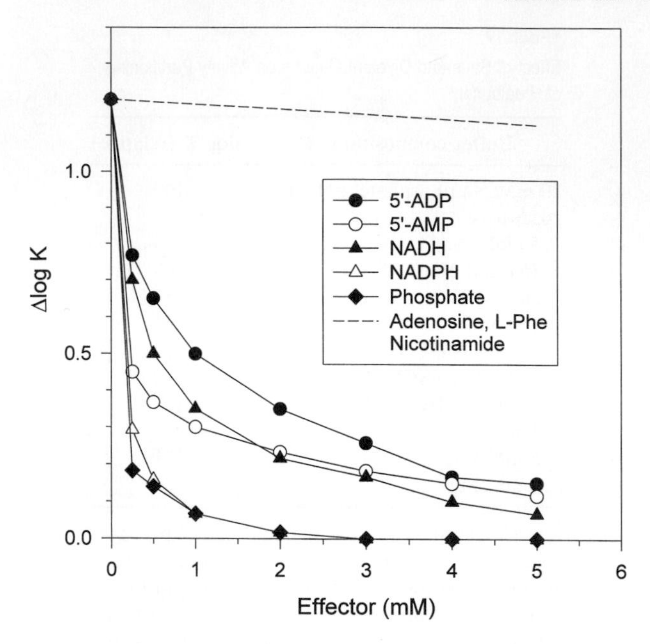

FIG. 6 Dependence of the affinity partition effect (Δlog K) of alkaline phosphatase on diverse effectors supplemented. Systems (4g) contain 6.5% (w/w) PEG 6000, 9.75g (w/w) dextran M70, 10 mM Tris/HCI buffer, pH 7.5, 2 mM $MgCl_2$, and 9 units enzyme. The concentration Red H-E3B-PEG was kept constant at 1%. The effector concentrations indicated refer to the whole system. The partition coefficients of the effectors were between 0.7 and 1.2. From Kopperschläger *et al.* (1988) with permission.

Another factor affecting affinity partitioning is the pH of the buffer. If the pK values of the polymer-bound ligands are outside of the pH range of the buffer used, which is often the case, the change in the net charge of the proteins will lead to an alteration of the binding strength. As a rule, an increase in the pH reduces the affinity partitioning effect.

The influence of the concentration of the phase-forming polymers on the effect of affinity partitioning was thoroughly studied with diverse yeast enzymes (Johansson and Andersson, 1984; Johansson *et al.,* 1983). Generally, with increasing concentrations of both polymers, the Δlog K rises leading to a maximal value.

The temperature has also been found to affect affinity partitioning. In PEG/dextran systems containing dye ligand, the effect of affinity partitioning is decreased with increasing temperature (Johansson and Andersson, 1984). Similar results were found at sub-zero temperatures (Johansson and Kopperschläger, 1987; Grimonprez and Johannson, 1996).

B. Conformational Changes

Since affinity partitioning is known to be a rather sensitive method for recognizing even small changes in binding forces between a ligand and a protein, conformation changes of proteins which involve the binding domain of the ligand should be recognized by alteration of the partition coefficient. Indeed, the conformation change in yeast phosphofructokinase after addition of the substrate fructose 6-phosphate can be followed in an aqueous two-phase system containing Cibacron Blue F3G-A-PEG (Kopperschläger and Birkenmeier, 1990). Similarly, the process of desensitization of ATP inhibition of this enzyme and its reversibility have been observed, as demonstrated in Fig. 7 (Kopperschläger and Johansson, 1985; Kopperschläger *et al.*, 1988).

The change of surface hydrophobicity of α_2-macroglobulin and pregnancy zone binding protein resulting in conformation changes after complexing

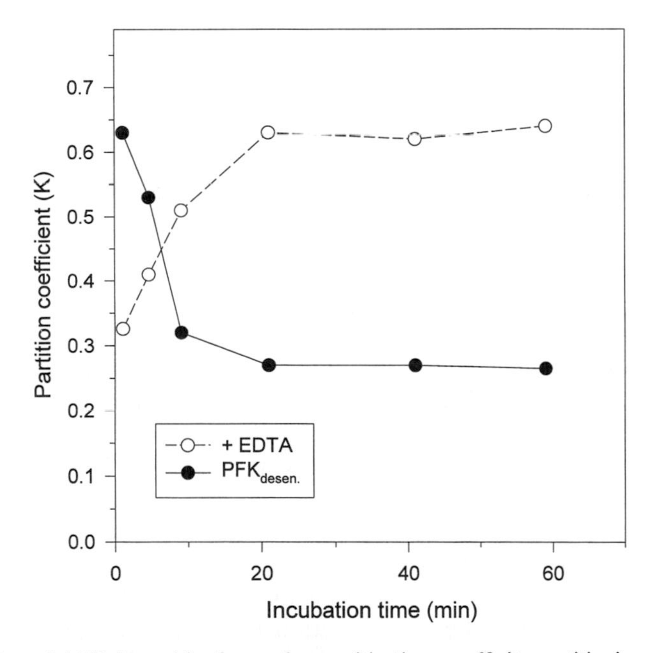

FIG. 7 Effect of ATP desensitization and resensitization on affinity partitioning of phosphofructokinase (PFK). The system contained 5% (w/w) PEG 6000 and 7.5% (w/w) dextran M 70 in 50 mM imidazole/HCl buffer, pH 6.8. 0.05% of the PEG were replaced by Procion Red H-E7B-PEG. For providing desensitization of the enzyme, $MgCl_2$ (1 mM), ADP (0.1 mM), fructose 6-phosphate (0.1 mM), NH_4Cl (10 mM), and NaF (10 mM) were added. The final concentration of PFK was 10 μg/ml. For resensitization, EDTA (5 mM) was added. From Kopperschläger *et al.* (1988) with permission.

with a proteinase or reaction of α_2-macroglobulin with methylamine yielding the transformed protein have been followed in aqueous two-phase systems containing palmitoyl-PEG (Jensen *et al.,* 1994; Birkenmeier *et al.,* 1989). With both proteins, a 2- to 3-fold increase in the degree of hydrophobicity has been observed as measured by the change of the partition coefficient. The change in the surface hydrophobicity parallels alterations in receptor binding properties of the derivatized forms of α_2-macroglobulin and could be a signal for binding to cell-surface receptors, followed by internalization.

Experiments using palmitoyl-PEG in aqueous two-phase systems have shown that binding steroidal and nonsteroidal ligands to the estrogen receptor or elevated temperature *in vitro* is associated with a conformation change in the protein followed by a reduction of the hydrophobic characteristics (Hansen and Gorski, 1986).

Changes in the hydrophobicity of α-lactalbumin were followd in aqueous two-phase systems containing various acyl-PEGs after addition of Ca^{2+}, and by variation of the pH (Shanbhag *et al.,* 1991). The Δlog K is strongly influenced by both parameters. The results are discussed in terms of conformational changes due to Ca^{2+}-binding and chelating of Ca^{2+}.

VII. Affinity Partitioning of Cell Membranes and Cells

A. General Remarks

More complex biomaterial such as membranes, organelles, and cells are often characterized by specific receptors present as integral proteins on the surfaces or attached as marker enzymes. The essential point in affinity partitioning of these particulates is to find an appropriate ligand which recognizes the marker molecules following complex formation with the ligand-polymer.

There are several factors to consider when adapting affinity partitioning to such complex structures. One is the partitioning behavior in conventional aqueous two-phase systems and how various parameters will influence this. They are the kind and the concentration of the phase-forming polymers and the ionic strength, whereby salt ions exhibiting different preferences for the two phases create an electrostatic potential difference and attract membranes in different ways. Another factor is the choice of the right ligand to achieve sufficient selectivity. A third factor to consider is the coupling procedure of the ligand to the phase while maintaining the biological function, and which kind of polymer should be used as carrier.

Biospecific ligands used so far in affinity partitioning of membrane fractions and cells include lectins, antibodies, receptor agonists, and biomimetic dyes. Typical ligands are listed in Table V. It is still unclear whether certain dyes react biospecifically with surface structures or interact as common hydrophobic ligands.

In comparison with proteins, affinity partitioning of complex particulates is less well understood and little is known about the requirements for successful separation in terms of number and density of binding sites required for transfer of these biomaterials from one phase into the other. Another factor concerns the reversibility of interaction without destroying the biological function of the particulate. This problem has not been solved satisfactorily.

B. Affinity Ligands for Plasma Membranes

The lectin wheat-germ agglutinin (WGA) has found application in the enrichment of plasma membranes (Persson *et al.,* 1991; Persson and Jergil, 1992). This affinity ligand recognizes specific carbohydrate residues and

TABLE V

Polymer-Ligand Combinations used in Affinity Partitioning of Cells and Membranes[a]

Polymer-ligand	Application	References
PEG-antibody	Erythrocytes	Karr *et al.* (1986); Sharp *et al.* (1986); Delgado *et al.* (1991)
PEG-secondary ligand-antibody	Erythrocytes	Karr *et al.* (1988); Stocks and Brooks (1988)
PEG-chelating agent	Erythrocytes, Mononuclear cells	Nanak *et al.* (1995); Botros *et al.* (1991); Laboureau *et al.* (1996)
PEG-amine-derivatives	Electroplax domains Synaptic membrane domains	Flanagan *et al.* (1976); Johansson *et al.* (1981); Olde and Johansson (1985, 1989)
PEG-naloxone	Synaptic membrane domains	Olde and Johansson (1985)
PEG-Procion Yellow H-E3G	Synaptic membrane domains	Muiño Blanco *et al.* (1986)
PEG-transferrin	Reticulocytes	Delgado *et al.* (1992)
Dextran-Procion Yellow H-E3G	Synaptic membrane domains	Muiño Blanco *et al.* (1991); Cebrian-Pérez *et al.* (1991)
Dextran-wheat germ agglutinin	Plasma membranes	Persson *et al.* (1991); Persson and Jergil (1992)

[a] From Persson and Jergil (1995).

therefore is capable of binding glycoproteins and glycolipids integrated in the bilayer of a cell membrane.

Fortunately, the lectin–glycoprotein complex can be dissociated by addition of a carbohydrate competing with the lectin binding sites. One example of membrane affinity partitioning is given in Fig. 8, in which rat liver membranes are partitioned in a PEG/dextran system containing the lectin-dextran in the bottom phase. In the absence of the affinity ligand, most of the membranes are concentrated in the top phase as indicated by the enzyme marker alkaline phosphodieserase (APDE) and 5′-nucleotidase (5′-N). With increasing WGA-dextran, the plasma membranes of both apical and basolateral domains were pulled into the bottom phase, resulting in a final to 30- to 40-fold purification (Persson and Jergil, 1992).

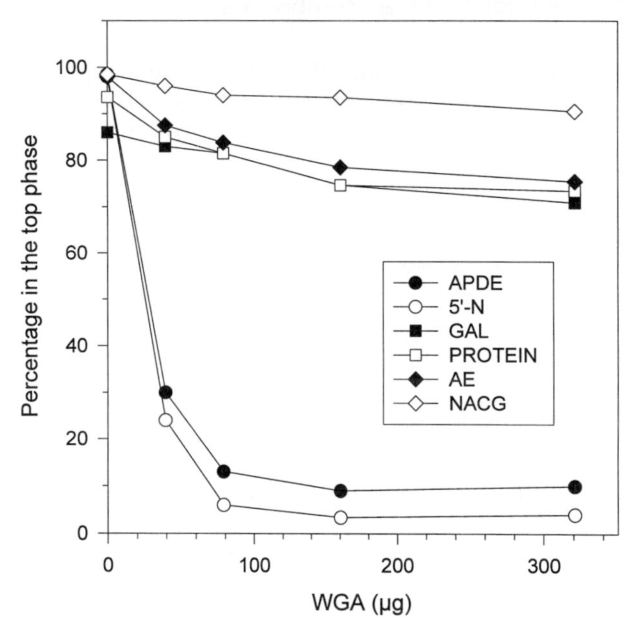

FIG. 8 Affinity partitioning of rat liver membranes. The system contained 6% (w/w) PEG 3350, 6% dextran, and increasing amounts of wheat-germ dextran. The protein concentration in the top phase before adding the bottom phase was 4 mg/g. After two reextractions of the wheat-germ agglutinin containing bottom phase, the protein content and the activities of alkaline phosphodiesterase (APDE), 5′-nucleotidase (5′-N), galactosyltransferase (GAL), arylesterase (AE), and N-acetylglucosaminidase (NACG) were measured in the combined top phases and the final bottom phase. They are expressed as percentages in the combined top phases of the total activity in the phases. Recoveries were 80–105% of the starting material. From Persson and Jergil (1992) with permission.

C. Affinity Ligands for Synaptic Membrane Domains

Affinity partitioning has been proven useful for separation of synaptic membranes obtained by fragmentation of synaptosomes from different sources (Johansson *et al.,* 1981, 1984; Olde and Johansson, 1985, 1989; Muiño Blanco *et al.,* 1986). The target membranes are characterized by specific receptor binding sites which can be recognized by natural or artificial ligands. For example, hexaethonium-PEG was used to study the heterogeneity of crude synaptosomal preparations (Olde and Johansson, 1989). Synaptic membrane domains from *Torpedo california* electroplax containing nicotinic cholinergic receptor molecules are separated by countercurrent distribution in the presence of diverse bis-quaternary amines coupled to PEG (Flanagan *et al.,* 1976). Naloxone and naltrexone, two ligands for the opiate receptor, have been applied after coupling to PEG for affinity partitioning of membranes from calf brain cortex yielding a selective extraction of membranes containing the receptor (Olde and Johansson, 1985).

In another approach to affinity partitioning, the reactive dye Procion Yellow H-E3G, coupled either to PEG or dextran, was applied to the separation of membrane fractions using a ficoll-dextran/PEG system (Muiño Blanco *et al.,* 1986, 1991; Cebrian-Pérez *et al.,* 1991). The dye is capable of extracting a part of the membranes into the upper phase (when bound to PEG) or into the lower phase (when bound to dextran). However, the chemical background of the interaction of the dye with the membranes still remains unclear. Nucleotides which frequently bind to membrane fragments due to the existence of specific binding domains do not compete with the polymer-bound dye.

D. Affinity Ligands for Separation of Cells

The introduction of aqueous two-phase systems for the separation of different cell populations is well documented (Walter, 1985, 1994). To increase selectivity in cell separation, affinity ligands recognizing surface receptors were coupled to one of the phase-forming polymers, causing a ligand-directed transfer of the cells into that phase where the ligand is present. However, the feasibility of extraction and separation of cells is often interfered with by certain factors which generally govern the distribution of cells in aqueous two-phase systems (see above).

The first experiment to separate erythrocytes obtained from different species by application of palmitoyl-PEG was reported by Eriksson *et al.* (1976). Although this hydrophobic ligand is believed to react unspecifically, two groups of cells could be distinguished, one consisting of cells from the dog, guinea pig, and rat, the other from human, sheep, and rabbit.

Immunoaffinity cell partitioning has been developed by several groups providing higher selectivity and specificity (Sharp *et al.,* 1986; Karr *et al.,* 1986; Delgado *et al.,* 1991, 1992). As a rule, the antibodies of choice are coupled to PEG, either directly after preceding activation (see Section III.C.), via a protein A linker (Karr *et al.,* 1988) or by using a monoclonal antibody–antibody bridge which is coupled first to the PEG polymer. One example is presented in Fig. 9 (Stocks and Brooks, 1988). A monoclonal mouse anti rabbit IgG-PEG, directed against the Fc fragment is used as affinity ligand in a system containing PEG and dextran to increase the partition of rabbit anti-*N,N*-glycophorin into the top phase specifically. The latter interacts with *N,N*-glycophorin on the surface of the human erythrocytes and results in a transfer of these cells into the top phase. The same system has no effect on the partition of rabbit erythrocytes. The advantage of this system for affinity partitioning is its versatility of use because only the rabbit bridge antibody has to be exchanged for other target cells.

Transferrin, bound to monomethoxy-PEG, was found to be an alternative ligand for affinity partitioning of reticulocytes (Delgado *et al.,* 1992). The interaction of the ligand polymer with transferrin receptors present on the

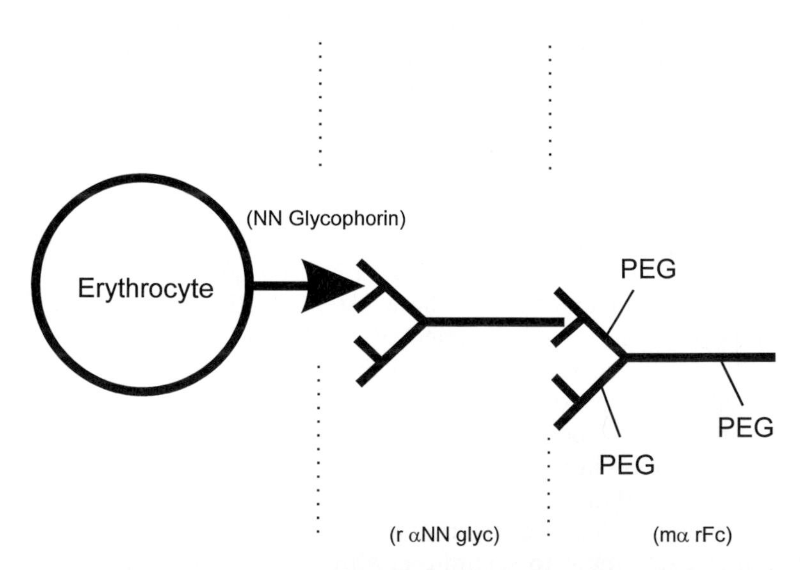

FIG. 9 A schematic representation of PEG 1900-modified monoclonal mouse anti-rabbit IgG, Fc fragment, IgG (mα rFc) bound to a human erythrocyte via rabbit anti-*N,N* glycophorin (r αNN glyc), thereby immunospecifically attaching PEG 1900 to a human erythrocyte. From Stocks and Brooks (1988) with permission.

surface of these cells leads to an enrichment of reticulocytes in the top phase. The specificity of this system was proven by adding unbound transferrin, which quenches the affinity partition effect. However, in comparison with immuno affinity partitioning, the system can be disturbed by other protein components such as albumin.

Metal-IDA-PEG has been proven effective for the segregation of normal and pathological human red blood cells, lymphocytes, and fibroblasts (Nanak *et al.,* 1995), and for separation of mononuclear cells from cord blood (Laboureau *et al.,* 1996). The partition of B cells, T cells, monocytes, and stem cells is varied by applying different metal chelate complexes coupled to PEG. Erythrocytes from different species can be also distinguished by metal chelate affinity partitioning (Botros *et al.,* 1991). A PEG/dextran system containing Cu^{2+}-IDA-PEG is quite effective in extraction of human and rabbit erythrocytes while systems containing Zn^{2+}-IDA-PEG display significant affinity only to the rabbit cells. There are some questions as to the specificity of these ligands as applied to cell separations, however (Walter *et al.,* 1993).

Another approach to affinity partitioning is provided by affinity-mediated modification of the electric charge on a cell surface (Mendieta and Johansson, 1992). Incubation of rat erythroblasts with a transferrin-polylysine derivative affects the partition of these cells in a charge-sensitive aqueous two-phase system composed of PEG and dextran. The change in the partition of the cells due to the conjugate causes a 15-fold decrease in the partition ratio. The polylysine part introduces a nonspecific influence on the partitioning that can be eliminated by preincubation of the cells with an excess of sialic acid.

Hybridoma cells have been separated from their IgG product by exploiting affinity partitioning in aqueous two-phase systems (Zijlstra *et al.,* 1996). Since systems composed of PEG and dextran are not capable of sufficiently separating hybridoma cells from thcir IgG, the introduction of biomimetic dyes coupled to PEG results in a significant separation, the cells being concentrated in the bottom phase, while the IgG, which interacts with the dye, transfers into the top phase. The partition coefficient of IgG was found to increase by three orders of magnitude, resulting in a 25-fold higher IgG concentration in the top phase than in the bottom phase.

VIII. Affinity Partitioning of Nucleic Acids

There are only few reports on using affinity partitioning for the separation of nucleic acids. A number of selective base pair-specific synthetic dyes of different structures have been proven suitable in various affinity separation techniques (Müller and Crothers, 1975; Müller and Gautier, 1975; Müller and Kütemeier, 1982; Müller *et al.,* 1982, Müller, 1985). DNA fractionation by affinity partitioning in systems containing PEG/dextran by introducing a GC-specific phenazinium dye covalently attached to PEG was reported by Müller and Eigel (1981). The partition coefficient of DNA increases as a function of its GC- content. However, the specificity of partitioning strongly depends on the ionic strength, the temperature, and on the length of the PEG chain attached to the ligand. Other DNA binding molecules, which are shown in Fig. 10, have been tested as ligands for the separation of DNA fragments, applying the technique of liquid–liquid affinity partition chromatography (Müller 1985). One example is given in Fig. 11, which demonstrates the separation of two DNA restriction fragments differing only by 7.7% in the AT-content. Although this study shows convincing results, a further extension of this approach has yet to be carried out.

Affinity partitioning of nucleic acids has been attempted by using diverse homologs of bis-quaternary amines coupled to PEG (Johansson *et al.,* 1985). The partition of RNA in an aqueous two-phase systems containing 11% ficoll 600, 7% dextran 500, and 0.125% PEG (with or without the ligand) is not affected by any of the ligands tested. On the other hand, the higher molecular weight DNA (calf thymus) shows higher affinity for the dextran-rich top phase. The ligand-PEG concentrated in the ficoll-rich bottom phase reduces the partition coefficient of DNA. As a result, about 40% of DNA from the top phase is pulled into the bottom phase.

A new approach to affinity partitioning was described by Jaschke *et al.* (1994), called hybridization affinity partitioning of nucleic acids using PEG-coupled oligonucleotides as ligands. This reaction principle is based on the hybridization of two single complementary nucleic acid strands. The oligonucleotide-PEG conjugate is included in a system of PEG/dextran, where the target nucleic acid is enriched in the absence of the conjugate in the dextran-rich bottom phase. As a result of hybridization of the two strands, a drastic increase in the K value of the DNA is observed. The system has been tested by hybridization of a labeled 18-mer oligonucleotide (sense DNA) with a 10- to 100-fold excess of PEG-coupled antisense strand. This led to a change of the partition coefficient from 0.08 to 0.81. However, the extent of the extraction depends on the chain length of the PEG. The utility of this system has been also proven by liquid–liquid affinity partition chromatography. The calculated number of the theoretical plates N was found to be 10,000 plates per meter (Jaschke *et al.,* 1994).

IX. Concluding Remarks

Affinity partitioning of proteins in aqueous two-phase systems has found comprehensive application in protein purification. Particularly, the applica-

Macroligand	Chemical structure	Color	Specific for	Affinity k (l/mol) I = 0.1, 37 °C	Molar extinction coefficient (nm)
I	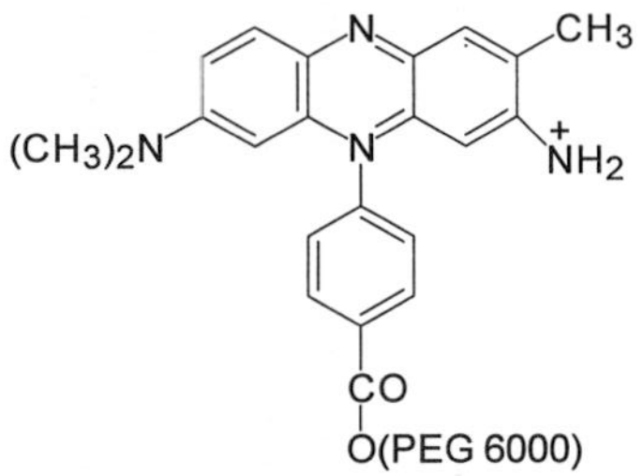 $(CH_3)_2N$, N, N, CH_3, $\overset{+}{N}H_2$, CO, O(PEG 6000)	Red	G C G C and G C C G	7.4×10^3	$44{,}700_{555}$
II	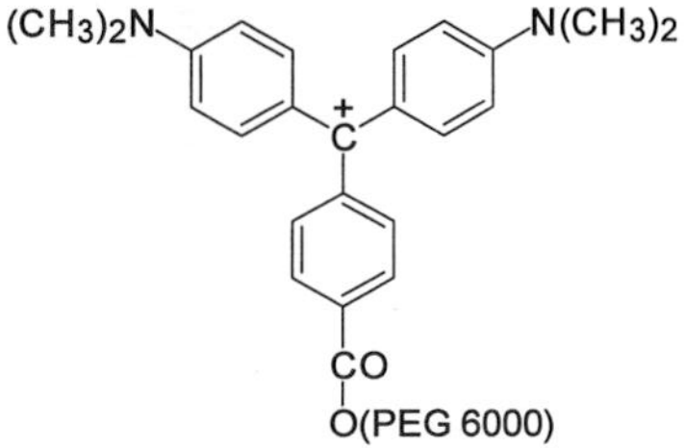 $(CH_3)_2N$, $N(CH_3)_2$, $\overset{+}{C}$, CO, O(PEG 6000)	Green	A T T A	6.2×10^3	$89{,}000_{637}$
III	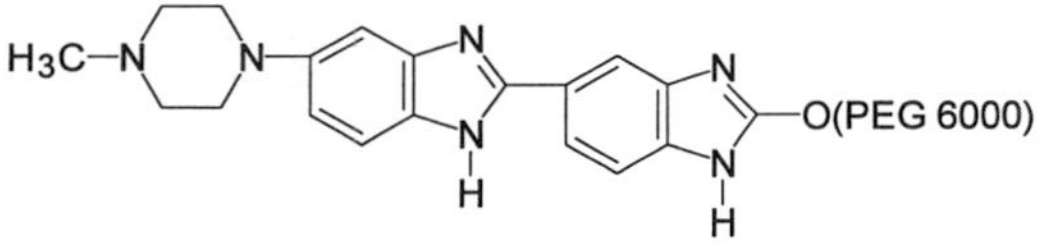 H_3C–N, N, N, N–H, N, N–H, O(PEG 6000)	Yellow	A T A T A T	2.5×10^5	$38{,}000_{338}$

FIG. 10 Structure and physical properties of DNA-binding macroligands. From Müller (1985).

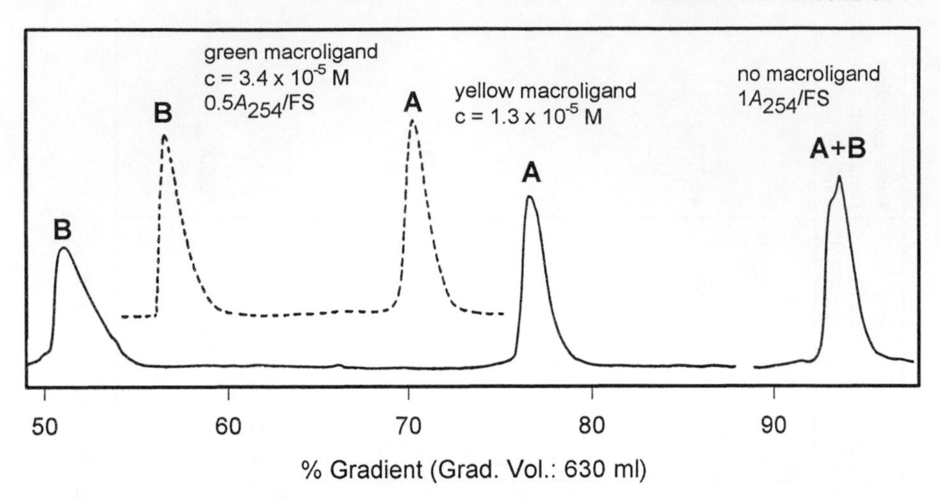

FIG. 11 Separation of the two halves of pBR322 plasmid on a PEG/dextran two-phase column according to (A+T) content by use of base- and sequence-specific macroligands (shown in Table VI). The two fragments were obtained by action of the restriction enzymes *Eco*RI and *Pvu*II. Fragment A is 2069 base pairs and fragment B is 2293 base pairs in size; A and B differ by 7% in (A+T) content. The separations were performed in the PEG/dextran system on a column of 35 × 1.5 cm (flow rate, 15 ml/h, 37°C) with cellulose CC31 (Whatman) as phase support, applying the same gradient from 0.2 M sodium acetate to 0.2 M lithium acetate for all three runs. The macroligand concentrations given in the figure indicate their concentrations in the mobile, PEG-rich phase. From Müller (1985) with permission.

tion of pseudo-biospecific dyes as general ligands in the liquid–liquid extraction of many proteins, including enzymes and plasma proteins, is well documented. Moreover, affinity partitioning has been elaborated as a sensitive tool for the investigation of ligand–protein interaction and for screening ligands. Small changes in the interaction caused by conformation changes in proteins are reflected by the alteration of the respective partition coefficients.

In addition, cell fragments, cells, and nucleic acids are also separated by applying affinity partitioning. In that case, the ligands of choice are antibodies or membrane receptors. The present theoretical approach to this principle does not make it possible to explain the partitioning of the macromolecules and other components comprehensively. A deeper consideration and the development of more selective ligands for various macromolecules will open up the great potential in analytic biochemistry and for practical biotechnology.

References

Abuchowski, T., Van Es, N., Palczuk, C., and Davis, F. F. (1977). Alteration of immunological properties of bovine serum albumin by covalent attachment of polyethylene glycol. *J. Biol. Chem.* **252,** 3578–3581.

Albertsson, P.-Å. (1971). "Partition of Cell Particles and Macromolecules." Wiley, New York.

Alred, P. A., Johansson, G., and Tjerneld, F. (1992). Interactions in affinity partition studied using fluorescence spectroscopy. *Anal. Biochem.* **205,** 351–358.

Arnold, F. H. (1991). Metal-affinity separations—a new dimension in protein processing. *Bio/Technology* **9,** 151–156.

Backman, L., and Johansson, G. (1976). Enzyme-enzyme interaction between aspartate amino-transferase and malate dehydrogenase from pig heart muscle. *FEBS Lett.* **65,** 39–43.

Baskir, J. N., Hatton, T. A., and Suter, U. W. (1989). Protein partitioning in two-phase aqueous polymer systems. *Biotechnol. Bioeng.* **34,** 541–558.

Birkenmeier, G., and Kopperschläger, G. (1987). Interaction of the dye Remazol Yellow GGL to prealbumin and albumin by affinity partition, difference spectroscopy and equilibrium dialysis. *Mol. Cell. Biochem.* **73,** 99–110.

Birkenmeier, G., Tschechonien, B., and Kopperschläger, G. (1984a). Affinity chromatography and affinity partitioning of human serum prealbumin using immobilized Remazol Yellow GGL. *FEBS Lett.* **174,** 162–166.

Birkenmeier, G., Usbeck, E., and Kopperschläger, G. (1984b). Affinity partitioning of albumin and alpha-fetoprotein in an aqueous two-phase system using poly(ethylene glycol)-bound triazine dyes. *Anal. Biochem.* **136,** 264–271.

Birkenmeier, G., Kopperschläger, G., and Johansson, G. (1986a). Separation and studies of serum proteins with aid of aqueous two-phase systems containing dyes as affinity ligands. *Biomed. Chromatogr.* **1,** 64–78.

Birkenmeier, G., Kopperschläger, U., Ehrlich, G., and Appelt, G. (1986b). Partition of purified human thyroxine-binding globulin in aqueous two-phase systems in response to reactive dyes. *J. Chromatogr.* **360,** 193–201.

Birkenmeier, G., Shanbhag, V. P., and Kopperschläger, G. (1986c). Comparison of the binding of poly (ethylene glycol)-bound fatty acids to human albumin and alpha-fetoprotein studied by affinity phase partitioning. *Biomed. Biochim. Acta* **45,** 285–289.

Birkenmeier, G., Kopperschläger, G., Albertsson, P.-Å., Johansson, G., Tjerneld, F., Åkerlund, H.E., Berner, S., and Wickstroem, H. (1987). Fractionation of proteins from human serum by counter-current distribution. *J. Biotechnol.* **5,** 115–129.

Birkenmeier, G., Carlsson-Bostedt, L., Kriegel, V. Shanbhag, T., Kopperschläger, G., Sottrup-Jensen, L., and Stigbrand, T. (1989). Differences in hydrophobic properties for human alpha2-macroglobulin and pregnancy zone protein as studied by affinity phase partitioning. *Eur. J. Biochem.* **183,** 239–243.

Birkenmeier, G., Vijayalakshmi, M. A., Stigbrand, T., and Kopperschläger, G. (1991). Immobilized metal ion affinity partitioning, a method combining metal-protein interaction and partitioning of proteins in aqueous two-phase systems. *J. Chromatogr.* **539,** 267–277.

Botros, H. G., Birkenmeier, G., Otto, A., Kopperschläger, G., and Vijayalakshmi, M. A. (1991). Immobilized metal ion affinity partitioning of cells in aqueous two-phase systems: Erythrocytes as a model. *Biochim. Biophys. Acta* **1074,** 69–73.

Brooks, D. E., Sharp, K. A., and Fisher, D. (1985). Theoretical aspects of partitioning. *In* "Partitioning in Aqueous Two-phase Systems: Theory, Methods, Uses, and Applications to Biotechnology" (H. Walter, D. E. Brooks, and D. Fisher, eds.), pp. 11–84. Academic Press, San Diego, CA.

Bückmann, A. F., Kula, M.-R., Wichmann, R., and Wandrey, C. (1981). An efficient synthesis of higher molecular weight NAD(H) derivatives suitable for continuous operation with coenzyme dependent enzyme systems. *J. Appl. Biochem.* **3,** 301–315.

Bückmann, A. F., Morr, M., and Kula, M.-R. (1987). Preparation of technical grade polyethylene glycol (Mr 20.000)-N6-(2-aminoethyl)-*N*ADH by a procedure adaptable to large-scale synthesis. *Biotechnol. Appl. Biochem.* **9,** 258–268.

Carlsson, M., Berggren, K., Linse, P., Veide, A., and Tjerneld, F. (1996). Effects of fused tryptophan-rich peptides to a recombinant protein A domain on the partitioning in polyethylene glycol-dextran and Ucon-dextran aqueous two-phase systems. *J. Chromatogr.* **A 756,** 107–117.

Cebrian Pérez, J. A., Muiño Blanco, M. T., and Johansson, G. (1991). Heterogeneity of synaptosomal membrane preparations from different regions of calf brain studied by partitioning and counter-current distribution. *Int. J. Biochem.* **23,** 1491–1495.

Chaabouni, A., and Dellacherie, E. (1979). Affinity partitioning of proteins in aqueous two-phase systems containing polyoxyethylene glycol-bound ligand and charged dextrans. *J. Chromatogr.* **171,** 135–143.

Chung, B. H., and Arnold, F. H. (1991). Metal-affinity partitioning of phosphoproteins in PEG dextran 2-phase systems. *Biotechnol. Lett.* **13,** 615–620.

Chung, B. H., Sohn, J. H., Rhee, S. K., Chang, Y. K., and Park, Y. H. (1994). Enhanced metal-affinity partitioning of genetically engineered hirudin variants in polyethylene glycol dextran 2-phase systems. *J. Ferment. Bioeng.* **77,** 75–79.

Clonis, Y. D., Atkinson, T., Bruton, C. J., and Lowe, C. R. (1987). "Reactive Dyes in Protein and Enzyme Technology." Stockton Press, New York.

Cordes, A., and Kula, M.-R. (1986). Process design for large-scale purification of formate dehydrogenase from *Candida boidinii* by affinity partitioning. *J. Chromatogr.* **376,** 375–384.

Cordes, A., Flossdorf, J., and Kula, M.-R. (1987). Affinity partitioning: Development of mathematical model describing behavior of biomolecules in aqueous two-phase systems. *Biotechnol. Bioeng.* **30,** 514–520.

Delgado, C., Tejedor, M. C., and Luque, J. (1990). Partial purification of glucose 6-phosphate dehydrogenase and phosphofructokinase from rat erythrocyte haemolysate by partitioning in aqueous two-phase systems. *J. Chromatogr.* **498,** 159–168.

Delgado, C., Anderson, R. J., Francis, G. E., and Fisher, D. (1991). Separation of cell mixtures by immunoaffinity cell partitioning: Strategies for low abundance cells. *Anal. Biochem.* **192,** 322–328.

Delgado, C., Sancho, P., Mendieta, J., and Luque, J. (1992). Ligand receptor interactions in affinity cell partitioning—studies with transferrin covalently linked to monomethoxypoly (ethylene glycol) and rat reticulocytes. *J. Chromatogr.* **594,** 97–103.

Elling, L., and Kula, M. R. (1991). Immunoaffinity partitioning—synthesis and use of polyethylene glycol oxirane for coupling to bovine serum albumin and monoclonal antibodies. *Biotechnol. Appl. Biochem.* **13,** 354–362.

Eriksson, E., Albertsson, P.-Å., and Johansson, G. (1976). Hydrophobic surface properties of erythrocytes studies by affinity partition in aqueous two-phases systems. *Mol. Cell. Biochem.* **10,** 123–128.

Flanagan, S. D., Barondes, S. H., and Taylor, P. (1976). Affinity partitioning of membranes. Cholinergic receptor-containing membranes from *Torpedo californica. J. Biol. Chem.* **251,** 858–865.

Flanagan, S. D., and Barondes, J. H. (1975). Affinity parition. *J. Biol. Chem.* **250,** 1484–1489.

Giuliano, K. A. (1991). Aqueous two-phase protein partitioning using textile dyes as affinity ligands. *Anal. Biochem.* **197,** 333–339.

Grimonprez, B., and Johansson, G. (1996). Liquid-liquid partitioning of some enzymes, especially phosphofructokinase from *Saccharomyces cerevisiae* at sub-zero temperature. *J. Chromatogr. B: Biomed. Appl.* **680,** 55–63.

Hansen, J. C., and Gorski, J. (1986). Conformational transitions of the estrogen receptor monomer. *J. Biol. Chem.* **261,** 13990–13996.

Hubert, P., Dellacherie, E., Neel, J., and Baulieu, E. E. (1976). Affinity partitioning of steroid binding proteins. The use of polyethylene oxide-bound estradiol for purifying $A_5 \rightarrow {}_4 3-$ oxosteroid isomerase. *FEBS Lett.* **65,** 169–174.

Huse, K., Himmel, M., Birkenmeier, G., Bohla, M., and Kopperschläger, G. (1983). A novel purification procedure for human alpha-fetoprotein. *Clin. Chim. Acta* **133,** 335–340.

Jaschke, A., Furste, J. P., Erdmann, V. A., and Cech, D. (1994). Hybridization-based affinity partitioning of nucleic acids using PEG-coupled oligonucleotides. *Nucleic Acids Res.* **22,** 1880–1884.

Jensen, P. E. H., Hagglof, E. M., Arbelaez, L. F., Stigbrand, T., and Shanbhag, V. P. (1993). Comparison of conformational changes of pregnancy zone protein and human alpha2-macroglobulin, a study using hydrophobic affinity partitioning. *Biochim. Biophys. Acta* **1164,** 152–158.

Jensen, P. E. H., Stigbrand, T., and Shanbhag, V. P. (1994). Use of hydrophobic affinity partitioning as a method for studying various conformational states of the human alpha2-macroglobulins. *J. Chromatogr.* **668,** 101–106.

Johansson, G. (1970). Studies on aqueous dextran-poly(ethylene glycol) two-phase systems containing charged poly(ethylene glycol). 1. Partitioning of albumin. *Biochim Biophys. Acta* **222,** 381–389.

Johansson G. (1984). Affinity partitioning. *In* "Methods in Enzymology" (W. B. Jakoby, ed.), Vol. 104, Part C, pp. 356–363. Academic Press, New York.

Johansson, G., and Andersson, M. (1984). Parameters determining affinity partitioning of yeast enzymes using polymer-bound triazine dye ligands. *J. Chromatogr.* **303,** 39–51.

Johansson, G., and Joelsson, M. (1985a). Preparation of Cibacron Blue F3G-A-poly(ethylene glycol) in large scale for use in affinity partitioning. *Biotechnol. Bioeng.* **27,** 621–625.

Johansson, G., and Joelsson, M. (1985b). Partial purification of D-glucose 6-phosphate dehydrogenase from baker's yeast by affinity partitioning using polymer-bound triazine dyes. *Enzyme Microb. Technol.* **7,** 629–634.

Johansson, G., and Joelson, M. (1986). Liquid-liquid extraction of lactate dehydrogenase from muscle using polymer-bound triazine dyes. *Appl. Biochem. Biotechnol.* **13,** 15–27.

Johansson, G., and Kopperschläger, G. (1987). Effects of organic solvents on partitioning of enzymes in aqueous two-phase systems. *J. Chromatogr.* **388,** 295–305.

Johansson, G., and Shanbhag, V. P. (1984). Affinity partitioning of proteins in aqueous two-phase systems containing polymer-bound fatty acids. I. Effect of polyethylene glycol palmitate on the partition of human serum albumin and alpha-lactalbumin. *J. Chromatogr.* **284,** 63–72.

Johansson, G., Hartmann, A., and Albertsson, P.-Å. (1973). Partition of proteins in two-phase systems containing charged poly(ethylene glycol). *Eur. J. Biochem.* **33,** 379–386.

Johansson, G., Gysin, R., and Flanagan, S. D. (1981). Affinity partitioning of membranes. *J. Biol. Chem.* **256,** 9126–9135.

Johansson, G., Kopperschläger, G., and Albertsson, P.-Å. (1983). Affinity partitioning of phosphofructokinase from baker's yeast using polymer bound Cibacron Blue F3G-A. *Eur. J. Biochem.* **131,** 589–594.

Johansson, G., Åkerlund, H.-E., and Olde, B. (1984). Liquid-liquid extraction of membranes from calf brain using conventional and centrifugal counter-current distribution techniques. *J. Chromatogr. B: Biomed. Appl.* **311,** 277–289.

Johansson, G., Joelsson, M., Olde, B., and Shanbhag, V. P. (1985). Affinity partitioning of biopolymers and membranes in ficoll-dextran aqueous two-phase systems. *J. Chromatogr.* **331,** 11–21.

Karr, L. J., Shafer, S. G., Harris, J. M., Van Alstine, J. M., and Snyder, R. S. (1986). Immunoaffinity partition of cells in aqueous polymer two-phase systems. *J. Chromatogr.* **354,** 269–282.

Karr, L. J., Van Alstine, J. M., Snyder, R. S., Shafer, S. G., and Harris, J. M. (1988). Cell separation by immunoaffinity partitioning with polyethylene glycol-modified protein A in aqueous polymer two-phase systems. *J. Chromatogr.* **442,** 219–227.

Kirchberger, J., Cadelis, F., Kopperschläger, G., and Vijayalakshmi, M. A. (1989). Interaction of lactate dehydrogenase with structurally related triazine dyes using affinity partitioning and affinity chromatography. *J. Chromatogr.* **483,** 289–299.

Kirchberger, J., Kopperschläger, G., and Vijayalakshmi, M. A. (1991). Dye-ligand affinity partitioning of lactate dehydrogenase isoenzymes. *J. Chromatogr.* **557,** 325–334.

Kirchberger, J., Domar, U., Kopperschläger, G., and Stigbrand, T. (1992). Interactions of human alkaline phosphatase isoenzymes with triazine dyes using affinity partitioning, affinity chromatography and difference spectroscopy. *J. Chromatogr.* **574,** 237–245.

Köhler, K., Ljungquist, T., Kondo, A., Veide, A., and Nilsson, B. (1991a). Engineering proteins to enhance their partition coefficient in aqueous two-phase systems. *Bio/Technology* **9,** 642–646.

Köhler, K., Veide, A., and Enfors, S. O. (1991b). Partitioning of beta-galactosidase fusion proteins in PEG/potassium phosphate aqueous 2-phase systems. *Enzyme Microb. Technol.* **13,** 204–209.

Kopperschläger, G., and Birkenmeier, G. (1990). Affinity partitioning and extraction of proteins. *BioSeparation* **1,** 235–254.

Kopperschläger, G., and Birkenmeier, G. (1993). Affinity partitioning of biomolecules in aqueous two-phase systems. *In* "Molecular Interactions in BioSeparations" (T. T. Ngo, ed.), pp. 499–509. Plenum, New York.

Kopperschläger, G., and Johansson, G. (1982). Affinity partitioning with polymer-bound Cibacron Blue F3G-A for rapid large-scale purification of phosphofructokinase from baker's yeast. *Anal. Biochem.* **124,** 117–124.

Kopperschläger, G., and Johansson, G. (1985). Studies on the ATP-sensitivity of yeast phosphofructokinase by means of affinity partitioning using polymer bound Cibacron Blue F3G-A. *Biomed. Biochim. Acta* **44,** 1047–1055.

Kopperschläger, G., and Lorenz, G. (1985). Interaction of yeast glucose 6-phosphate dehydrogenase with diverse triazine dyes. A studies by means of affinity partitioning. *Biomed. Biochim. Acta* **44,** 517–525.

Kopperschläger, G., Böhme, H.-J., and Hofmann, E. (1982). Cibacron Blue F3G-A and related dyes as ligands in affinity chromatography. *Adv. Biochem. Eng.* **25,** 101–138.

Kopperschläger, G., Lorenz, G., and Usbeck, E. (1983). Application of affinity partitioning in an aqueous two-phase system to the investigation of triazine dye-enzyme interactions. *J. Chromatogr.* **259,** 97–105.

Kopperschläger, G., Kirchberger, J., and Kriegel, T. (1988). Studies on triazine dye-enzyme interaction by means of affinity partitioning. *Macromol. Chem., Macromol. Symp.* **17,** 373–385.

Kroner, K. H., Cordes, A., Schelper, A., Morr, M., Bückmann, A. F., and Kula, M. R. (1982). Affinity partition studies with glucose 6-phosphate dehydrogenase in aqueous two-phase systems in response to triazine dyes. *In* "Affinity Chromatography and Related Techniques" (T. C. J. Gribneau, J. Visser, and R. J. F. Nivard, eds.), pp. 491–501. Elsevier, Amsterdam.

Laboureau, E., Capiod, J. C., Dessaint, C., Prin, L., and Vijayalakshmi, M. A. (1996). Study of human cord blood lymphocytes by immobilized metal ion affinity partitioning. *J. Chromatogr. B: Biomed. Appl.* **680,** 189–195.

Lowe, C. R., Burton, S. J., Burton, N. P., Alderton, W. K., Pitts, J. M., and Thomas, J. A. (1992). Designer dyes: "biomimetic" ligands for the purification of pharmaceutical proteins by affinity chromatography. *Trends Biotechnol.* **10,** 442–448.

Lu, M., and Tjerneld, F. (1997). Interaction between tryptophan residues and hydrophobically modified dextran, effect on partitioning of peptides and proteins in aqueous two-phase systems. *J. Chromatogr. A,* **766,** 99–108.

Lu, M., Tjerneld, F., Johansson, G., and Albertsson, P.-Å. (1991). Preparation of benzoyl dextran and its use in aqueous two-phase systems. *BioSeparation* **2,** 247–255.

Lu, M., Albertsson, P.-Å., Johansson, G., and Tjerneld, F. (1996). Ucon-benzoyl dextran aqueous two-phase systems: Protein purification with phase component recycling. *J. Chromatogr. B: Biomed. Appl.* **680,** 65–70.

Mendieta, J., and Johansson, G. (1992). Affinity-mediated modification of electrical charge on a cell surface: A new approach to the affinity partitioning of biological particles. *Anal. Biochem.* **200,** 280–285.

Muiño Blanco, M. T., Cebrian-Pérez, J. A., Olde, B., and Johansson, G. (1986). Effect of dextran- and poly(ethylene glycol)-bound Procion Yellow HE-3G on the paritition of membranes from calf brain synaptosomes within an aqueous two-phase system. *J. Chromatogr.* **358,** 147–158.

Muiño Blanco, M. T., Cebrian-Pérez, J. A., Olde, P., and Johansson, G. (1991). Subfractions of membranes from calf brain synaptosomes obtained and studied by liquid-liquid partitioning. *J. Chromatogr.* **547,** 79–87.

Müller, W. (1985). Partitioning of nucleic acids. *In* "Partitioning in Aqueous Two-phase Systems: Theory, Methods, Uses, and Applications to Biotechnology" (H. Walter, D. E. Brooks, and D. Fisher, eds.), pp. 227–266. Academic Press, San Diego, CA.

Müller, W., and Crothers, D. M. (1975). Interaction of heteroaromatic compounds with nucleic acids. I. The influence of heteroatoms and polarizability on the base specifity of intercalating ligands. *Eur. J. Biochem.* **54,** 267–277.

Müller, W., and Eigen, A. (1981). DNA fractionation by two-phase partitioning with the aid of a base-specific macroligand. *Anal. Biochem.* **118,** 269–277.

Müller, W., and Gautier, F. (1975). Interaction of heteroaromatic compounds with nucleic acids. AT-specific non-intercalating ligands. *Eur. J. Biochem.* **54,** 385–394.

Müller, W., and Kütemeier, G. (1982). Size fractionation of DNA fragments ranging from 20 to 30000 base pairs by liquid-liquid chromatography. *Eur. J. Biochem.* **128,** 231–238.

Müller, W., Hatteshohl, I., Schuetz, H., and Meyer, G. (1981). Polyethylene glycol derivatives of base and sequence specific DNA ligands: DNA interaction and application for base specific separation of DNA fragments by gel electrophoresis. *Nucleic Acids Res.* **9,** 95–119.

Müller, W., Bünemann, H., Schuetz, H.-J., and Eigel, A. (1982). Nucleic acid interacting dyes suitable for affinity chromatography, partitioning and electrophoresis. *In* "Affinity Chromatography and Related Techniques" (T. C. J. Gribneau, J. Visser, and R. J. F. Nivard, eds.), pp. 437–444. Elsevier, Amsterdam.

Nanak, E., Vijayalakshmi, M. A., and Chadha, K. C. (1995). Segregation of normal and pathological human red blood cells lymphocytes and fibroblasts by immobilized metal ion affinity partitioning. *J. Mol. Recognition* **8,** 77–84.

Nguyen, A. L., and Luong, J. H. T. (1990). Development and application of a new affinity partitioning system for enzyme isolation and purification. *Enzyme Microb. Technol.* **12,** 663–668.

Nishimura, H., Munakata, N., Hayashi, K., Hayakawa, M., Iwamoto, H., Terayama, S., Takahata, Y., Kodera, Y., Tsurui, H., Shirai, T., and Inada, Y. (1995). Polyethylene glycol-modified avidin: A novel agent for the selective extraction of biotinylated immune-complex in an aqueous two-phase system. *J. Biomater. Sci., Polym. Ed.* **7,** 289–296.

Olde, B., and Johansson, G. (1985). Affinity partitioning and centrifugal counter-current distribution of membrane-bound receptors using naloxone-poly(ethylene glycol). *Neuroscience* **15,** 1247–1253.

Olde, B., and Johansson, G. (1989). Heterogeneity of a crude synaptosomal preparation, studied by affinity partitioning using hexaethonium-poly(ethylene glycol). *Mol. Cell. Biochem.* **87,** 153–160.

Otto, A., and Birkenmeier, G. (1993). Recognition and separation of isoenzymes by metal chelates—immobilized metal ion affinity partitioning of lactate dehydrogenase isoenzymes. *J. Chromatogr.* **644,** 25–33.

Patricia A., Tjerneld, V., Kozlowski, A., and Harris, M. (1992). Synthesis of dye conjugates of ethylene oxide-propylene oxide copolymers and application in temperature-induced phase partitioning. *BioSeparation* **2,** 363–373.

Persson, A., and Jergil, B. (1992). Purification of plasma membranes by aqueous two-phase affinity partitioning. *Anal. Biochem.* **204,** 131–136.

Persson, A., and Jergil, B. (1995). The purification of membranes by affinity partitioning. FASEB J. 1304–1370.

Persson, A., Johansson, B., Olsson, H., and Jergil, B. (1991). Purification of rat liver plasma membranes by wheat-germ-agglutinin affinity partitioning. *Biochem. J.* **273,** 173–177.

Persson, L. O., and Olde, B. (1988). Synthesis of ATP-polyethylene glycol and ATP-dextran and their use in the purification of phosphoglycerate kinase from spinach chloroplasts using affinity partitioning. *J. Chromatogr.* **457,** 183–193.

Pesliakas, H., Zutautas, V., and Baskeviciute, B. (1994). Immobilized metal ion affinity partitioning of NAD(+)- dependent dehydrogenases in poly(ethylene glycol) dextran two-phase systems. *J. Chromatogr. A* **678,** 25–34.

Pesliakas, J.-H. J., Zutautas, V. D., and Glemža, A. A. (1988). Affinity partitioning of yeast and horse liver alcohol dehydrogenase in polyethylene glycol-dyes/dextran two-phase systems. *Chromatographia* **26,** 85–90.

Pillai, V. N. R., Mutter, M., Bayer, E., and Gatfield, I. (1980). New, easily removable poly (ethylene glycol) supports for the ligand-phase method of peptid synthesis. *J. Org. Chem.* **45,** 5364–5370.

Poráth, J., Carlsson, J., Olsson I., and Belfrage, G. (1975). Metal chelate affinity chromatography, a new approach to protein fractionation. *Nature* (*London*) **258,** 598–599.

Radzicka, A., and Wolfenden, R. (1988). Comparing the polarities of the amino acids: Side-chain distribution coefficients between the vapor phase cyclohexane, 1-octanol and neutral aqueous solutions. *Biochemistry* **27,** 1664–1670.

Schustolla, D., Deckwer, W. D., Schugerl, K., and Hustedt, H. (1992). Enzyme purification by immobilized metal ion affinity partitioning—application to D-hydroxyisocaproate dehydrogenase. *BioSeparation* **3,** 167–175.

Shanbhag, V. P., and Johansson, G. (1974). Specific extraction of human serum albumin by partition in aqueous biphasic systems containing PEG bound ligand. *Biochem. Biophys. Res. Commun.* **61,** 1141–1146.

Shanbhag, V. P., Johansson, G., and Ortin, A. (1991). Ca^{2+} and pH dependence of hydrophobicity of alpha-lactalbumin: Affinity partitioning of proteins in aqueous two-phase systems containing poly(ethylene glycol) esters of fatty acids. *Biochem. Int.* **24,** 439–450.

Sharp, K. A., Yalpani, M., Howard, S. J., and Brooks, D. E. (1986). Synthesis and application of a poly(ethylene glycol)-antibody affinity ligand for cell separations in aqueous two-phase systems. *Anal. Biochem.* **154,** 110–117.

Singh, M., and Clark, W. M. (1994). Partitioning of vancomycin using poly(ethylene glycol)-coupled ligands in aqueous two-phase systems. *Biotechnol. Prog.* **10,** 503–512.

Stocks, S. J., and Brooks, D. E. (1988). Development of a general ligand for immunoaffinty partitioning in two phase aqueous polymer systems. *Anal. Biochem.* **173,** 86–92.

Suh, S.-S., and Arnold, F. H. (1990). A mathematical model for metal affinity protein partitioning. *Biotechnol. Bioeng.* **35,** 682–690.

Sulkowski, E. (1985). Purification of proteins by IMAC. *Trends Biotechnol.* **3,** 1–7.

Takerkart, G., Segard, E., and Monsigny, M. (1974). Partition of trypsin in two-phase systems containing a diamino-alpha,omega-diphenyl carbamyl PEG as competitive inhibitor of trypsin. *FEBS Lett.* **42,** 218–220.

Tejedor, M. C., Delgado, C., Grupeli, M., and Luque, J. (1992). Affinity partitioning of erythrocytic phosphofructokinase in aqueous two-phase systems containing poly(ethylene glycol)-bound Cibacron Blue. Influence of pH, ionic strength and substrates/effectors. *J. Chromatogr.* **589,** 127–134.

Vlatakis, G., and Bouriotis, V. (1991). Affinity partitioning of restriction endonucleases. Application to the purification of EcoR I and EcoR V. *J. Chromatogr.* **538,** 311–321.

Walsdorf, A., Forciniti, D., and Kula, M. R. (1990). Investigation of affinity partition chromatography using formate dehydrogenase as a model. *J. Chromatogr.* **523,** 103–107.

Walter, H. (1985). Surface properties of cells reflected by partitioning: Red blood cells as a model. *In* "Partitioning in Aqueous Two-phase Systems: Theory, Methods, Uses, and Applications to Biotechnology" (H. Walter, D. E. Brooks, and D. Fisher, eds.), pp. 328–377. Academic Press, San Diego, CA.

Walter, H. (1994). Analytical applications of partitioning: detection of differences or changes in surface properties of mammalian cell populations. *In* "Methods in Enzymology" (H. Walter and G. Johansson, eds.), Vol. 228, pp. 299–320. Academic Press, San Diego, CA.

Walter, H., Widen, K. E., and Birkenmeier, G. (1993). Immobilized metal ion partitioning of erythrocytes from different species in dextran-poly(ethylene glycol) aqueous phase systems. *J. Chromatogr.* **641,** 279–289.

Wuenschell, G. E., Naranjo, E., and Arnold, E. H. (1990). Aqueous two-phase metal affinity extraction of heme proteins. *Bioprocess. Eng.* **5,** 199–202.

Wuenschell, G. E., Wen, E., Todd, R., Shnek, D., and Arnold, F. H. (1991). Chiral copper-chelate complexes alter selectivities in metal affinity protein partitioning. *J. Chromatogr.* **543,** 345–354.

Zijlstra, G. M., Michielsen, M. J. F., Degooijer, C. D., Vanderpol, L. A., and Tramper, J. (1996). Hybridoma and CHO cell partitioning in aqueous two-phase systems. *Biotechnol. Prog.* **12,** 363–370.

Properties of Interfaces and Transport across Them

Heriberto Cabezas
U.S. Environmental Protection Agency, Office of Research and Development, National Risk Management Research Laboratory, Sustainable Technology Division, Systems Analysis Branch, Cincinnati, Ohio 45268

Much of the biological activity in cell cytoplasm occurs in compartments some of which may be formed, as suggested in this book, by phase separation, and many of the functions of such compartments depend on the transport or exchange of molecules across interfaces. Thus, a fundamentally based discussion of the properties of phases, interfaces, and diffusive transport across interfaces has been given to further elucidate these phenomena. An operational criterion for the width of interfaces is given in terms of molecular and physical arguments, and the properties of molecules inside phases and interfaces are discussed in terms of molecular arguments. In general, the properties of the interface become important when the molecules diffusing across are smaller than the width of the interface. Equilibrium partitioning, Donnan phenomena, and electrochemical potentials at interfaces are also discussed in detail. The mathematical expressions for modeling transport across interfaces are discussed in detail. These describe a practical and detailed model for transport across interfaces. For molecules smaller than the width of the interface, this includes a detailed model for diffusion inside the interface. Last, the question of the time scale for phase formation and equilibration in biological systems is discussed.

KEY WORDS: Interfaces, Transport, Phases, Diffusion, Partitioning, Cytoplasm

I. Introduction

The possibility of phase separation occurring in cytoplasm that is examined in this book has many implications (Walter and Brooks, 1995). One of these relates to the effects any phase separation would exert on the cell

0074-7696/00 $30.00

via the properties of the interfaces which delimit the phases. As is discussed below, the properties of such interfaces would be expected to play a significant role in diffusion and distribution of solutes, for instance. Hence, the topic is examined in some detail in the following sections.

The best described examples of phase separated aqueous mixtures of macromolecules are the two polymer/water mixtures which are discussed, for example, in the chapters by Tolstoguzov and by Johansson and Walter in this volume. The properties of the interfaces which form between the phases in these systems have been studied to some extent. Interfaces of protein-based systems have received less attention, however. In the absence of an extensive experimental literature, the approach taken in this chapter is to provide a basic treatment of the topic from the perspective of fundamental thermodynamics and transport theory. For this reason, the language is perhaps more mathematical than cell biologists would normally encounter. The ideas are developed without providing mathematical details of the derivations associated with the statements, however, so it is hoped the results and conclusions will be of interest to those curious about the topic but who have not been exposed previously to more mathematically rigorous treatments of the issues raised. Because a fundamental approach is taken, the conclusions should apply to a wide variety of liquid–liquid interfaces, including any which might form in the complex milieu of the cell cytoplasm.

II. Interfaces

A. What Are Interfaces?

A phase is a region of space which is occupied by matter whose measurable properties are slowly changing functions with space over the extent of the phase. For example, the measurable mass density within a body of water is nearly constant with space. Even in nonequilibrium situations, the change in measurable properties with space is finite within a phase. For instance, the temperature difference between the bottom and the top of the water in a vessel full of water being heated on a flame may be large but it is still finite. Therefore, a phase can be defined as a region of space occupied by matter over which the first derivative of any measurable property of matter P with respect to any spatial coordinate x is finite as shown in Eq. (1):

$$\frac{\partial P}{\partial x} \ll \infty \tag{1}$$

In general, the properties of matter will vary from one phase to another and no two phases have identical values for all the properties. Because these material properties do not vary much within a phase, when two phases come in physical contact, the measurable properties of matter will change from those characteristic of one phase to those characteristic of the other phase over a very small measure of space. The region of space over which the measurable properties of matter change from those characteristic of one phase to those characteristic of another phase is called an interface. Within an interface, the first derivative of any measurable property of matter P with respect to any spatial coordinate x can approach infinity as shown in Eq. (2):

$$\frac{\partial P}{\partial x} \rightarrow \infty \tag{2}$$

The reason is that the interface is rather a thin region of width $L_l + L_r$ as shown in Fig. 1 in very simplified and exaggerated form. In fact, the interface is frequently approximated as a surface located at the heavy dotted line ($X = 0$) with no width ($L_l + L_r = 0$) and no volume. Exactly where the hypothetical interface surface should be located is not simple. According to Gibbs, the surface should be located such that the two semi-triangular areas under the curve of property P should be equal. Statistical mechanics (Croxton, 1975) tells us that for a one component fluid with two phases, this condition can be represented by

$$\int_{-\infty}^{0} \rho_1^{(l)}[1 - g_1(x)]dx = \int_{0}^{+\infty} \rho_1^{(r)}[1 - g_1(x)]dx \tag{3}$$

where $\rho_1^{(l)}$ and $\rho_1^{(r)}$ are the bulk number density or concentration of molecules of the one component *1* on the left (l) and the right (r) side phase, respec-

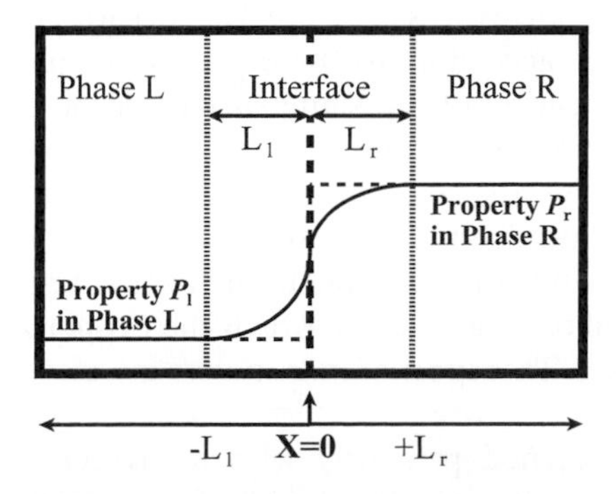

FIG. 1 Representation of the macroscopic structure of an interface separating a phase L from another phase R. Note the steep change in the value of a property P with position in going from phase L through the interface to phase R.

tively, g_1 (x) is the single particle distribution function for the one component l, and x is the spatial coordinate perpendicular to the interface surface. The single particle distribution function g_1 (x) is the probability of finding a molecule at given distance x from the interface surface. Far away from the interface ($x \rightarrow \infty$), the probability approaches one ($g_1(x) \rightarrow 1$) while near the interface this probability is not equal to one ($g_1(x) \neq 1$). Thus, the integrants $\rho_1^{(l)}[1 - g_1(x)]$ and $\rho_1^{(r)}[1 - g_1(x)]$, respectively, represent the deviation of the density of molecules of l from the bulk densities of molecules of l ($\rho_1^{(l)}$ and $\rho_1^{(r)}$) at a position x away from the interface surface for the left and right side phases in Fig. 1. Therefore, according to Gibbs, the interface surface should be located at the point where the excess or depletion of molecules of l is the same in both phases. For a system that contains n components rather than one, one could generate several criteria such as equating the excess or depletion of the dominant species or minimizing the difference in the excess or depletion of all the species, taking each one individually. Possibly, the most appropriate and practical criterion for placing the interface surface is

$$\sum_{i=1}^{n} \int_{-\infty}^{0} \rho_i^{(l)}[1 - g_i(x)]dx - \sum_{i=1}^{n} \int_{0}^{+\infty} \rho_i^{(r)}[1 - g_i(x)]dx = 0 \tag{4}$$

where the sum is taken over all components n. This criterion insures that the sum total of the excess or depletion of molecules of all components is the same in both phases. However, this criterion is not without some ambiguity. To illustrate the problem, one can consider a system with one solvent and two very dilute solutes ($n = 3$) forming two separate liquid phases. In this case, it could be argued that the important excess is that of the solutes and not the solvent, and that Eq. (4) should be replaced by two expressions like Eq. (3) for the two solutes only. Although, one could also argue that there should also be an expression like Eq. (3) for the solvent. But the placing of the surface according to the equation for the solvent would be different from the location of the same surface according to the equation for the solutes—a dilemma. One must, therefore, conclude that replacing the interface region by a surface is a serious approximation which is not readily extended to multicomponent systems. Thus, this approximation is likely to be most useful for those phenomena that are dominated by the properties of the bulk phase and for which the properties of the interface are not too important. Transport of mass across interfaces is, unfortunately, not is this category.

Since we are concerned primarily with interfaces and since we have discussed how replacing the interface by a surface is not entirely satisfactory, it is then logical to ask that if the interface is to be considered a region of finite width, where does the interface region start and where does it end?

That is, what is the value of L_l and what is the value of L_r? After all, the dilemma regarding the placing of the interface surface arose partly out of an effort to avoid having to define the beginning and the end of the interface. For a one component system, it is reasonable to propose that L_l and L_r should be placed at that point where the local density of molecules of the one component ($\rho_1^{(l)} g_1(x)$ and $\rho_1^{(r)} g_1(x)$) becomes negligibly different from the bulk density in the respective phase ($\rho_1^{(l)}$ and $\rho_1^{(r)}$), i.e., the local composition of the phase at the microscopic level becomes nearly equal to the overall composition of the phase at the macroscopic level. For a one component (1) system, this can be represented in terms of the single particle distribution function as

$$g_1 (L_l) = 1 \tag{5}$$

$$g_1 (L_r) = 1 \tag{6}$$

These criteria assure that the excess or depletion of molecules of *1* with respect to the bulk densities in their respective phases is negligible, i.e., they set the width $L_l + L_r$ around the interface at which the effect of the interface vanishes. For a multicomponent system with n components, the equivalent criterion involves one expression for each component i as shown in Eqs. (7) and (8):

$$g_i (L_l^{(i)}) = 1 \qquad i = 1,2 \ldots n \tag{7}$$

$$g_i (L_r^{(i)}) = 1 \qquad i = 1,2 \ldots n \tag{8}$$

where $L_l^{(i)}$ and $L_r^{(i)}$ set the distance where the local density of species i becomes equal to the bulk density of species i in the left and right phases, respectively. Then one can set L_l equal to the largest $L_l^{(i)}$ and L_r equal to the largest $L_r^{(i)}$ to define the extent of the interface. The reason for choosing the largest $L_l^{(i)}$ and $L_r^{(i)}$ as the interface boundaries is that these are the distances around the interface at which the local densities of each and every component can be guaranteed to be approximately equal to its bulk density. At distances less than these, there is at least one component whose local density is not equal to its bulk density. These criteria avoid some of the artificialities and ambiguities associated with Eq. (4). The reason is that the model is closer to physical reality and allows us to bring some physical reasoning in terms of molecules to bear on the problem.

Some theoretical methods by which one might compute the single particle distribution function ($g_i(x)$) include Monte Carlo or Molecular Dynamic Simulation using computers (Allen and Tildesley, 1989). The application of these methods to systems of small molecules is fairly well established. However, applying them when macromolecules are present is very difficult because the computing power required can exceed the capacity of even the fastest available machines unless major simplifications are introduced. Since the biological systems of interest in the present context contain high concentrations of macromolecules, this limitation can present a very serious problem. One should not, however, confuse the relatively simpler simulation of individual proteins or other macromolecules with the simulation of an entire system including solvents, small molecules, and macromolecules, which is a far more taxing problem. The single particle distribution function can also be calculated from experimental neutron scattering measurements (Skold and Price, 1987). Unfortunately, the signal normally obtained from such experiments is dominated by the bulk phase unless a very thin film is employed in the experiment. This introduces the difficulty of extracting a faint signal from the surrounding noise. Nonetheless, these developments will be used in the following section to outline a method for determining when the properties of interfaces are important and when they are not in interface transport.

B. The Size and Volume of Interfaces

Since there are no simple, readily available means of determining the single particle distribution function [$g_1(x)$] and, thus, the size of the interface region, it is then fair to propose the heuristic use of the theoretical construct to approximately estimate the extent of the interface. In a liquid phase, molecules are as tightly packed together as their size will allow, i.e., the closest distance of approach between centers of mass of two small molecules or parts, say monomers, of two large molecules is determined by the distance of separation at which the molecular orbitals start to overlap. For a one-component, one-phase system, the single particle distribution function is, therefore, a smooth, unchanging, and uninteresting function since essentially every point along a given coordinate x has a molecule. For a one-component system with two phases, the single particle distribution function will track the change in the probability of finding a molecule at a point x as it changes from the probability characteristic of one phase to that of the other phase, as shown in the caricature of Fig. 2. The reason is that every point along the coordinate x will have a molecule of component i associated with it but the probability of finding the molecule at that point will be different because the density of molecules is, in general, different in different phases. For this case, the single particle distribution function is, thus, only interesting near the interface. For a system with n components and more than one phase, the single particle distribution function $g_i(x)$ for a

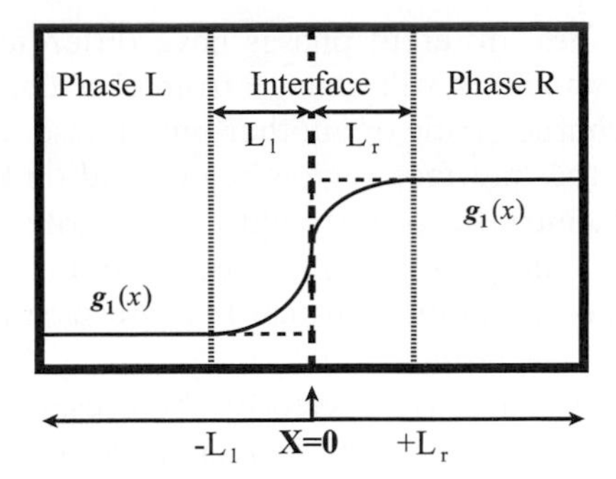

FIG. 2 Representation of the single particle distribution function $g_1(x)$ in a fluid consisting of one component as it changes with position in going from a phase L through an interface into another phase R. Note that, according to Gibbs, the surface at $X = 0$ representing the interface should be placed such that the semitriangular areas under and above the curve of $g_1(x)$ are equal.

component i has an oscillating quality, as shown in Fig. 3. The reason is that not every point along the coordinate x has a molecule of component i. Rather, it could just as easily have a molecule of some other component j where $j \neq i$, and one would generally expect to find molecules of i and other components alternating with a certain periodicity along the coordinate x. This periodicity correlates with the sizes of the molecules such that a peak in the single particle distribution function $g_i(x)$ for a component i roughly corresponds to the location along the x coordinate of the center of a molecule of this component i. This periodicity may be different for

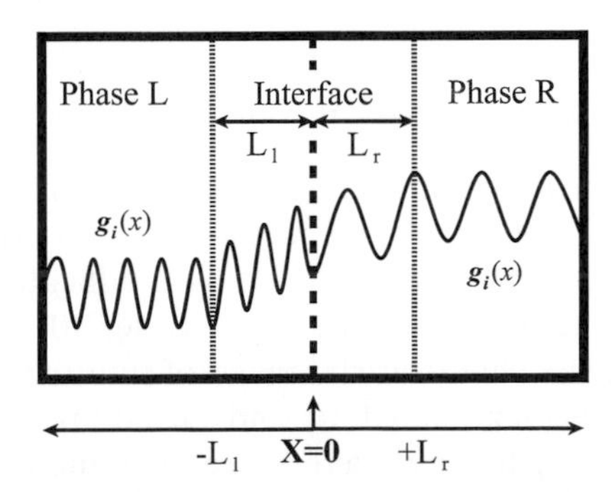

FIG. 3 Representation of the single particle distribution function $g_i(x)$ for a component i in a mixture of several other components. Note that the periodicity of $g_i(x)$ changes in going from a phase L to another phase R because the structures of the two phases are not equal.

different phases because different phases have different structures. Near the interface, the periodicity will change from that characteristic of one bulk phase to that characteristic of another bulk phase. In trying to define the points at which the interface region begins and ends, it is worth noting that the point at which the local structure of the solution, i.e., the aforementioned periodicity in $g_i(x)$, becomes negligibly different from that of either bulk phase sets the point at which the interface region begins and ends.

The structure of all phases is affected by the presence of the interface, and for a one-component system, the proper length scale for this effect is a multiple of the characteristic size parameter l_1 of the molecules of component *1*, i.e.,

$$L_l + L_r = \alpha^{(l)} l_1 + \alpha^{(r)} l_1 \approx 2 \alpha l_1 \tag{9}$$

where $\alpha^{(l)}$ is an number characteristic for phase *l*, $\alpha^{(r)}$ is an number characteristic for phase *r*, and where one can approximately assume that for two liquid phases with one component $\alpha^{(l)} \approx \alpha^{(r)} \approx \alpha$. The parameter l_1 can be taken to be the equivalent diameter of a small molecule, the end-to-end distance of a polymer coil, or the effective diameter of a protein molecule. The parameter α represents the distance in terms of characteristic molecular sizes l_1 over which the interface is able to affect the structure of the phase. The parameter α, therefore, represents the effect of intermolecular forces to the extent that the longer the range of the intermolecular forces, the larger the value of α. For a multicomponent system, the effect of the interface will extend to at least a multiple of the size of the largest molecule present in the system. Therefore, the proper length scale should be a multiple of the characteristic size parameter of the largest molecule present in the system. If the largest molecules in the system belong to a component *i* with a size parameter l_i, then the proper criterion for the extent of the interface should be

$$L_l + L_r = \alpha^{(l)} l_i + \alpha^{(r)} l_i \approx 2 \alpha l_i \tag{10}$$

where one can again approximately assume that for two liquid phases $\alpha^{(l)} \approx \alpha^{(r)} \approx \alpha$, since molecules of all components are always presumed to be present in all phases. The exact value of α is difficult to determine, but one would speculate that it should be of the order 10. The reason is that, for small molecules, the characteristic range of intermolecular forces is of the order of 10 molecular diameters, and, in addition, pair distribution functions typically decay to 1 over a distance of 10 molecular diameters or so (Friedman, 1985). Whether this argument can also be made for the case of macromolecules can be debated because it could easily lead to phases that consist only of interface in small compartments such as those inside a cell. This could, nevertheless, be an accurate picture of the physical reality. Therefore, the appropriate number for macromolecules may well be less than 10, although 10 is likely to be a reasonable order of magnitude estimate. Thus, the width of the interface should be about 10 molecular diameters for the largest species present. However, one should exercise some judgment in choosing the largest component i to avoid very dilute species that play a negligible role in determining the structure of the interface. For instance, it would seem inappropriate to choose very large species found in trace quantities. If such a species should collect at an interface, then the collection should be treated as another phase with an interface between it and the bulk phases. One should, therefore, look for large species found in relatively high concentrations.

For a solution of macromolecules in water with other smaller components such as salts, the appropriate length (l_i) scale is the radius (R) or the end-to-end distance (R) for the macromolecules. For linear macromolecules such as unfolded DNA or RNA or linear polymers, the end-to-end distance is approximately related to the radius of gyration by

$$l_i = R = 2.5105\ R_g \tag{11}$$

where R is the end-to-end distance and R_g is the radius of gyration. The radius of gyration is readily measurable from light-scattering experiments (Van Holde, 1985). For generally spherical macromolecules, the radius can be set equal to the length scale ($l_i = R$). The width of the interface region is then

$$L_l + L_r \approx 2\ \alpha l_i \tag{12}$$

where all the symbols have their previously defined interpretation.

One can also calculate the ratio of interface volume to total system volume ($R_{is} = V_i/V_s$), including all phases as a function of interface width and the ratio of interface area to total system volume according to the expression

$$R_{is} = (L_l + L_r)\frac{A}{V_s} = 2\alpha l_i \frac{A}{V_s} \tag{13}$$

where A is the area of the interface, V_s is the total volume of the system, and where all the other symbols have their previously assigned interpretation. In addition, the ratio of interface volume to the phase volume (R_{ik}) for a given phase k ($k = l$ or r) can be calculated from

$$R_{ik} = L_k \frac{A}{V_k} = \alpha l_i \frac{A}{V_k} \tag{14}$$

where L_k is the width of the interface on the side of phase *k*, and V_k is the volume of phase *k*, and all the other symbols have their previously assigned interpretation. Both of these are important because, in many experimental situations, it is possible to measure the area of the interface and the total volume of the system, but it is generally difficult or impossible to measure the ratio of interface volume to phase volume or to total system volume.

To illustrate the significance of the ratio of the interface volume to total system volume ($R_{is} = V_i/V_s$), we consider a simple example involving an aqueous two-phase system formed by mixing polyethylene glycol of 8000 molecular mass and dextran of 2,700,000 molecular mass in water (Albertsson, 1986; Walter *et al.*, 1985). The radius of gyration (R_g) for polyethylene glycol can be estimated from the correlations (Cabane and Duplessix, 1982), and the radius of gyration (R_g) for dextran can also be estimated from the correlations (Senti *et al.*, 1955). The length scale (l_i) is estimated from the radii of gyration as already discussed. The ratio of interface area to total system volume (A_i/V_s) can be readily estimated from geometric arguments and can range widely from 0 to 10,000 or larger. For example, for a common benchtop laboratory experiment done with test tubes, the ratio of interface area to system volume is of the order of 0 to 1. For a dispersion or emulsion of two phases of, say, a milliliter each with droplets of, say, a 1 millimeter diameter, this same ratio is of the order 1000, and for smaller droplets of, say, 1/3 mm in diameter, the ratio quickly rises to about 5000. Thus, if the aforementioned aqueous two-phase system is agitated and mixed so that the two phases are dispersed into droplets of about 1/3 mm diameter ($A_i/V_s \approx 5000$), the ratio of interface volume to system volume ($R_{is} = V_i/V_s$) is of the order of 10^{-2}, which is not negligible. However, for two phases in a test tube, the ratio of interface area to system volume (A_i/V_s) is of the order of 1, and the ratio of interface volume to system volume ($R_{is} = V_i/V_s$) is of the order 10^{-6}, which is probably negligible in many situations.

C. When and Why Are Interfaces Important

The presence of interfaces in a system is always important in the transport of molecules across interfaces. The reason is that transporting a molecule across an interface is roughly equivalent to undissolving a molecule from one phase, dissolving it in the interface, undissolving it from the interface, and finally, redissolving it in the other phase. This is energetically costly. The more interesting and important question is whether the properties of the bulk phases or the properties of the interface or both are the important factors in interface transport. This question can be addressed by some simple criteria using the theory developed above.

First, when the characteristic size of the molecules being transported is smaller than the width of the interface as determined from Eq. (10), then, during transport across the interface, the transported molecules will, at some point, be completely immersed in the interface, and the properties of the interface will affect the transport. This can be expressed by

$$L_l + L_r > l_t \tag{15}$$

where l_t is the characteristic size of the chemical species being transported and the other symbols have their previous interpretation. In this case, one has to consider transport from one bulk phase to the interface and from the interface to the other bulk phase, each one separately. This means that the transport properties associated with the transported molecule, e.g., the diffusivity, etc., would be different in the interface than they would be in either bulk phase. Small ionic species being transported across the interface in almost any biological system fall in this category.

Second, when the characteristic size of the molecules being transported is larger than the width of the interface as determined by Eq. (10), then, during transport across the interface, the transported molecules will never be completely immersed in the interface, and the properties of the interface are unlikely to play a major role in the transport. This can be expressed by

$$L_l + L_r < l_t \tag{16}$$

where all the symbols have their previously given interpretation. In this case, one can usually describe all the transport phenomena in terms of transport from one bulk phase to another bulk phase and the properties of the transported molecules in the bulk phases. Very large protein aggregates such as albumins or viruses being transported across the interface in almost any biological system fall in this category. Accumulations of large protein-based structures (e.g., aggregates) at an interface are an extreme case of this category, and the collected mass at the interface needs to be treated as a separate phase.

The aforementioned considerations indicate when the transport of molecules across an interface can be treated in terms of the properties of the transported molecules in the bulk phases, and when the properties of the transported molecules in the interface need to be included. They do not, however, indicate how important the effect of the interface is likely to be, i.e., they do not give a quantitative sense of the importance of the effect. How important the effect of the interface is depends on how many molecules are in the interface at any given time during transport. Molecules generally are not evenly distributed across phases and interfaces because of the various intermolecular interactions. However, the probability (P_t) that a given molecule of a component t is found within the interface at a given time is defined by

$$P_t \equiv \frac{N_t^{(i)}}{N_t^{(s)}} = \frac{N_t^{(i)}}{N_t^{(l)} + N_t^{(r)} + N_t^{(i)}} \tag{17}$$

where $N_t^{(i)}$ is the number of molecules of component t in the interface i, $N_t^{(s)}$ is the total number of molecules of component t in the system s, and where for a system consisting of two phases, l and r, and an interface i, $N_t^{(s)}$ is equal to the sum of the number of molecules of t found in phase l ($N_t^{(l)}$), phase r ($N_t^{(r)}$), and the interface i ($N_t^{(i)}$). Equation (17) is not very practical in this form but it can be rewritten in terms of measurable component concentrations and interface volume to phase volume ratios using the results of Eq. (14) to give

$$P_t = \frac{1}{1 + \frac{1}{C_t^{(i)}}\left[\frac{C_t^{(l)}}{R_{il}} + \frac{C_t^{(r)}}{R_{ir}}\right]} \tag{18}$$

where $C_t^{(i)}$, $C_t^{(l)}$, and $C_t^{(r)}$ are the concentrations of component t in the interface i, the left-side phase l, and the right-side phase r, respectively, and where R_{il} and R_{ir} are the interface volume-to-phase volume ratios given by Eq. (14) for the left-side and right-side phase, respectively. It worth noting that as R_{il} and R_{ir} become large, the probability P_t of finding a molecule of t within the interface approaches one, and that as R_{il} and R_{ir} become small, the probability P_t of finding a molecule of t within the interface approaches zero. Therefore, for a system that meets the criterion of Eq. (15) so that the molecules being transported are smaller than the width of the interface, the larger the value of P_i, the more significant the effect of the interface. As a rough guideline, if 10% or less of the molecules of a species t are within the interface, i.e., $P_t \leq 1/10$, then the interface is likely to play a minor role in the transport of component t across the interface. One could, in addition, consider the ratio of $P_t^{(\alpha)}$ in a system α to $P_t^{(\beta)}$ for a different system β to gauge the relative importance of the interface in different transport problems. Finally, it is important to acknowledge that the measurement of the quantities in Eqs. (14) and (18) is not trivial for the case of biological systems. However, estimates can be obtained which may afford by use of Eqs. (14), (15), (16), and (18) at least qualitative judgment and guidance on the importance of the interface for a particular transport problem. In general, the transport properties of a component within the interface need to be carefully considered when the size of the interface is larger than the characteristic size of the molecules being transported, Eq. (15), and when more than 10% of the molecules are within the interface, $P_t \leq 1/10$ at any given time. If either of these criteria are not met, then it is likely that the interface can be neglected.

III. Properties of Interfaces and Phases

The properties of Interfaces can be classified into equilibrium properties and dynamic properties. The equilibrium properties can be further classified into bulk properties and component specific properties. Bulk properties are those that are associated with a particular phase but not with a specific component in the phase. The equilibrium bulk property that will be treated here is the density. Component specific properties are those that can be associated with a particular component in the phase. The equilibrium component specific properties that will be treated here are the component chemical potentials and the component activity coefficients. The dynamic properties can also be further classified into bulk properties and component specific properties following the definitions already given. The most important dynamic bulk properties are the viscosity and the thermal conductivity, but they are not particularly important in the transport of mass across interfaces in biological systems so they will not be further discussed here. The dynamic component specific property that will be treated here is the diffusivity. In addition, the effect of the molecular interactions, i.e., the nonidealities, on the equilibrium and the dynamic properties will also be discussed.

A. Equilibrium Properties

The equilibrium properties of a phase are those that characterize the equilibrium state. They do not change with time at any given point in space within the phase. If the phase is homogeneous, the equilibrium properties of the phase do not change with space within the confines of the phase, i.e., any given equilibrium property has a single value throughout the phase. For a heterogeneous phase, the equilibrium properties of the phase vary from point to point within the phase. Interfaces, being the boundary between two phases are like heterogeneous phases.

1. Bulk Properties

The density of a phase is a measure of the number of molecules or the number of moles per unit volume within the phase. This is the number density which can easily be related to mass density by multiplying by molecular mass. Since molecules are generally packed as closely as possible in liquid phases, the density of liquid phases primarily depends on the size of the molecules, i.e., the distance between molecular centers at which orbitals begin to overlap. As result, the density of liquid phases does not change significantly within a phase even for heterogeneous phases. The number density of a liquid phase containing n chemical components can be calculated from

$$\frac{1}{\rho^{(p)}} = x_1 \overline{V}_1(T,\underline{x}) + x_2 \overline{V}_2(T,\underline{x}) \ldots + x_n \overline{V}_n(T,\underline{x}) \approx x_1\underline{V}_1(T) + x_2\underline{V}_2(T) \ldots + x_n\underline{V}_n(T) = \frac{x_1}{\rho_1(T)} + \frac{x_2}{\rho_2(T)} \ldots + \frac{x_n}{\rho_n(T)} \quad (19)$$

where $\rho^{(p)}$ is the number density of phase p, x_i is the mole fraction of component i, $\overline{V}_i$ is the partial molar volume of component i, $\underline{V}_i$ is the specific molar volume of pure component i, and ρ_i is the number density of pure component i. Equation (19) tacitly assumes that the liquid phase is independent of pressure which is a reasonable assumption for biological systems, and it further assumes that the partial molar volume of all components ($\overline{V}_i$) is approximately equal to the pure component molar volume ($\underline{V}_i$). The latter assumption, setting $\overline{V}_i \approx \underline{V}_i$, implies that there is negligible volume change on mixing of the chemical components of the phase. This is usually reasonable but there are exceptions. For example, if one mixes very large and rigid molecules with very small molecules, the small molecules can fit in the space between the large molecules so that there is an apparent reduction in volume upon mixing, i.e., 1 milliliter of large molecules plus 1 milliliter of small molecules yields less than 2 milliliters of solution.

The density $\rho^{(i)}$ associated with an interface will rapidly vary from that associated with one phase to that associated with the other phase. As a result, the density of an interface needs to be treated in a pointwise fashion in space so that Eq. (19) becomes

$$\left.\frac{1}{\rho^{(i)}}\right|_x \approx x_1|_x\underline{V}_1(T) + x_2|_x\underline{V}_2(T) \ldots + x_n|_x\underline{V}_n(T) = \frac{x_1|_x}{\rho_1(T)} + \frac{x_2|_x}{\rho_2(T)} \ldots + \frac{x_n|_x}{\rho_n(T)} \quad (20)$$

where x is a spatial coordinate, $\underline{x}$ is the vector of all component mole fractions, ρ_i is the density of pure component i, and where it has been assumed that the density of a liquid phase depends only on the temperature and not the pressure. For cases where the largest molecules involved in the phenomena of interest are smaller than the size of the interface, i.e., the criterion of Eq. (15), the molecules will experience the density of the interface in a pointwise fashion as they move across the interface. Thus, the density must be treated in a pointwise fashion as expressed by Eq. (20). To a good degree of approximation, one can treat the density inside the interface as a linear function interpolating between the density of the two adjacent phases. By using Eq. (19), one can, therefore, give the density $\rho^{(i)}$ of the interface as

$$\left.\frac{1}{\rho^{(i)}}\right|_x \approx \frac{1}{\rho^{(l)}(T)} + \frac{1}{L_r - L_l}\left[\frac{1}{\rho^{(r)}(T)} - \frac{1}{\rho^{(i)}(T)}\right](x - L_l) + \ldots \tag{21}$$

where $\rho^{(k)}$ is the number density of phase k ($k = r$ or l), which can be estimated using Eq. (19). For cases where the largest molecules of interest are larger than the size of the interface, i.e., the criterion of Eq. (16), the density of the interface can be treated as an average over the interface because the large molecules experience the entire interface at a any given time. Then, Eq. (20) becomes

$$\rho^{(i)} \approx \frac{\rho^{(l)} + \rho^{(r)}}{2} \tag{22}$$

where $\rho^{(k)}$ is the number density of phase $k(k = r$ or $l)$, which can be estimated using Eq.(19).

The density of many components can be estimated from various readily available literature sources. Thus, the density of pure water is approximately 0.99 kg/L. The density ρ_p of polymers in aqueous solution can be estimated from

$$\rho_p \approx \frac{1}{\left(\frac{M_p}{M_m}\right)\overline{\mathrm{V}}^{\infty}{}_m} \tag{23}$$

where $\overline{\mathrm{V}}_m^{\infty}$ is the formal partial molar volume at infinite dilution of a monomer m of polymer p in water, M_p is the molecular mass of polymer p, M_m is the molecular mass of a monomer m of polymer p. For example, for polyethylene glycol $\overline{\mathrm{V}}_m^{\infty} = 37\ \mathrm{m}^3/\mathrm{kmol}$ (Zana, 1980). The density ρ_s of salts in aqueous solution can be estimated from

$$\rho_s \approx \frac{1}{\overline{\mathrm{V}}_s^{\infty}} \tag{24}$$

where $\overline{V}_s^{\infty}$ is the partial molar volume of salts at infinite dilution in water that can be estimated from the work of Millero and coworkers (Millero, 1972). The partial molar volumes of proteins can be obtained from various widely available sources (Durchschlag, 1989; Durchschlag and Rainer, 1982, 1983).

2. Component Specific Properties

The chemical potential of a component i in a phase is roughly that portion of the free energy of the phase that can be assigned to the molecules of component i. The chemical potential is an abstract quantity that cannot be directly measured, but it can be calculated from related measurements. For a homogeneous phase, the chemical potential is defined by

$$\mu_i^{(k)} = \left.\frac{\partial G^{(k)}}{\partial N_i^{(k)}}\right|_{T,P,N_{j\neq i}} \tag{25}$$

where $\mu_i^{(k)}$ is the chemical potential of component i in phase k, $G^{(k)}$ is the total Gibbs free energy of phase k, $N_i^{(k)}$ is the number of moles of component i in phase k, and where the temperature, the pressure, and the number of moles of all components j not equal to i are held constant during differentiation. The force driving the diffusive transport of molecules across an interface is the difference in the chemical potential of the transported component across the interface, i.e., the difference between the chemical potential of the transported component in the left and the right phases. The chemical potential is not the most convenient function for calculations because it approaches minus infinity as the concentration goes to zero. In practice, it is frequently related to a measure of concentration and other better-behaved functions, activity coefficients, by

$$\begin{aligned}\mu_i^{(k)} &= \mu_i^{(ko)}(T) + RT\ln[x_i^{(k)}\gamma_i^{(kx)}(T,x^{(k)})] \\ &= \mu_i^{(ko)}(T) + RT\ln[C_i^{(k)}\gamma_i^{(kc)}(T,C^{(k)}] \\ &= \mu_i^{(ko)}(T) + RT\ln[m_i^{(k)}\gamma_i^{(km)}(T,m^{(k)}]\end{aligned} \tag{26}$$

where $\mu_i^{(k)}$ is the chemical potential of component i in phase k, $\mu_i^{(ko)}$ is the reference chemical potential of component i in phase k, R is the gas constant, T is the temperature, $x_i^{(k)}$ is the mole fraction of component i in phase k, and $\gamma_i^{(kx)}$ is the activity coefficient on the mole fraction scale for component i in phase k, $C_i^{(k)}$ is the molar concentration of component i in phase k, and $\gamma_i^{(kc)}$ is the activity coefficient on the concentration scale for component i in phase k, $m_i^{(k)}$ is the molality of component i in phase k, $\gamma_i^{(km)}$ is the activity coefficient on the molality scale for component i in phase k, and $x^{(k)}$, $C^{(k)}$, and $m^{(k)}$, respectively, represent the set of mole fractions, molar concentrations, and molalities for all components in phase k. One can, in fact, simply write an activity coefficient for any composition scale of interest. However, the three aforementioned activity coefficients are the most commonly used ones. It should be noted that the activity coefficient on the mole fraction scale, $\gamma_i^{(kx)}$, cannot be combined with any composition measure other than the mole fraction, that the activity coefficient on the concentration scale, $\gamma_i^{(kc)}$, cannot be combined with any measure of composition other than the molar concentration, and that the activity coefficient on the molality scale, $\gamma_i^{(km)}$, cannot be combined with any composition measure other than the molality. It is worth noting that the different activity coefficients are related by

$$x_i^{(k)}\gamma_i^{xk)} = C_i^{(k)}\gamma_i^{(kc)} = m_i^{(k)}\gamma_i^{(km)} \tag{27}$$

so that if one knows the mole fraction, the molar concentration, and the molality of component i in phase k and one of the activity coefficients, then they can all be readily calculated.

The force driving the diffusive transport of molecules across an interface can also be expressed in terms of mole fractions or molar concentration and the corresponding activity coefficient, as well as in terms of chemical potentials. The activity coefficients measure the extent to which a solution of chemicals deviates from ideal behavior. One very important point that should always be emphasized is that the activity coefficient of any component i depends on the composition of all the components in the solution and not just component i. There are numerous methods for estimating the activity coefficients and the chemical potential for systems of interest here (Prausnitz et al., 1986). Many of these have been recently reviewed (Cabezas, 1996).

B. Dynamic Properties

The dynamic properties are those that characterize the dynamic behavior of a phase. In general, they change with time and space within the phase, but to a good degree of approximation, they can often be considered constant. For example, they can be considered approximately constant with space and time within a homogeneous phase for unsteady state transport. For steady state transport within a homogeneous phase, the dynamic properties can be considered exactly constant with time and approximately constant with space to a good degree of approximation. For heterogeneous phases, the dynamic properties vary with space as the composition varies from point to point within the phase. Again, interfaces being the boundary between two phases are like heterogeneous phases. Since dynamic bulk properties are not being treated as already mentioned, the discussion will then focus on the diffusivity.

The diffusivity is the characteristic property for the transport of mass across a phase or an interface. The diffusivity D_{ip} of a component i in phase p is defined by Fick's first law which gives

$$\mathrm{J}_{ix} = -D_{ip}\frac{\partial C_i}{\partial x} \tag{28}$$

where J_{ix} is the flux of component i in the x direction, C_i is the concentration of component i, and x is a spatial coordinate. The standard model indicates that the diffusivity of a component depends on the viscosity of the phase, the size of the diffusing species, the intermolecular forces, and the temperature. This is approximately summarized by the modified Einstein-Stokes expression

$$\mathrm{D}_{ip} = \frac{RT}{6\pi\eta R_i}\left(1 + \mathrm{C}_i \frac{\partial \ln \gamma_i^{(pc)}}{\partial C_i}\bigg|_{T,P}\right) \tag{29}$$

where R is the gas constant, T is the temperature, π is the number pi, η is the viscosity of the phase, R_i is the effective radius of a molecule of component i, C_i is the concentration of component i, and $\gamma_i^{(pc)}$ is the activity coefficient on the concentration scale of component i in phase p. The fact that the diffusivity increases with temperature and decreases with viscosity and molecular size is intuitive and generally known. The activity coefficient, however, incorporates the effect of the forces between molecules and their effect on the diffusivity and this is far from obvious.

As illustrated in Fig. 4, the forces between molecules can be classified as repulsive or attractive. At molecular separations less than the radius of the molecules, the prevalent forces are repulsive because the molecules collide, orbitals start to overlap, and for macromolecules, major changes in conformation would be caused. All of this requires a great deal of energy so that a very strong repulsive force is generated between the molecules. This force is qualitatively proportional to the negative of the slope of the potential energy curve in Fig. 4. At short distances of separation, this slope is quite steep and negative so that the force is large and positive, i.e., strongly repulsive. At distances of separation greater than the size of the molecules, the slope of the potential energy curve becomes smaller and

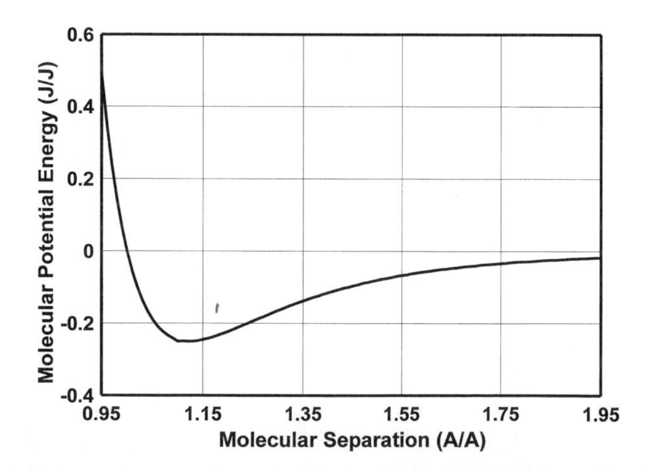

FIG. 4 Representation of dimensionless molecular energy of interaction between two small spherical molecules. Note that as the molecules get very close, they strongly repel and that as they move further apart, they weakly attract.

positive so that a relatively weaker attractive force is generated. All molecules have strong repulsive forces simply because they have finite sizes but attractive forces of varying strengths. Some molecules, such as normal hydrocarbons, have relatively weak attractive forces dominated by dispersion. The strongest attractive forces are generally found in molecules that have electrostatic charges and dipoles, such as proteins and salts.

The function that connects molecular forces to macroscopic quantities such as the diffusivity is the statistical mechanical partition function Q. To a fair approximation, one can separate the statistical mechanical partition function into the product of a partition function for repulsive forces $Q^{(R)}$ and a partition function for attractive forces $Q^{(A)}$. The rationale for this factorization is again that when the distance of separation between molecules is approximately equal to the molecular size, i.e., molecules are touching, the repulsive forces completely dominate the molecular force field, and when the molecules are separated from each other, the attractive forces dominate. The partition function is related to the chemical potential by

$$\mu_i = -RT\frac{\partial \ln Q}{\partial N_i}\bigg|_{T,V,N_{j\neq i}} = -RT\frac{\partial \ln Q^{(R)}}{\partial N_i}\bigg|_{T,V,N_{j\neq i}} - RT\frac{\partial \ln Q^{(A)}}{\partial N_i}\bigg|_{T,V,N_{j\neq i}} = \mu_i^{(R)} + \mu_i^{(A)} \tag{30}$$

where R is the gas constant, T is the temperature, N_i is the number of moles of component i, and V is the phase volume, and where the partition function has been factored into contributions from repulsive and attractive forces. As a consequence of Eq. (30), the chemical potential ($\mu_i^{(R)}$ and $\mu_i^{(A)}$), and through Eq. (26), the activity coefficient, can both be resolved into additive contributions from repulsive and attractive intermolecular forces as shown by

$$\ln\gamma_i = \ln\gamma_i^{(R)} + \ln\gamma_i^{(A)} \tag{31}$$

where $\gamma_i^{(R)}$ and $\gamma_i^{(A)}$ are, respectively, the contributions from repulsive and attractive molecular forces to the activity coefficient of a component i.

The activity coefficient generally increases with concentration for molecules that have weak attractive forces such that the activity coefficient is dominated by repulsive forces. The reason is that the activity coefficient represents the effect of the intermolecular forces on the tendency of the molecules to leave the solution, i.e., the higher the activity coefficient, the easier it is for the molecules to leave the solution. Thus, the stronger the repulsive forces and the weaker the attractive forces, the easier it is for the molecules to leave the solution and the bigger the activity coefficient. In such cases, the derivative of the activity coefficient in Eq. (29) is always positive because the more molecules present, the easier it is for them to leave the solution so that the diffusivity tends to increase with concentration. The relative importance of attractive forces, however, depends on the composition of the solution. In general, attractive forces are strongest at low concentration of charged species, i.e., low ionic strength, with the strength of attractive forces decreasing with concentration, i.e., increasing ionic strength. The reason is that charged species tend to shield each other so that, at high concentration, the mixture acts like a solution of molecules with weak attractive forces. Therefore, at low concentration of charged species, the derivative of the activity coefficient in Eq. (31) is negative and decreasing with concentration so that the diffusivity actually decreases with increasing concentration for a charged diffusing species. All known biologically related phases consist of solutions of charged species such as ions and proteins and dipolar species such as water and macromolecules often at low ionic strength. These effects, therefore, cannot be generally neglected in biological solutions. In summary, the diffusivity is proportional to the concentration gradient of the activity coefficient,

$$D_i \propto \left(1 + C_i \frac{\partial \ln \gamma_i^{(R)}}{\partial C_i}\bigg|_{T,P,C_{j\neq 1}} + C_i \frac{\partial \ln \gamma_i^{(A)}}{\partial C_i}\bigg|_{T,P,C_{j\neq i}}\right) \tag{32}$$

which for biological systems at low ionic strength is dominated by electrostatic, i.e., attractive forces, so that

$$D_i \propto \left(1 + C_i \frac{\partial \ln \gamma_i^{(A)}}{\partial C_i}\bigg|_{T,P,C_{j\neq i}}\right) \tag{33}$$

and since the derivative of the activity coefficient is negative and increasing with concentration, the diffusivity coefficient decreases with concentration. At high ionic strength, the charges shield each other so the gradient of the activity coefficient is dominated more by repulsive forces such that

$$D_i \propto \left(1 + C_i \frac{\partial \ln \gamma_i^{(R)}}{\partial C_i}\bigg|_{T,P,C_{j\neq 1}}\right) \tag{34}$$

and the gradient of the activity coefficient is now positive and increasing with concentration so the diffusivity also increases with concentration. This is important because in going from one phase to another in biological systems, one can completely change ionic strength regimes so that the diffusivity of a given species can significantly increase or decrease. In fact, in transport cases where the interface region is important, the diffusivity can significantly change from the beginning to the end of the interface.

IV. Equilibrium Partitioning across Interfaces

When any species is present in a system that is not too far from equilibrium and that embodies more than one phase, the species typically exhibits a preference for some phases more than others, i.e., the concentration of the species is highest in the most preferred phase and lower in the less preferred phase and so on. This observation gives rise to what is generally known as equilibrium partitioning across interfaces, which we shall henceforth refer to as partitioning phenomena.

A. Partitioning Phenomena

The equilibrium partitioning of a component i across an interface between two homogeneous phases occurs such that the chemical potential of each component is the same inside the interface and on the left and right phases as prescribed by the First and Second Laws of Thermodynamics. When molecules of component i are larger than the size of the interface, the criterion of Eq. (16), these molecules are never entirely inside the interface, and the partitioning can be simply treated in terms of the chemical potential of the component in the left and right phases as given by

$$\mu_i^{(l)}(T,P,\underline{x}^{(l)}) = \mu_i^{(r)}(T,P,\underline{x}^{(r)}) \qquad (35)$$

where $\mu_i^{(k)}$ is the chemical potential of component or species i in phase k ($k = l$ or r) and where the chemical potentials depend on the temperature T, the pressure P, and the composition $\underline{x}^{(k)}$ of phase k. For practical calculations, the series of expressions above is usually recast in terms of activity coefficients and concentrations to give

$$C_i^{(l)}\gamma_i^{(lc)}(T,P,\underline{x}^{(l)}) = C_i^{(r)}\gamma_i^{(rc)}(T,P,\underline{x}^{(r)} \; n = 1,2 \ldots n \qquad (36)$$

where $C_i^{(k)}$ is the molar concentration of component i in phase k ($k = l$ or r) and the activity coefficients $\gamma_i^{(kc)}$ for phase k on the molar concentration scale depend on the molar concentration of all components in the phase. Eq. (36) can be written in terms of mole fraction or molality scale activity coefficients as well. It is written here using molar concentration scale activity coefficients because biological systems are most often analyzed using molar concentration. From Eq. (36), one can further obtain expressions for the partition coefficient K_i of any component i in terms of activity coefficients. The expression is

$$K_i = \frac{C_i^{(l)}}{C_i^{(r)}} = \frac{\gamma_i^{(rc)}(T,P,\underline{x}^{(r)})}{\gamma_i^{(lc)}(T,P,\underline{x}^{(l)})} \; i = 1,2 \ldots n \qquad (37)$$

which relates partition coefficients to molecular interactions. It is important to note that the activity coefficients $\gamma_i^{(kc)}$ are functions of the concentration of all the components in the phase k and not just the concentration of component i.

For the case when the molecules of component i are smaller than the size of the interface, the criterion of Eq. (15), one needs to consider the interface as a separate phase and treat the equilibrium between the left phase and the interface and the interface and the right phase. Further, since the interface is an inhomogeneous phase, one must also consider that the composition and, consequently, the chemical potential and the activity coefficients will vary with space within the interface. This problem can be treated by assuming that the interface can be divided into a series of m slices which are small enough so that composition is constant within any given slice, and by further assuming that each of the slices is at equilibrium with the other slices. One, therefore, extends the method of Eqs. (35), (36), and (37) by writing a series of equilibrium expressions,

$$\begin{aligned}\mu_i^{(l)}(T,P,\underline{x}^{(l)}) &= \mu_i^{(1)}(T,P,\underline{x}^{(1)}) = \mu_i^{(2)}(T,P,\underline{x}^{(2)}) \\ &= \ldots = \mu_i^{(m)}(T,P,\underline{x}^{(m)}) = \mu_i^{(r)}(T,P,\underline{x}^{(r)}) \qquad (38)\end{aligned}$$

where all the symbols have their previously assigned interpretation and where the chemical potentials $\mu_i^{(1)}$ to $\mu_i^{(m)}$ refer to the hypothetical slices inside the interface. It is also quite simple to write equivalent expressions for Eqs. (36) and (37) in terms of activity coefficients and concentrations after the fashion of Eq. (38). In addition, one could, in the interest of simplicity but with a much greater degree of approximation, assume that the interface can be treated in an averaged fashion so that Eq. (38) reduces to

$$\mu_i^{(l)}(T,P,\underline{x}^{(l)}) = \mu_i^{(i)}(T,P,\underline{x}^{(i)}) = \mu_i^{(r)}(T,P,\underline{x}^{(r)}) \qquad (39)$$

where $\mu_i^{i)}$ represents an average chemical potential for the interface. The approximations embodied in Eq. (39) are likely to be most accurate when the difference between the left- and right-side phases are small so the property gradients are small, and they are likely to be less accurate when the left and right phases are quite different so large property gradients exist across the interface.

B. Donnan Potentials

When equilibrium partitioning across an interface involves charged molecules such as proteins and ions, the partitioning phenomena are further complicated by the requirement that all of the phases remain electrically neutral. The reason is that, in general, bulk phases are electrically neutral at equilibrium, and, therefore, charged molecules are found in concentrations such that the number of opposite charges cancel, e.g., proteins always have a number of counter ions associated with them whose sum total charge must be equal to the charge on the protein. Biological solutions normally consist of complex mixtures of various ions (Na^+, K^+, Cl^-, etc.) and charged macromolecules along with neutral molecules in water. As already discussed in the previous section, the equilibrium partitioning of a component *i* needs to be treated according to Eqs. (35) to (39), depending on the size of the molecule and the degree of approximation deemed appropriate. But when charged molecules are present, the additional condition of electrical neutrality needs to be imposed on each phase. Thus, for a system with *n* components, the expressions for electrical neutrality for the left side (l) and right side (r) phases are

$$C_1^{(l)}Z_1 + C_2^{(l)}Z_2 \ldots + C_n^{(l)}Z_n = 0 \tag{40}$$

$$C_1^{(r)}Z_1 + C_2^{(r)}Z_2 \ldots + C_n^{(r)}Z_n = 0 \tag{41}$$

and for the interface treated as an average phase, the expression is

$$C_1^{(i)}Z_1 + C_2^{(i)}Z_2 \ldots + C_n^{(i)}Z_n = 0 \tag{42}$$

and for the interface treated as series of *m* slices each at equilibrium, the expression are

$$C_1^{(i)}Z_1 + C_2^{(i)}Z_2 \ldots + C_n^{(i)}Z_n = 0 \quad i = 1,2, \ldots m \tag{43}$$

where Z_i is the valence of species *i*, $C_i^{(k)}$ is the concentration of species *i* in phase *k*, and where neutral species obviously have a valence of zero.

The additional requirement of electroneutrality causes the concentration of charged molecules in each phase at equilibrium to deviate from the concentrations that would be observed if all the molecules were neutral. To further illustrate the point, we consider a system of two bulk phases, *l* and *r*, in equilibrium containing macromolecules *m* of negative charge $-Z_m$ and ions, Na^+ and Cl^-. Neglecting the interface for the moment, let us also assume that the macromolecules *m* preferentially partition to the left-side phase *l* so that $C_m^{(l)} \ggg C_m^{(r)}$. Because the charges $-Z_m$ on the macromolecules *m* need to be balanced by the charges on the ions Na^+ for each phase, the concentration of Na^+ in the left-side phase at equilibrium is going to be greater than it would otherwise be if the macromolecules were neutral. This effect was first described by F.G. Donnan and is generally known as the Donnan effect.

If two electrodes are respectively inserted into the left and right phases of a two-phase system containing charged molecules (Bamberger *et al.*, 1985), one will generally find that a small electrical potential difference can be measured between the left and right electrodes. To explore the origin of this electrical potential difference, we construct a hypothetical electrochemical cell with two identical electrodes inserted into the left and right phases respectively. The reversible electrodes are not directly inserted into the phases. Rather, metal electrodes are inserted into two matched salt bridges which are then inserted into the phases. A schematic for this cell is given by Fig. 5 in very exaggerated form. Since the salt bridges are matched, i.e., their electrical conductivity is essentially the same, they are canceled out in the potential calculations and need not further concern us here. One should note, however, that the salt bridges contact the phases through capillaries that restrict the diffusion of ions, and that this insures that the composition of the phases does not significantly change over the short period of time required to do the measurements. For any particular charged species *i*, the measurable electrical potential (V) across the electrodes in this cell is given by

$$V = Z_i[\Phi^{(l)} - \Phi^{(r)}] = \frac{\mu_i^{(l)} - \mu_i^{(r)}}{F} \tag{44}$$

where Z_i is the charge of species *i*, F is Faraday's constant, $\Phi^{(k)}$ is the electrochemical potential of either the left ($k = l$) or the right ($k = r$) electrode, and $\mu_i^{(k)}$ is the chemical potential of a charged species *i* in phase *k*. It is worth noting that because both phases must be electrically neutral, the electrochemical potential difference, $\Phi^{(l)} - \Phi^{(r)}$, between the two phases is independent of the species *i* (Newman, 1991). It is important to emphasize that the electrochemical potential difference, $\Phi^{(l)} - \Phi^{(r)}$, and the measured electrical potential, V, depend on the total chemical potential difference of any charged species *i*. If the left- and right-side phases were to be at

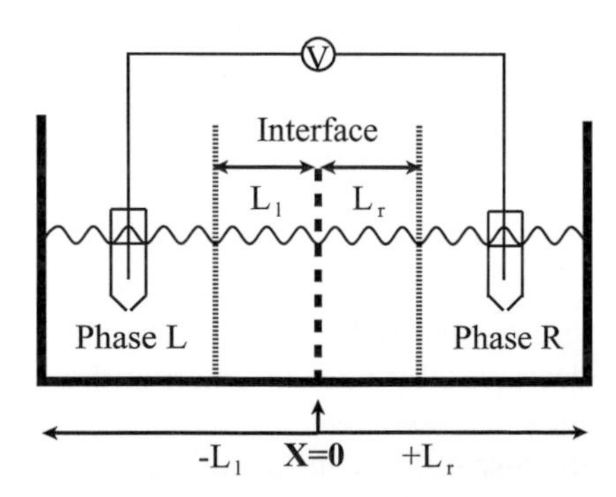

FIG. 5 Sketch of a hypothetical electrochemical cell used to measure the potential between two phases. Note that the metal electrodes are inserted into salt bridges to create a nonequilibrium system, making the measurements possible.

equilibrium, the chemical potentials of any species i would be the same in both phases as already discussed, and the electrochemical potential difference, $\Phi^{(l)} - \Phi^{(r)}$, would have to be zero. But because salt bridges are introduced, a nonequilibrium is created which allows the measurements to be performed.

Nonetheless, the electrochemical potential difference between the phases, $\Phi^{(l)} - \Phi^{(r)}$, is a measure of the reversible work required to transfer electrical charge from the left-side to the right-side phase and vice versa. It is, therefore, quite useful in constructing models for the equilibrium partitioning of charged species between two phases, provided one is willing to ignore the deviation from equilibrium that the existence of a nonzero electrochemical potential difference implies. For example, if a mathematical model exists for the nonelectrostatic contribution to the chemical potential ($\mu_i^{(ne)}$) of a charged species i, one can use the separation of molecular forces arguments from Eq. (30) to construct an expression for the chemical potential and the partitioning of species i between the left and right phases by use of Eq. (35). The expression is

$$\mu_i^{(ne,l)} = \mu_i^{(ne,r)} + FZ_i[\Phi^{(r)} - \Phi^{(l)}] \tag{45}$$

where $\mu_i^{(ne,k)}$ is the nonelectrostatic contribution to the chemical potential of species i and where $Z_iF[\Phi^{(l)} - \Phi^{(r)}]$ contains the electrostatic contribution to the chemical potential of species i. However, one should note that the existence of the electrochemical potential difference $\Phi^{(l)} - \Phi^{(r)}$ does not imply that one phase is positive or negative with respect to another phase since the phases must be electrically neutral near equilibrium. Rather, $F[\Phi^{(l)} - \Phi^{(r)}]$ is simply the work required to transfer a unit charge from one phase to another. The difference $\Phi^{(l)} - \Phi^{(r)}$ is actually a function of the ionic atmospheres of the two phases, i.e., the ionic strength, the charge and size of the ions, the dielectric constants of the phases, etc.

C. Particles at Interfaces

Biological systems normally include particles with a wide range of sizes from small ions with characteristic sizes of the order of 1 Å, to molecules with characteristic sizes of the order of 100 Å, to macromolecules and viruses with characteristic sizes of the order of 1000 Å, and to particles such as organelles with characteristic sizes of the order of 10,000 Å. In general, particles of characteristic size l_t up to 1000 Å (macromolecules and viruses) partition between the phases and interface as do small molecules. These are particles that generally fit within the interface, Eq. (15), because their characteristic size is smaller than the width of the interface, and they experience the interface as just another phase. However, if small particles agglomerate or if large particles exist which are larger than the width of the interface, Eq. (16), these large particles then experience the interface more like a surface than a phase. These large particles may be too few in number to take part in phase formation. In this case, one frequently observes adsorption of these large particles at the interface. In reality, whether adsorption occurs or not also depends on the surface tension between the phases. The reason that the particles collect at the interface instead of somewhere else in the phases is because of the interfacial tension. Reduction of the interfacial area by particle adsorption reduces the free energy of the system. The criterion is that one would generally expect to see adsorption of large particles at the interface if the characteristic particle size l_t is such that

$$l_t \geq \left[\frac{\kappa T}{\sigma}\right]^{1/2} \tag{46}$$

where κ is Boltzmann's constant, T is the temperature, and σ is the surface tension between the two phases. For particles that are smaller than this critical size, one would not expect to observe significant adsorption at the interface.

Once a significant accumulation of particles occurs at the interface, it needs to be treated as another phase, albeit a fairly inhomogeneous one. One can then, in principle, apply the theoretical treatment previously developed for partitioning across interfaces and the treatment for transport across an interface to be developed below. The difficulty here is going to be that of either experimentally measuring or theoretically estimating properties such as diffusivities and activity coefficients inside this inhomogeneous phase. It is simple enough to write a phase equilibrium relation or Fick's laws inside the layer of particles adsorbed at the interface, but obtaining the necessary properties to model the system adequately is not trivial.

V. Transport across Interfaces

The most important and most difficult issue in the transport of chemical species across interfaces in biological systems is that of defining the problem of interest itself. The reason is that biological systems are so complicated that there is little hope of being able to treat all of the phenomena present in detail. Rather, a sequence of decisions must be made with the objective of simplifying the problem and focusing attention on the most important phenomena. Thus, one must, for example, decide which species are to be treated in detail and which are to be treated approximately as background, which species are bigger and which are smaller than the interface, and when the interface is to be treated as a plane and when it is to be treated as phase. An effort has been made in the span of this chapter to provide the reader with some guidance on these issues.

The purpose of this section is to outline the relevant theory that must be used in computations of transport across interfaces, in as general a fashion as possible. The basic equations governing transport are well known and there is an extensive literature on this subject (Bird *et al.,* 1960). Essentially, the treatment consists of postulating a mass balance in the form of a differential equation within each phase and the interface and then adding constraints such as boundary and initial concentrations, either phase equilibrium or some other assumption at interfaces, and when appropriate, electrical neutrality within the phases. Many of these mathematical constructs have no analytical solutions, but the advent of modern computers and numerical methods have adequately addressed this limitation.

A. Mathematical Model for the Phases

We start constructing the mathematical model for the one-dimensional transport of a species i within the left-side and the right-side phase by use Fick's Second Law which states that

$$\frac{\partial C_i^{(l)}}{\partial t} = D_{il}\frac{\partial^2 C_i^{(l)}}{\partial x^2} \quad x \leq - L_l \tag{47}$$

$$\frac{\partial C_i^{(r)}}{\partial t} = D_{ir}\frac{\partial^2 C_i^{(r)}}{\partial x^2} \quad x \geq + L_r \tag{48}$$

where $C_i^{(k)}$ is the concentration of species i in phase k, t is time, D_{ik} is the diffusivity of component i in phase k, and x is a spatial coordinate. Each of these two equations simply states in mathematical form that the concentration of species i within the phase changes with time because of diffusion of i in or out of the phase. Equations (47) and (48) assume that all transport is driven by diffusion so thermal and forced convection are neglected. However, this is a reasonable approximation in biological systems because thermal convection requires significant temperature gradients and forced convection requires a pumping device, and neither of these is normally found in interface transport in this particular situation.

In addition, initial and boundary conditions are also imposed on Eqs. (47) and (48) for the left- and right-side phases. These establish some of the physical characteristics of the system in mathematical form for use with Eqs. (47) and (48). Typically, the initial conditions state that

$$C_i^{(l)} = C_i^{(l,0)} \text{ @ } t = 0; x \leq - L_l \tag{49}$$

$$C_i^{(r)} = C_i^{(r,0)} \text{ @ } t = 0; x \geq + L_r \tag{50}$$

which implies that at some point in time which we call $t = 0$ the concentration of species i had fixed values, $C_i^{(l,0)}$ and $C_i^{(r,0)}$, throughout the left- and right-side phases, respectively. If the phases are very large or if there is some reservoir of species i so that the concentration of species i in the phases does not change, then it is possible to assume that far away from the interface, the concentration of species i is constant with time and space and equal to the initial values so the boundary conditions become

$$C_i^{(l)} = C_i^{(l,0)} \text{ @ } x \leq - \xi L_l \tag{51}$$

$$C_i^{(r)} = C_i^{(r,0)} \text{ @ } x \geq + \xi L_r \tag{52}$$

where ξL_l and ξL_l are the distances away from the interface at which the concentrations reach their initial values, $C_i^{(l)} = C_i^{(l,0)}$ and $C_i^{(r)} = C_i^{(r,0)}$, and where L_l and L_r define the size of the interface, and ξ is a multiplicative factor of the order of 100. If the phases are small or if there is no reservoir for maintaining the concentration of species i constant, then other more complicated boundary conditions must be used. For example, if the system consisting of the left-side and the right-side phases plus the interface can be assumed to be a closed system, i.e., no significant amounts of species i are entering or leaving the system, then a conservation relation can be imposed so that

$$\begin{aligned} V_l C_i^{(l)} + V_r C_i^{(r)} + A_i(L_l + L_r) C_i^{(i)} \\ = V_l C_i^{(l,0)} + V_r C_i^{(r,0)} + A_i(L_l + L_r) C_i^{(i,0)} \end{aligned} \tag{53}$$

where V_k is the volume of phase k, $C_i^{(k)}$ is the concentration of species i in phase k, A_i is the area of the interface, $C_i^{(i)}$ is the concentration of species i in the interface, and the superscript 0 indicates an initial concentration at $t = 0$. Equation (53) simply states that although the concentration of species i can vary among the phases and the interface, the total amount of species i in the system is constant. Equation (53) can be used to generate the boundary conditions,

$$C_i^{(l)} = \frac{V_l C_i^{(l,0)} + V_r[C_i^{(r,0)} - C_i^{(r)}] + A_i(L_l + L_r)[C_i^{(i,0)} - C_i^{(i)}]}{V_l} \quad \text{@ } x \leq - \xi L_l \tag{54}$$

$$C_i^{(r)} = \frac{V_r C_i^{(r,0)} + V_l[C_i^{(l,0)} - C_i^{(l)}] + A_i(L_l + L_r)[C_i^{(i,0)} - C_i^{(i)}]}{V_r} \quad \text{@ } x \geq + \xi L_r \tag{55}$$

where the term involving the interface, $A_i(L_l + L_r)[C_i^{(i,0)} - C_i^{(i)}]$, may well be negligible compared to the other terms since the volume of the interface is often rather small compared to the volume of the phases. For these boundary conditions, however, the concentrations $C_i^{(k)}$ are going to be changing with time which is a manageable but significant complication.

In addition to the diffusion model of Eqs. (47) and (48) and the initial and boundary conditions of Eqs. (49) to (55), the condition of electrical neutrality within the left- and the right-side phases must also be imposed. Since both phases are nonequilibrium phases, the concentration will change with space throughout the phase, so the most appropriate electroneutrality condition is that of Eq. (43). This implies that electrical equilibration is fast compared to transport, and that the phase can be divided into a series of very thin slices so that electrical neutrality is maintained within each slice. Since the sum of the slices is equal to the phase and each slice is electrically neutral, the phase as a whole is electrically neutral. This is generally thought to be the case in biological systems where there are always significant amounts of small mobile ions available to maintain electrical neutrality. Note, however, biological systems also include many large molecules such a proteins that are slow-moving so that equilibrium compositions can take much longer to get established than electrical equilibration.

Last, if the composition of the phases are presumed to be changing with space throughout the phase, then the diffusivity must also be assumed to be changing. The reason is that the diffusivity of a species i is a function of not only the concentration of i but of the concentration of all other species present. There are at least three possible means of incorporating a changing diffusivity into the model. The most rigorous way of treating the diffusivity is to again divide the phase into a series of thin slices and to assume that within each slice the diffusivity has a characteristic value consistent with the composition in that slice. The next approach in descending terms of rigor is to assume that the diffusivity is some function of space within the phase as given by

$$D_{il}(x) = f_l(x) \approx D_{il}|_{x=-\xi L_l} - \frac{D_{il}|_{x=-\xi L_l} - D_{il}|_{x=-L_l}}{\xi L_l - L_l}(\xi L_l + x) \ @ \ -L_l \geq x \geq -\xi L_l \tag{56}$$

$$D_{ir}(x) = f_r(x) \approx D_{ir}|_{x=\xi L_r} - \frac{D_{ir}|_{x=L_r} - D_{ir}|_{x=\xi L_r}}{\xi L_r - L_r}(\xi L_r - x) \ @ \ L_r \leq x \leq \xi L_r \tag{57}$$

where $D_{il}|_{x=-\xi L_l}$ and $D_{il}|_{x=-\xi L_r}$ are the diffusivity of i in the left- and right-side phase far away from the interface and $f_k(x)$ is some function of x for phase k which is nonlinear but could be approximated as a linear function.

The last method is to assume that the diffusivity has some average and constant value throughout the phase

$$D_{il} \approx \frac{D_{il}|_{x=-\xi L_l} + D_{il}|_{x=-L_l}}{2} \tag{58}$$

$$D_{ir} \approx \frac{D_{ir}|_{x=-\xi L_r} + D_{ir}|_{x=-L_r}}{2} \tag{59}$$

which could be taken to be the arithmetic average of the diffusivity of species i just outside the interface and far away from the interface. There are many other means of computing an approximate average value for the diffusivity, however.

B. Mathematical Model for the Interface: Small Diffusing Molecules

The mathematical treatment of the interface depends first on the relative sizes of the molecules of the species of interest i and the interface. As discussed already, if the characteristic size of the molecules of species i is smaller than the width of the interface, Eq. (15), then it is prudent to treat the interface as a separate phase and write Fick's Second Law for diffusive transport inside the interface.

$$\frac{\partial C_i^{(i)}}{\partial t} = D_{ii} \frac{\partial^2 C_i^{(i)}}{\partial x^2} \quad L_l \leq x \leq L_r \tag{60}$$

where $C_i^{(i)}$ is the concentration of species i in the interface and D_{ii} is the diffusivity of species i in the interface. The initial conditions for the interface state that the concentration of species i inside the interface must be equal to the initial concentration of species i in the left-side phase at the boundary between the left-side phase and the interface at $t = 0$ so that

$$C_i^{(i)} = C_i^{(l,0)} \ @ \ t = 0; x =- L_l \tag{61}$$

and that the same criteria apply to the right-side phase so that

$$C_i^{(i)} = C_i^{(r,0)} \ @ \ t = 0; x =+ L_r \tag{62}$$

The boundary conditions for the interface state that the time-dependent concentration of species i inside the interface must always match the time-dependent concentration of species i in the left-side phase at the boundary between the two at all times other than $t = 0$ so that

$$C_i^{(i)} = C_i^{(l)} \ @ \ x =- L_l \tag{63}$$

and that again the same criteria apply to the right-side phase so that

$$C_i^{(i)} = C_i^{(r,0)} \ @ \ x = + L_r \tag{64}$$

Additionally, one more boundary condition must be imposed on both phases and the interface which requires that the flux of species i must be constant across the boundary between the interface and each of the two phases. This simply states that molecules must come out of the boundary at the same rate at which they came in, i.e., molecules cannot be created or consumed at the boundary between the interface and either phase. This is expressed by

$$D_{il} \frac{\partial C_i^{(l)}}{\partial x} = D_{ii} \frac{\partial C_i^{(i)}}{\partial x} \ @ \ x = - L_l \tag{65}$$

$$D_{ir} \frac{\partial C_i^{(r)}}{\partial x} = D_{ii} \frac{\partial C_i^{(i)}}{\partial x} \ @ \ x = + L_r \tag{66}$$

As done for the two phases, electrical neutrality is also imposed on the interface. Because concentrations are changing rather steeply with space and time throughout the interface, the only appropriate electroneutrality condition is that of Eq. (43). Again, this implies that the interface is divided into a series of thin slices and that electrical neutrality is maintained within each slice. Thus, the interface as a whole is electrically neutral but the concentration of the various species will be changing with time and space throughout the interface.

Since the interface has, by definition, rather steep composition gradients with space, the diffusivity is going to be changing steeply throughout the interface. The reason is again that the diffusivity of species i is a function of the concentration of not only i but of all the other species present. The strategy for modeling the diffusivity within the interface follows the treatment used for the two phases. The most rigorous way is to divide the interface into a series of thin slices and to assign a diffusivity to each. The next most rigorous method is to assume that the diffusivity is some function of space within the interface that can, perhaps, be approximated by a linear interpolation between the two phases as given by

$$D_{ii}(x) = f_i(x) \approx D_{il}|_{x=-L_l} + \frac{D_{ir}|_{x=L_r} - D_{il}|_{x=-L_l}}{L_l + L_r}(L_l + x) \ @ \ -L_l \leq x \leq + L_r \tag{67}$$

where f_i is some function of distance x, and $D_{il}|_{x=-L_l}$ and $D_{ir}|_{x=-L_r}$ are the diffusivities of species i just outside of the interface in the left-side and right-side phases, respectively. The last and least exact approach is to assign an average value to the diffusivity throughout the interface. This average diffusivity can be approximated as the arithmetic average of the diffusivities of species i in the surrounding phases just outside the interface as given by

$$D_{ii} \approx \frac{D_{il}|_{x=-L_l} + D_{ir}|_{x=+L_r}}{2} \tag{68}$$

C. Mathematical Model for the Interface: Large Diffusing Molecules

Since the mathematical modeling of transport across the interface is intrinsically tied to the relative sizes of the diffusing species and the interface, the treatment for the diffusive transport of a species i whose molecules are larger than the width of the interface, Eq. (16), will differ from that developed in Section B above. One should note, however, that since the width of the interface is about 10 times the characteristic size of the molecules of the largest nontrace species, there are relatively few situations to which this is applicable with the exception of protein agglomerations, viruses, etc. In this case, the molecules are never completely immersed in the interface during their transport between the two phases. The properties of these molecules inside the interface are, therefore, unlikely to be important as far as interface transport is concerned and, for most practical purposes, the interface can be approximately treated as a surface. It is, thus, unnecessary to set up Fick's Second Law and initial and boundary conditions inside the interface. The mathematical model then consists of that developed for two phases in Section A above, i.e., Eqs. (47) to (59), as appropriate. However, one must also require that the flux of species i across the interface must match on both sides as already postulated, Eqs. (65) and (66), so that

$$D_{il} \left. \frac{\partial C_i^{(l)}}{\partial x} \right|_{x=-L_l} = D_{ir} \left. \frac{\partial C_i^{(r)}}{\partial x} \right|_{x=+L_r} \tag{69}$$

and that some type of relationship be postulated between the concentrations of species i on both sides of the interface. If the two phases are at equilibrium, then the partition coefficient of Eq. (37) relates the concentration as given by

$$C_i^{(l)}|_{x=-L_l} = K_i C_i^{(r)}|_{x=+L_r} \tag{70}$$

where the partition coefficient can be either measured experimentally or estimated from the activity coefficient of species i. If the two phases are not at equilibrium, usually because transport of species i across the interface is relatively slow (Levine *et al.*, 1992), then a mass transfer expression is postulated, giving

$$J_{ix}|_{x=-L_l} = J_{ix}|_{x=+L_r} = \alpha_i[K_i C_i^{(r)}|_{x=+L_r} - C_i^{(l)}|_{x=-L_l}] \tag{71}$$

which again states that all molecules entering the interface at $x = -L_i$ must exit at $x = +L_r$ and vice versa, and where α is a mass transfer coefficient, usually fitted empirically. Choosing between Eqs. (70) and (71) is not simple, but as a guide one can use the characteristic time t_c for transport of species i across the interface given by

$$t_c = \frac{(L_l + L_r)^2}{D_{ii}}. \tag{72}$$

Thus, if the time period of interest Δt, i.e., characteristic time of the phenomenon of interest or, perhaps, the experiment, is much larger than t_c, then the interface is equilibrated and Eq. (70) should provide an adequate approximation. If the time period of interest Δt is much smaller than t_c, then the interface is unlikely to be at equilibrium and Eq. (71) would provide a better approximation.

VI. The Time Scale for Phase Formation

The challenges posed by transport across interfaces in biological systems are mainly due to the tremendous complexity of biological systems and the difficulty of obtaining physical measurements in these systems. Rather than attempt to catalog all of these problems, this section will focus on the very important questions of the time scale for phase formation and equilibration in biological systems.

When two phases start to spontaneously form from a single phase inside a biological system, the process involves various species diffusing toward or away from the forming phases. Thus, the rate at which the phases form is dependent on the speed of diffusive transport in the system. The slower the diffusion, the longer it takes to form phases, and the faster the diffusion, the faster that phases form. Thus, if one has a measure of the dimensions of the two phases, then it may be possible to estimate how long it might take to form the phases. The reasonably exact answer to this question can be obtained by solving the model equations developed in the previous section, starting with one homogeneous phase and progressively adding the interface and the other phase. This calculation, however, is well beyond the scope of this chapter. A much rougher estimate can be obtained by assuming that the rate of phase formation is set by the slowest diffusing species required for phase formation, and then assuming that the time scale for phase formation, t_p, is equal to the characteristic time for the diffusion of the slowest species across the region comprising the two phases. If the system consists of two phases of characteristic size L_p each, so that the characteristic size of the system is $2L_p$, then t_p is given by

$$t_p = \frac{(2L)^2}{D_s} \tag{73}$$

where D_s is the diffusion coefficient of the slowest diffusing species required for phase formation. This excludes species found in such small concentrations that they would not play a significant role in phase formation. Unfortunately, there is no simple criterion for determining whether any given species is required for phase formation.

Whether any two phases inside a biological system are equilibrated or not is another question. This is an important consideration because one would suspect that phases inside biological systems are equilibrated with respect to small molecules but not large ones such as proteins. The reason that this is important is that the mechanisms by which nonequilibrium multiphase systems maintain their stability are dynamic, and these mechanisms may be more fragile than the equilibrium mechanisms by which multiphase equilibrium systems exist. The equilibration of phases in biological systems can probably be studied by solving the model equations discussed in the previous section and comparing the results to experiment, but this is also well beyond the scope of this chapter. A rough estimate of whether a given system of two phases is or is not equilibrated with respect to a given species i can be obtained. This can be done by looking at the time the system has existed since formation, Δt_e, and the characteristic times t_i for diffusion of species i. This is given by

$$t_i = \frac{(2L)^2}{D_i} \tag{74}$$

Thus, if$\Delta t_e > t_i$, then the system is likely to be equilibrated with respect to species i. But if $\Delta t_e \leq t_i$, then the system is unlikely to be equilibrated with respect to species i because it would not have been in existence long enough for species i to have diffused throughout the system.

VII. Summary

Understanding the nature of interfaces and the transport of molecules across them in biological systems is an important but rather difficult problem. In the span of this chapter, an effort has been made to outline the known theory in both of these areas and to suggest some directions for further scientific inquiry. Thus, the nature of interfaces has been discussed, starting with fundamental theory and ending with some of the practical consequences of the theory. These results tell us that the width of interfaces and the characteristic sizes of the molecules in the system are critical parameters in determining when interfaces need to be treated in detail and when they do not. Roughly, small molecules see the interface as another phase and large molecules see it more as a surface. The properties of molecules in the phases surrounding an interface and inside the interface have also been discussed starting from fundamental theory. This discussion tells us that the properties of phases are consequences of molecular properties and interactions, and that inside interfaces phase properties can be steeply changing functions of position. It is, therefore, unwise to treat them as constant. The formal equations used to mathematically model transport from a phase across an interface and on to another phase have also been presented and discussed. The treatment again depends on the relative dimensions of the width of the interface and the characteristic molecular sizes. Roughly, when the molecules are smaller than the width of the interface, the properties and transport inside the interface need to be treated in detail, and when the molecules are larger than the width of the interface, then these can be largely ignored. Because much of the activity of the various compartments in cytoplasm may consist of the transport of molecules across the interface between compartments, neglecting the interface or treating it as a surface would likely miss important behavior. Last, the questions of the time scale for phase formation and equilibration in biological systems have been discussed. These are important questions because phases can form and disappear in the nonequilibrium environment inside biological systems. Since it is suggested that at least part of the structure in cytoplasm is due to multiphase formation, this question merits further scientific inquiry.

Acknowledgments

The author wishes to acknowledge the very generous assistance with the writing of this chapter from Professor Donald Elliott Brooks from the Department of Pathology and Laboratory Medicine at the University of British Columbia. The author also acknowledges the very steadfast encouragement of Dr. Harry Walter, retired from the Laboratory of Chemical Biology, Veterans Affairs Medical Center, Long Beach, California, who has been the very soul of this entire project. The author also wishes to acknowledge his managers, Mr. Gregory

Carroll and Dr. Subhas K. Sikdar, for creating and maintaining the spirit of scientific inquiry which made the writing of this chapter possible.

References

Albertsson, P.-A. (1986). "Partition of Cell Particles and Macromolecules," 3rd ed. Wiley, New York.

Allen, M. P., and Tildesley, D. J. (1989). "Computer Simulation of Liquids." Oxford University Press, Oxford.

Bamberger, S., Brooks, D. E., Sharp, K. A., van Alstine, J. M., and Webber, T. J. (1985). Preparation of phase systems and measurement of their physicochemical properties. *In* "Partitioning in Aqueous Two Phase Systems: Theory, Methods, and Applications to Biotechnology" (H. Walter, D. E. Brooks, and D. Fisher, eds), Chapter 3, pp. 113–119. Academic Press, Orlando, FL.

Bird, R. B., Steward, W. E., and Lightfoot, E. N. (1960). "Transport Phenomena." Wiley, New York.

Cabane, B., and Duplessix, R. (1982). Organization of surfactant micelles adsorbed on a polymer molecules in water: A neutron scattering study. *J. Phys.* (*Paris*) **43,** 1529–1542.

Cabezas, H. (1996). Theory of phase formation in aqueous two phase systems. *J. Chromatogr. B* **680** 3–30.

Croxton, C. A. (1975). The liquid surface. *In* "Introduction to Liquid State Physics," Chapter 7, pp. 170–190. Wiley, New York.

Durchschlag, H. (1989). Determination of the partial specific volume of conjugated proteins. *Colloid Polym. Sci.* **267**, 1139–1150.

Durchschlag, H., and Rainer, J. (1982). Partial specific volumes changes of proteins: Densimetric studies. *Biochem. Biophys. Res. Commun.* **108,** 1074–1079.

Durchschlag, H., and Rainer, J. (1983). Partial specific volume changes of proteins: Ultracentrifugal and viscometric studies. *Int. J. Biol. Macromol.* **5,** 143–148.

Friedman, H. L. (1985). Simulation and diffraction. *In* "A Course in Statistical Mechanics," Chapter 5, pp. 93–109. Prentice-Hall, Englewood Cliffs, NJ.

Levine, M. L., Cabezas, H., and Bier, M. (1992). Transport of solutes across aqueous phase interfaces by electrophoresis: mathematical modeling. *J. Chromatogr.* **607,** 113–118.

Millero, F. J. (1972). The partial molar volume of electrolytes in aqueous solution. *In* "Water and Aqueous Solutions: Structure, Thermodynamics, and Transport Properties" (R. A. Horne, ed.), p. 519. Wiley-Interscience, New York.

Newman, J. S. (1991). "Electrochemical Systems." Prentice-Hall, Englewood Cliffs, NJ.

Prausnitz, J. M., Lichtenthaler, R. N., and Gomes de Azevedo, E. (1986). "Molecular Thermodynamics of Fluid Phase Equilibria," 2nd ed. Prentice-Hall, Inc. Englewood Cliffs, NJ.

Senti, F. R., Hellman, N. N., Ludwig, N. H., Babcock, G. E., Tobin, R., Glass, C. A., and Lamberts, B. L. (1955). Viscosity, sedimentation, and light-scattering properties of fractions of an acid hydrolyzed dextran. *J. Polym. Sci.* **17,** 527–546.

Skold, K., and Price, D. L. (1987). Neutron scattering. *In* "Methods of Experimental Physics" (R. Celotta and J. Levine, eds.), Vol. 23, Academic Press, Orlando, Florida.

Van Holde, K. E. (1985). Scattering. *In* "Physical Biochemistry," Chapter 9. pp. 209–234. Prentice-Hall, Englewood Cliffs, NJ.

Walter, H., and Brooks, D. E. (1995). Phase separation in cytoplasm, due to macromolecular crowding, is the basis for microcompartmentation. *FEBS Letts* **361,** 135–139.

Walter, H., Brooks, D. E., and Fisher, D., eds. (1985). "Partitioning in Aqueous Two-Phase Systems: Theory, Methods, Uses, and Applications to Biotechnology." Academic Press, Orlando, FL.

Zana, R. (1980). Partial molal volumes of polymers in aqueous solutions from partial molal volume group contributions. *J. Polym. Sci.* **18,** 121.

Compartmentalization of Enzymes and Distribution of Products in Aqueous Two-Phase Systems

Folke Tjerneld and Hans-Olof Johansson
Department of Biochemistry, University of Lund, S-22100 Lund, Sweden

---

Phase separation is a common phenomenon in water solutions of polymers due to "polymer incompatibility." Polymeric aqueous two-phase systems are much used for separations in biochemistry and cell biology. When macromolecules are included in a phase system, it is often possible to obtain a one-sided distribution to one of the phases, i.e., the macomolecule is compartmentalized within one aqueous phase. This chapter describes the thermodynamic forces which govern the partitioning of molecules in aqueous two-phase systems. For a high molecular weight macromolecule, e.g., an enzyme, both enthalpic and entropic effects contribute to a one-sided partitioning. Molecules of low molecular weight will be more evenly distributed between the phases. These mechanisms are significant in biological systems and can be used for enzyme reactors in bioconversions. Enzymatic reactions can take place with enzyme and substrate compartmentalized in one of the phases. A low-molecular weight product which is evenly partitioned between the phases can be continuously removed from the enzyme-substrate compartment. These principles are described in the enzymatic conversion of cellulose in an aqueous two-phase system.

KEY WORDS: Aqueous two-phase systems, Protein partitioning, Thermodynamics, Flory-Huggins theory, Bioconversion, Enzyme reactions.

I. Introduction

Aqueous polymer two-phase systems are obtained in water solutions of two polymers with differing chemical structure and above certain concentrations. In biochemistry and cell biology studies, an aqueous phase system composed of poly(ethylene glycol) (PEG) and dextran has been much used for separations of macromolecules, membranes, organelles, and cells (Walter *et al.,* 1985; Albertsson, 1986; Walter and Johansson, 1994). The physical chemical basis for phase separation in water solutions of polymers has been analyzed in considerable detail (Brooks *et al.,* 1985; Gustafsson *et al.,* 1986; Abbott *et al.,* 1992). Phase separation in polymer solutions is a general phenomenon and is due to "polymer incompatibility." The entropy increase upon mixing of molecules in a solution will favor the formation of a single phase. However, the mixing entropy is much reduced in solutions of macromolecules and, for long polymer chains, the entropy of mixing is a relatively small term in the total free energy of mixing. A weak repulsive enthalpic interaction between monomer units on the different polymers is sufficient to dominate the mixing entropy. The solution will then separate into two phases where each of the two polymers is enriched in one of the phases.

In the well-known PEG-dextran-water system, PEG is enriched in the top phase and dextran in the bottom phase. Both phases have a water content of 80–90%. In the phase diagram (see chapter by Johansson/Walter, this volume), the polymer concentrations needed to obtain phase separation is shown by the binodal curve which separates the one- and two-phase regions. The tie-lines show the composition of the two phases in equilibrium. The degree of polymerization, i.e., the polymer molecular weight, has a strong influence on phase separation and, by increasing the molecular weight of the polymers, the phase separation will occur at lower polymer concentrations. This will also lead to more one-sided distributions of the two polymers between the phases.

Substances can be partitioned in aqueous two-phase systems and their partitioning behavior will depend on their size and surface properties. Several different mechanisms govern the partitioning and can be used to influence the distribution of a substance. The ionic composition will determine the nature and magnitude of the electrical potential difference between the phases. The charge of a partitioned substance will then have a strong influence on the partition coefficient. Direct interactions, e.g., hydrophobic, between the partitioned substance and the phase-forming polymers also contribute to the partitioning. These interactions, and thus the partition coefficient, can be influenced by polymer concentrations and polymer molecular weights (see chapter by Johansson/Walter, this volume).

As will be shown, macromolecules and particles can be made to favor one of the phases in an aqueous two-phase system. The reason for this can be found in the high molecular weights for these substances which, due to both entropic and enthalpic effects, strongly contribute to extreme partitioning to one of the phases. Thus, enzymes (high molecular weight) can be partitioned to one of the phases in a two-phase system while substrate or product molecules (low molecular weight) will be much more evenly partitioned. The enzyme is thus compartmentalized within one aqueous phase with access to substrate molecules partitioned to the same phase. Depending on its properties, the product molecules can be partitioned to the other phase and removed from the enzyme reaction. The nature of the thermodynamic mechanisms in macromolecular solutions may thus lead to different degrees of compartmentalization of macromolecules such as enzymes, depending on surface properties and to distribution of substrates and products between the phases. These mechanisms will be significant in biological systems and can be used for enzyme reactors in bioconversions.

II. Thermodynamic Forces for Partitioning of Proteins and Solutes between Two Aqueous Polymer Phases

With the Flory-Huggins theory for polymer solutions (Flory, 1953), it is possible to get insight into the thermodynamic driving forces for partitioning in aqueous two-phase systems. In this theory, the solution is regarded as an incompressible regular lattice. Each component in this lattice is divided into "monomers," where each monomer occupies a lattice site. Furthermore, each monomer experiences the resultant interaction from the mean composition in the solution; thus, the theory is a mean-field theory. In this model, proteins are treated as flexible polymers (Brooks *et al.,* 1985; Johansson *et al.,* 1998). This approximation may seem crude. However, by choosing a suitable energetic interaction between the protein and the other components and by choosing a suitable polymerization degree for the protein, it is possible to qualitatively reproduce the partitioning behavior of the protein (Johansson *et al.,* 1996, 1998).

In a lattice with p components, the total moles of lattice sites are

$$N = \sum_{i=1}^{p} M_i n_i \tag{1}$$

where n_i is the number of moles of component i and M_i is its degree of polymerization. Furthermore, each component is homogeneous, i.e., a polymeric component is a homopolymer. The volume fraction of component i in the system is

$$\phi_i = M_i n_i / N$$

The enthalpy of mixing can be expressed by introducing an effective pairwise interaction parameter w_{ij} defined through

$$w_{ij} = w'_{ij} - \frac{w'_{ii} + w'_{jj}}{2} \tag{2}$$

where the prime denotes an absolute interaction parameter. All w_{ii} are put equal to zero in the following discussion. This can be done without loss of generality for the derived effects below. Thus, the enthalpy of mixing is then

$$\Delta H_{mix} = N \sum_{i=1}^{p-1} \sum_{j=i+1}^{p} \phi_i \phi_j w_{ij} \tag{3}$$

The entropy of mixing is

$$\Delta S_{mix} = -NR \sum_{i=1}^{p} (\phi_i / M_i) \ln \phi_i \tag{4}$$

By minimizing the free energy $\Delta G_{mix} = \Delta H_{\mathrm{mix}} - T\Delta S_{mix}$ for a given composition, it is possible to reproduce the composition of an aqueous two-phase system (for more detail, see Gustafsson *et al.,* 1986; Johansson *et al.,* 1998).

A relatively simple expression can be derived for an uncharged solute, e.g., a protein at the isolelectric point, which relates the partition coefficient of the solute with parameters such as polymerization degree, interaction parameters, and temperature (Johansson *et al.,* 1998). The partition coefficient K of a component in this model is a function of the components' partial molar mixing enthalpy and entropy, respectively. In this discussion, the effects of entropy and enthalpy will be discussed separately.

By substituting N^t, n^t, ϕ^t_i into Eqs. (1), (3), and (4), where superscript t stands for top phase, the free energy of the top phase is obtained. The chemical potential for component k (the protein) in the top phase is $\mu^t_k = \partial \Delta G^t_{mix} / \partial n^t_k$. In a similar way, the chemical potential for the bottom phase (superscript b) can be obtained. At equilibrium

$$\mu^t_k - \mu^b_k = 0 \tag{5}$$

A. The Entropic Contributions in the Partition Coefficient

Consider now the special case in which enthalpic contributions in Eq. (4) are canceled. Then Eq. (5) yields

$$-\mathrm{T}\Delta\left(\frac{\partial \Delta S_{mix}}{\partial n_k}\right) = M_k RT\left(\sum_{i=1, i\neq k}^{p} - \left(\frac{\phi^t_i}{M_i} - \frac{\phi^b_i}{M_i}\right)\right) + RT(\ln\phi^t_k - \ln\phi^b_k) - RT(\phi^t_k - \phi^b_k) = 0 \tag{6}$$

For the case in which component k (the protein) is very dilute, the last term in Eq. (6) can be considered to be zero. By writing the partition coefficient K as the ratio of the volume fraction of component k in the two phases, Eq. (6) yields

$$\ln K_k = M_k\left(\sum_{i=1,i\neq k}^{p} \frac{\phi_i^t}{M_i} - \sum_{i=1,i\neq k}^{p} \frac{\phi_i^b}{M_i}\right) \tag{7}$$

Since $\frac{\phi_i^t}{M_i} = \frac{n_i^t}{N^t} = \frac{n_i^t}{\rho V^t}$, where ρ is the number of lattice sites per volume and V^t is the top phase volume (analogous expressions for the bottom phase), Eq. (7) can be further simplified

$$\ln K_k = \frac{M_k}{\rho}\left(\frac{n^t}{V^t} - \frac{n^b}{V^b}\right) \tag{8}$$

where n^t and n^b are the total number of moles in the top and bottom phases, respectively. This expression indicates that the protein k will partition to the phase which has the higher number of molecules per volume unit. There will be an entropic driving force for a component to partition to this phase, and by increasing the polymerization degree of the partitioned component (the protein molecular weight M_k), this driving force will increase (Johansson *et al.*, 1997). An even partitioning is obtained if the number of molecules per volume unit is the same in both phases. This entropic driving force leads to an exclusion from the phase with the lowest number of molecules per volume unit. The effect is also called the excluded volume effect.

B. Enthalpic Contributions to the Partition Coefficient

Consider now a case where entropic contributions in Eq. (5) are canceled in such a way that the number of molecules per volume unit is the same in both phases, but the enthalpic contributions are not canceled. Furthermore, the concentration of component k (the protein) is very low, which means that the transfer of k between the phases does not significantly change the composition of the underlying system. Equation (4) then yields

$$\Delta\left(\frac{\partial \Delta H_{mix}}{\partial n_k}\right) = M_k \sum_{i=1,i\neq k}^{p-1} \sum_{j=i+1,j\neq k}^{p} - (\phi_i^t\,\phi_j^t - \phi_i^b\,\phi_j^b)\,w_{ij} + M_k \sum_{i=1,i\neq k}^{p} (\phi_i^t - \phi_i^b)\,w_{ik} + RT(\ln\phi_k^t - \ln\phi_k^b) = 0 \tag{9}$$

Equation 9 can be written in a more compact form.

$$\ln K_k = -\frac{M_k}{RT}\left(\sum_{i=1,i\neq k}^{p-1} \sum_{j=i+1,j\neq k}^{p} -(\phi_i^t\phi_j^t - \phi_i^b\phi_j^b)w_{ij} + \sum_{i=1,i\neq k}^{p} (\phi_i^t - \phi_i^b)w_{ik} \right) \quad (10)$$

The last sum on the right side of Eq. (10) is the difference between the average interaction of component k and the top phase components and corresponding average interaction of component k and the bottom phase components. This term can be written as $-M_k/RT\ (w_{tk} - w_{bk})$ where w_{tk} and w_{bk} are interaction parameters between component k and an average lattice site in the top and bottom phases, respectively. The first sum in Eq. (10) can be further simplified by identifying this as the difference of the self energies (E_t and E_b) of the top and bottom phases, respectively. The self energy is the total enthalpy of formation of the phase from the pure components (except the protein) per lattice site (Johansson *et al.,* 1998). Thus one gets

$$\ln K_k = -\frac{M_k}{RT}[w_{tk} - w_{bk} - (E_t - E_b)] \quad (11)$$

From this equation, it is clear that if the protein has strong attractive interaction with the top phase components, the interaction parameter w_{tk} will be more negative, which will lead to an increased lnK value, i.e., partitioning to the top phase, and vice versa for w_{bk}. Also, there is an underlying driving force which depends on the self energies E_t and E_b of the two phases. The protein k will partition into the phase which has the highest self energy, i.e., to the phase where the components have the strongest repulsive interactions with each other. This can be understood so that the protein locally breaks the repulsive interactions between the phase components and this results in a driving force toward the phase where this effect is strongest. From Eq. (11) it can be seen that the preference of the protein for one of the phases due to enthalpic interactions increases linearly with the molecular weight M_k.

C. Entropic and Enthalpic Contributions

For the case that neither enthalpic nor entropic contributions are canceled in Eq. (5), the partition coefficient of protein k can be written as the sum of Eqs. (8) and (11).

$$\ln K_k = \ln K_k^H + \ln K_k^S \quad (12)$$

Superscripts H and S stand for enthalpic and entropic contributions, respectively. Again, this expression is valid for low concentrations of k. The partition coefficient of the protein ($\ln K_k$) in Eq. (12) is linearly dependent on the polymerization degree (molecular weight) of k, which is clearly seen in Eqs. (8) and (11) (Johansson *et al.,* 1997). Thus, Eq. (12) predicts an increased partitioning to the preferred phase with increasing polymerization degree of k. Proteins with high molecular weights will have stronger tendencies for one-sided partitioning between the phases compared to smaller proteins, based on both entropic and enthalpic effects in the phase systems. This has been experimentally observed for high molecular weight proteins (Albertsson *et al.,* 1987).

D. Electrostatic Potential Difference Generated by Salts

It has been demonstrated experimentally that some salts partition unequally between the two phases in an aqueous phase system (Johansson, 1974). This is due to the different affinities of the ions for the two phases and the resulting macroscopic partition coefficient for the salt is a result of the electroneutrality condition, which forces one ion to partition together with a counterion. The differing tendencies of ions to partition between the phases lead to a driving force for partitioning of charged species in the system. Such differing affinities of ions for the phases lead to the creation of an electrostatic potential difference over the interface between the phases. This has been explained theoretically by Albertsson (1986). He was able to show that the potential difference could be written as a function of the hypothetical partition coefficients of the individual ions, where the hypothetical individual partition coefficient for an ion could be taken as a measure of the affinity of this ion for one of the phases. The electrical potential difference caused by the addition of a salt will thus influence the partitioning of a charged solute, and the partitioning (log K) of a macromolecule is linearly related to the net charge of the macromolecule. This can be seen in Eq. (13)

$$\ln K_k = \ln K_k^o + \frac{z_k}{z_C - z_A}(\ln K_A^o - \ln K_C^o) \quad (13)$$

where $K^o{}_k$ is the hypothetical partition coefficient of the charged protein (charge z_k). $K^o{}_A$ and $K^o{}_C$ are the hypothetical partition coefficients of the salt anion and cation, respectively (Albertsson, 1986; Pfennig and Schwerin, 1995). For highly charged macromolecules, e.g., nucleic acids, the salt effects are very strong and lead to extreme partitioning. For proteins, this electrostatic effect is strongly dependent on the net charge of the protein and, thus, on the number of titratable amino acids on the protein surface and the solution pH (Johansson, 1974). Depending, then, also on the actual ionic composition, the electrostatic potential difference between the phases may constitute a driving force, in addition to the entropic and enthalpic effects described above, to compartmentalize an enzyme in one of the aqueous phases.

III. Enzyme Reactions in Aqueous Phase Systems

The consequences of the thermodynamic principles for partitioning between two aqueous polymeric phases are that high molecular weight species can be made to partition one-sidedly between the phases and that low molecular weight solutes partition more evenly. Enzymatic reactions can thus take place in an aqueous phase system with the enzyme partitioned to one phase, i.e., compartmentalized, and the substrates and products partitioning depending on their molecular weights. With a high molecular weight substrate for the enzyme, e.g., a macromolecule, it is possible to have an enzymatic reaction system where both enzyme and substrate are compartmentalized in one aqueous phase and a low molecular product is continuously removed from the enzyme compartment due to the spontaneous partitioning to the other phase. If the product is further converted, the product concentration can be maintained at a low level in the enzyme-substrate compartment, thus leading to reduced product inhibition of the enzyme catalyzed reaction. This is shown schematically in Fig. 1.

One example which describes these principles will be presented here. It is the enzymatic hydrolysis of cellulose. However, the system described can be realized with other macromolecular substrates for enzymes, e.g., hydrolysis of starch (Larsson *et al.,* 1989), proteins, and nucleic acids.

Aqueous two-phase systems composed of the polymers PEG and dextran have been used in laboratory-scale enzymatic cellulose hydrolysis processes (Tjerneld *et al.,* 1985a,b,c). In a biotechnical conversion of cellulose, these systems offer several advantages. The enzymes can be partitioned to one of the phases together with substrate. Due to the low interfacial tension in the aqueous phase system, there is rapid equilibrium of the formed product across the interface, which facilitates mass transfer of produced sugar. After phase separation, the phase containing the product is removed while the enzyme is retained in the reactor in the second phase and can be resupplied with substrate. The phase system thus makes it possible to recycle the enzymes in the bioconversion process.

The three enzyme activities involved in cellulose hydrolysis are cellobiohydrolase (CBH), endo-β-glucanase (EG), and β-glucosidase. The most studied enzyme system has been the cellulolytic enzymes produced by the fungus *Trichoderma reesei,* where the main enzymes are CBH I and II, EG I and II. CBH and EG act in synergism to break down cellulose particles

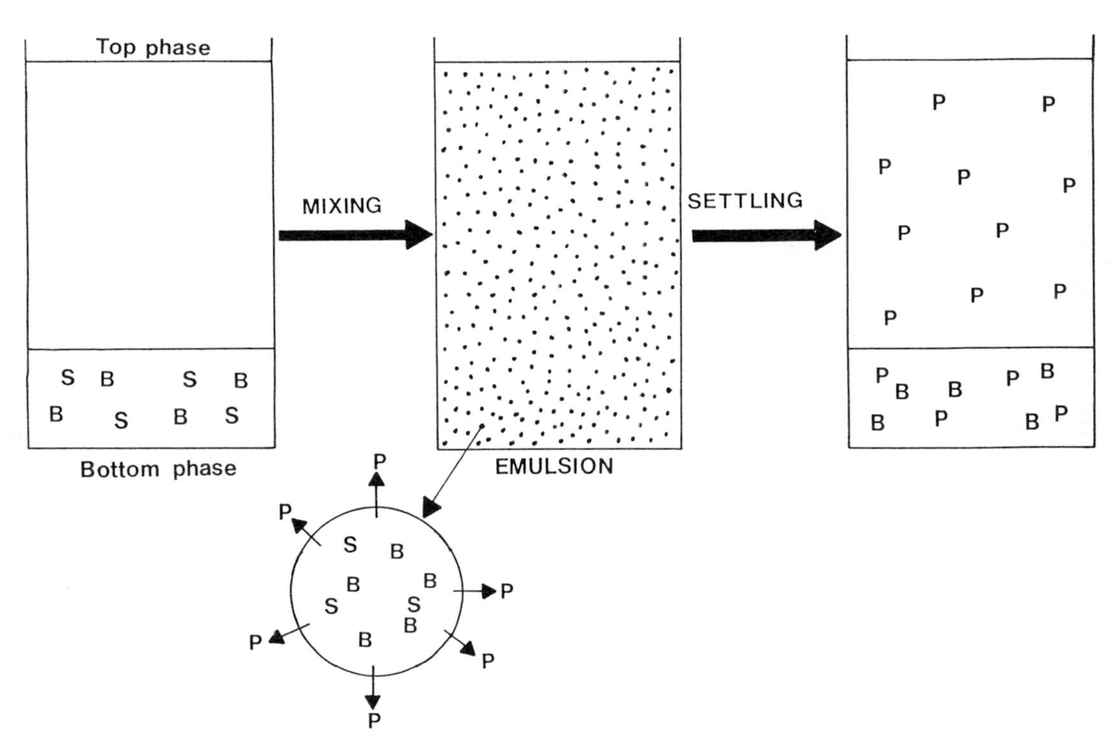

FIG. 1 A common arrangement when aqueous two-phase systems are applied in bioconversion processes. The high-molecular weight substrate (S) is partitioned together with the biocatalyst (B, e.g., an enzyme) to the bottom phase. The low-molecular weight product (P) is partitioned equally between the phases. The conversion takes place during mixing of the phases and equilibrium is established. After settling, the top phase can be removed with the product and a new top phase can be added.

and the products are glucose and cellobiose (Tomme *et al.,* 1995). In order to get a total conversion of cellulose to glucose, it is necessary to add β-glucosidase, which is often obtained from *Aspergillus niger.*

The optimal conditions for cellulose hydrolysis with *Trichoderma* enzymes are at pH values of 4.5–5.0 in 0.05M acetate or citrate buffers at 40–50°C. In a PEG-dextran phase system, the low molecular weight sugar products, glucose and cellobiose, partition evenly between the phases (Tjerneld *et al.,* 1985a). This is understandable from the described thermodynamics, where the low molecular sugars are relatively hydrophilic, and thus, for enthalpic reasons, will not prefer either phase, and the entropy of mixing dominates in the free energy leading to an equal partition between the phases for these low molecular weight products.

Cellulose particles are partitioned to the bottom phase in dextran-PEG systems. Here, the enthalpic interactions dominate in the partitioning and the polysaccharide particles prefer the dextran phase. In the partitioning of macroscopic particles, it is necessary to consider the influence of the surface energies, i.e., between cellulose particle and PEG phase, dextran phase, or the interface between the phases, which here also contribute to partition the cellulose to the dextran phase.

The cellulolytic enzymes have a strong affinity for particulate cellulose due to the domain structure of the enzymes with a catalytic domain and a cellulose binding domain separated by a linker (Tomme *et al.,* 1995). Figure 2 shows the partition coefficients for the cellulase enzymes at increasing

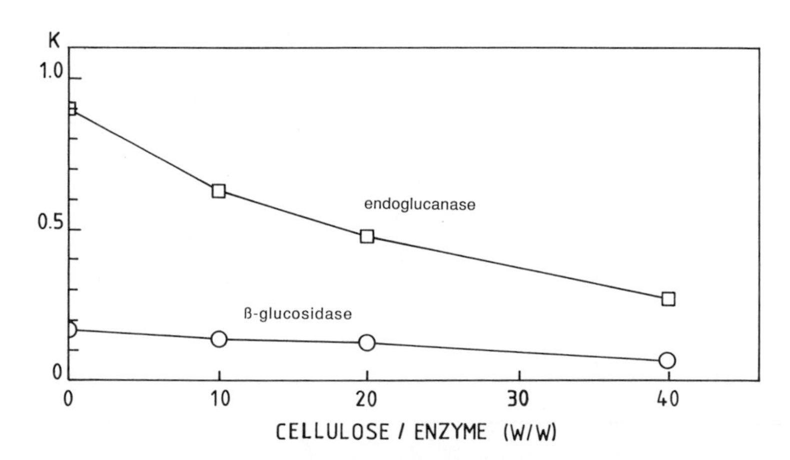

FIG. 2 The partition coefficient (K) for endo-β-glucanase and β-glucosidase from *Trichoderma reesei* as a function of the cellulose-to-enzyme ratio (w/w). The phase system was composed of 8% (w/w) Dextran T40 (Mw 40,000), 6% (w/w) PEG 8000 (Mw 8,000), 10 mM sodium acetate buffer, pH 4.8, with 2% Avicel PH 105 cellulose. The cellulase enzymes were from NOVO Celluclast 2.0L (NOVO A/S, Bagsvaerd, Denmark). [Data from Tjerneld *et al.,* 1985a.]

ratios of particulate cellulose to enzyme (w/w) in dextran T40-PEG 8000 phase systems (Tjerneld *et al.,* 1985a). The distribution of endo-β-glucanase activity to the bottom (dextran) phase at higher cellulose/enzyme ratios is due to the strong adsorption of endo-β-glucanase to the cellulose particles. β-glucosidase strongly partitions to the bottom phase in the PEG-dextran two-phase system. This enzyme does not specifically bind to cellulose particles and only slightly decreased K values at high cellulose/enzyme ratios are observed.

Polymer molecular weight influences partitioning of substances. Generally, if the bottom phase polymer molecular weight is increased, the substance partitions more to the top phase; increasing the top phase polymer molecular weight, the substance partitions more to the bottom phase (Albertsson *et al.,* 1987). Dextran of low molecular weight (40,000) was found to be optimal for partitioning of the enzymes to the dextran phase (Tjerneld *et al.,* 1985a). The K values of cellulases were also influenced by the molecular weight of PEG. Both endo-β-glucanase and β-glucosidase were distributed more to the bottom phase when a high-molecular-weight fraction of PEG is used. By combining the effects of polymer molecular weight and adsorption to cellulose, very low K values could be obtained. In Fig. 3, the K values for endo-β-glucanase and β-glucosidase are shown in phase systems with a dextran of low molecular weight (dextran T40) and a high

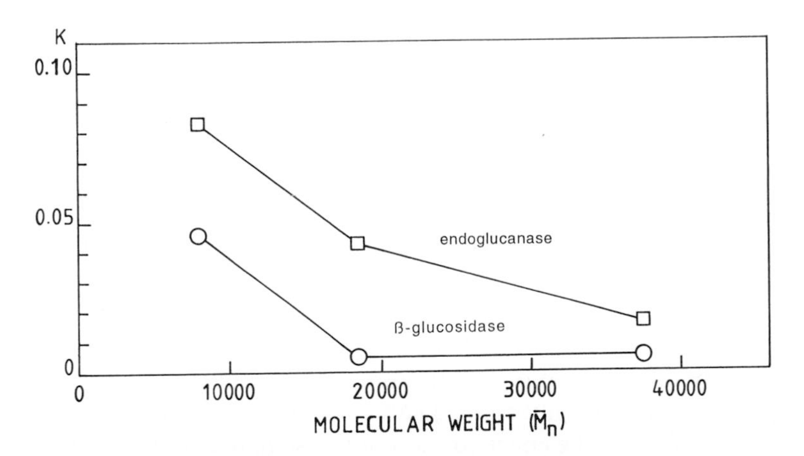

FIG. 3 The partition coefficient (K) for endo-β-glucanase and β-glucosidase from *Trichoderma reesei* as a function of the molecular weight of PEG with cellulose present. The phase system was composed of 8% (w/w) Dextran T40, 6% (w/w) PEG, 10 mM sodium acetate buffer, pH 4.8, 5% Solka Floc BW 200 cellulose. The cellulase enzymes were from NOVO Celluclast 2.0L (NOVO A/S, Bagsvaerd, Denmark). The cellulose-to-enzyme weight ratio was 40. [Data from Tjerneld *et al.,* 1985a.]

cellulose/enzyme weight ratio (C/E = 40) as a function of PEG molecular weight (Tjerneld *et al.,* 1985a)

Thus, when the cellulases are used in a process for cellulose hydrolysis the most important experimental parameters for the partitioning of the enzymes to the bottom phase are polymer molecular weight and enzyme adsorption to the cellulose particles. A one-sided partitioning of both enzyme and substrate can be obtained, which makes it possible to run an enzyme reactor for cellulose hydrolysis in which the low molecular products are continuously removed from the reactor.

The enzyme reactor is a mixer-settler in which the hydrolysis of particulate cellulose takes place in a PEG (MW 20,000)-crude dextran system at 50°C in the mixer unit (Tjerneld *et al.,* 1985b) The hydrolysis mixture is pumped to the settler unit where the phases are separated at room temperature. The sugar products are continuously withdrawn with the top phase from the settler. The bottom phase, containing the unhydrolyzed cellulose and the cellulolytic enzymes, is recycled back to the mixer unit. Although the cellulases strongly partition to the bottom phase, there is a low concentration of enzymes in the top phase, which leads to a slow washout of enzymes. In order to reduce enzyme losses during prolonged running of the enzyme reactor, and for recycling of the top phase polymer, an ultrafiltration unit was added to the mixer-settler (Tjerneld *et al.,* 1985c). After phase separation in the settler unit, the top phase was passed over an ultrafiltration membrane (cut off 10,000 MW) and the polymer (PEG 20,000) is recycled to the mixer together with any enzymes present in the top phase. The sugars produced are recovered in the filtrate.

A cellulose concentration of approximately 80 g/liter is maintained in the reactor throughout the hydrolysis. Cellulose particles are then present in excess over enzymes, which is of advantage for enzyme adsorption and partitioning to the bottom phase containing cellulose. The ultrafiltration of the top phase is started after 50 h when the concentration of reducing sugars is above 60 g/liter. The sugar concentration is then kept at 80 g/liter by adjusting the withdrawal of sugars with the UF unit. A steady-state situation is thus reached, i.e., new sugars are produced in the bottom phase, which are rapidly equilibrated between the phases, and the production is balanced by the removal of sugars from the top phase with the UF unit. The concentrations of glucose and cellobiose in the UF filtrate are shown in Fig. 4a. In Fig. 4b, the concentration of xylose in the UF filtrate is shown. The semicontinuous cellulose hydrolysis can be run for 1200 h with only the initially added enzymes (Tjerneld *et al.,* 1985c).

Trichoderma cellulases are very stable in solution and the use of polymeric phase systems in cellulose hydrolysis further increases cellulase stability. Semicontinuous hydrolysis for 50 days using only initially added en-

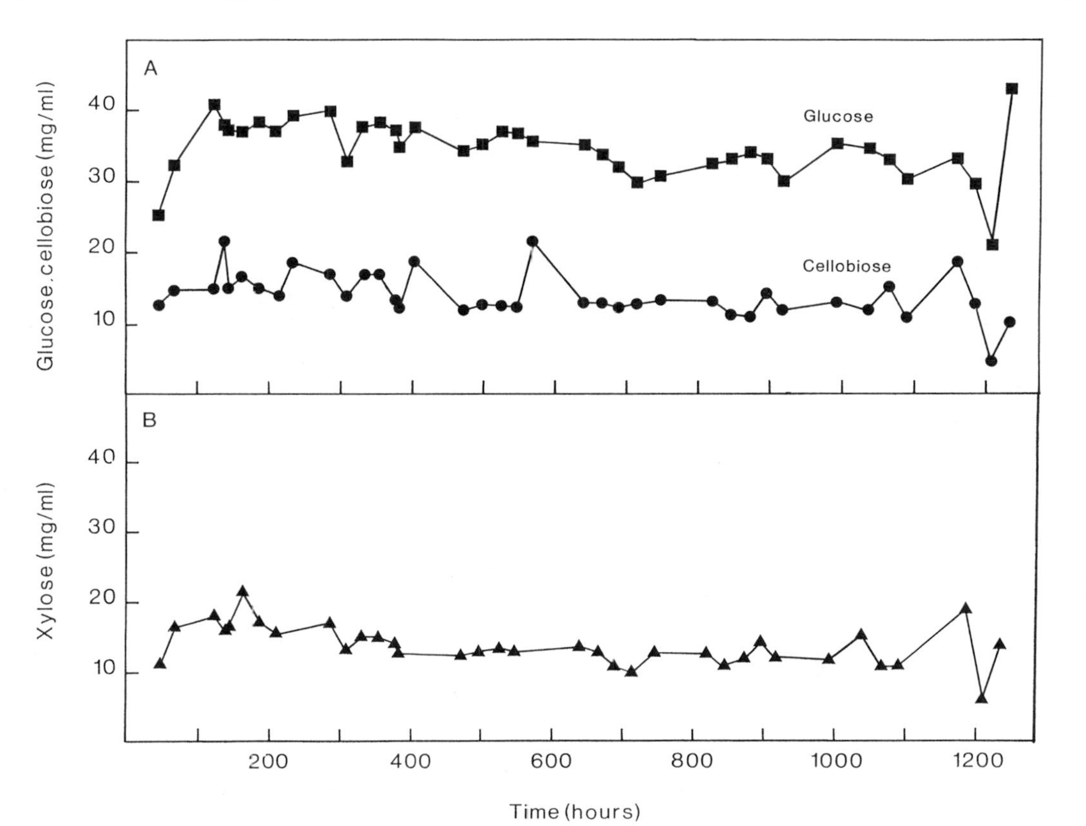

FIG. 4 Sugar concentrations in the UF filtrate from semicontinuous hydrolysis of cellulose (Solka Floc BW 200) in an aqueous two-phase system combined with ultrafiltration. The phase system was composed of 5% (w/w) crude dextran, 3% (w/w) PEG 20000, 50 mM sodium acetate buffer, pH 4.8, and 0.03% sodium azide. The temperature in the mixer was 50°C. The sugars were analyzed by HPLC. (A) Concentrations of glucose and cellobiose. (B) Xylose concentration. [Data from Tjerneld *et al.,* 1985c.]

zymes demonstrates excellent enzyme stability in aqueous two-phase systems.

The principles of an enzyme reactor where the enzyme and the substrate are compartmentalized in one phase in an aqueous phase sytem have also been used in other reaction systems. Enzymatic hydrolysis of starch using amylases has been carried out in a PEG/dextran system with the use of a mixer-settler unit (Larsson *et al.,* 1989). The conversion of penicillin with penicillin acylase in PEG/salt aqueous two-phase systems has also been described (Andersson *et al.,* 1984).

IV. Concluding Remarks

A thermodynamic analysis of the partitioning of solutes between two aqueous polymer phases shows the underlying forces governing the compartmentalization of enzymes in phase systems. High molecular weight solutes have much stronger tendency to one-sided partitioning due to entropic reasons. The enthalpic interactions can be weak for monomer units in macromolecules. However, when added up for the whole molecule, these interactions will contribute to a one-sided partitioning. The presence of electrolytes will generate an electrostatic potential difference between the phases, which affects partitioning of charged solutes. Also, here there is a strong dependence on molecular weight and highly charged macromolecules display extreme partitioning to one of the phases. Thus, enzymes can be partitioned one-sidedly in a phase system and a compartmentalization can be achieved. For enzymes and other macomolecules, this has been demonstrated both in theory and experimentally. Substrates and products will distribute depending on surface properties, e.g., charge and molecular weight. Enzyme reactors have been run with enzyme and substrate located in the same phase and the products removed with the other phase. By using the principles of phase partitioning, an advantageous system for enzymatic reactions can be set up which allows continuous removal of products. This has been shown to work for hydrolysis of cellulose and starch with enzymes. The described principles are of general nature and will have a role in systems with mixtures of macromolecules. Inside the cell, these principles may provide the cell with methods for compartmentalization of enzymes and for the channeling of substrates and products in metabolic pathways.

Acknowledgments

The research on thermodynamic driving forces for partitioning in aqueous two-phase systems is supported by a grant from the Swedish Research Council for Engineering Sciences (TFR).

References

Abbott, N. L., Blankschtein, D., and Hatton, T. A. (1992). Protein partitioning in two-phase aqueous polymer systems. 4. Proteins in solutions of entangled polymers. *Macromolecules* **25,** 5192.

Albertsson, P.-Å. (1986). "Partition of Cell Particles and Macromolecules," 3rd ed. Wiley, New York.

Albertsson, P.-Å., Cajarville, A., Brooks, D. E., and Tjerneld, F. (1987). Partition of proteins in aqueous polymer two-phase systems and the effect of molecular weight of the polymer. *Biochim. Biophys. Acta* **926,** 87.

Andersson, E., Mattiasson, B., and Hahn-Hägerdal, B. (1984). Enzymatic conversions in aqueous two-phase systems: Deacylation of benzylpenicillin to 6-aminopenicillanic acid with penicillin acylase. *Enzyme Microb. Technol.* **6,** 301.

Brooks, D. E., Sharp, K. A., and Fisher, D. (1985). Theoretical aspects of partitioning. *In* "Partitioning in Aqueous Two-Phase Systems: Theory, Methods, Uses, and Applications to Biotechnology" (H. Walter, D. E. Brooks, and D. Fisher, eds.), pp. 11–85. Academic Press, New York.

Flory, P. J. (1953). "Principles of Polymer Chemistry." Cornell University Press, Ithaca, NY.

Gustafsson, Å., Wennerström, H., and Tjerneld, F. (1986). The nature of phase separation in aqueous two-polymer systems. *Polymer* **27,** 1768.

Johansson, G. (1974). Effects of salts on the partition of proteins in aqueous polymeric biphasic systems. *Acta Chem. Scand.* **B28,** 873.

Johansson, H.-O., Lundh, G., Karlström, G., and Tjerneld, F. (1996). Effects of ions on partitioning of serum albumin and lysozyme in aqueous two-phase systems containing ethylene oxide-propylene oxide copolymers. *Biochim. Biophys. Acta* **1290,** 289.

Johansson, H.-O., Karlström, G., and Tjerneld, F. (1997). Temperature-induced phase partitioning of peptides in water solutions of ethylene oxide and propylene oxide random copolymers. *Biochim. Biophys. Acta* **1335,** 315.

Johansson, H.-O., Karlström, G. and Tjerneld, F., and Haynes, C. A. (1998). Driving forces for phase separation and partitioning in aqueous two-phase systems. *J. Chromatogr. B* **711,** 3.

Larsson, M., Arasaratnam, V., and Mattiasson, B. (1989). Integration of bioconversion and downstream processing: Starch hydrolysis in an aqueous two-phase system. *Biotechnol. Bioeng.* **33,** 758.

Pfennig, A., and Schwerin, A. (1995). Analysis of the electrostatic potential difference in aqueous polymer two-phase systems. *Fluid Phase Equilib.* **108,** 305.

Tjerneld, F., Persson, I., Albertsson, P.-Å., and Hahn-Hägerdal, B. (1985a). Enzymatic hydrolysis of cellulose in aqueous two-phase systems. I. Partition of cellulases from Trichoderma reesei. *Biotechnol. Bioeng.* **27,** 1036.

Tjerneld, F., Persson, I., Albertsson, P.-Å., and Hahn-Hägerdal, B. (1985b). Enzymatic hydrolysis of cellulose in aqueous two-phase systems. II. Semi-continuous conversion of a model substrate, Solka Floc BW 200. *Biotechnol. Bioeng.* **27,** 1044.

Tjerneld, F., Persson, I., Albertsson, P.-Å., and Hahn-Hägerdal, B. (1985c). Enzyme recycling in cellulose hydrolysis by combined use of aqueous two-phase systems and ultrafiltration. *Biotechnol. Bioeng. Symp.* **15,** 419.

Tomme, P., Warren, R. A. J., and Gilkes, N. R. (1995). Cellulose hydrolysis by bacteria and fungi. *Adv. Microb. Physiol.* **37,** 1.

Walter, H., and Johansson, G., eds. (1994). "Methods in Enzymology," Vol. 228. Academic Press, San Diego, CA.

Walter, H., Brooks, D. E., and Fisher, D., eds. (1985). "Partitioning in Aqueous Two-Phase Systems: Theory, Methods, Uses, and Applications to Biotechnology." Academic Press, New York.

Part II

Physicochemical Properties of Cytoplasm

Macromolecular Crowding and Its Consequences

H.-O. Johansson,* D. E. Brooks,† and C. A. Haynes‡
*Biotechnology Laboratory, University of British Columbia, Vancouver, B.C., Canada V6T 1Z3; †Department of Pathology and Laboratory Medicine and Department of Chemistry, University of British Columbia, Vancouver, B.C., Canada V6T 2B5; and ‡Biotechnology Laboratory and Department of Chemical and Bioresource Engineering, University of British Columbia, Vancouver, B.C., Canada V6T 1Z3

Incompatible pairs of polymers separate into two phases in aqueous solution above a few percentage points total concentration. Protein pairs can also produce phase separation, but at somewhat higher concentrations. In this chapter, we explore the effect of high background concentrations of macromolecules on phase separation of pairs of species which would not be at sufficiently high concentration to separate in the absence of the uninvolved species. Effects produced by such high background concentrations are known as macromolecular crowding. Dramatic enhancements in various association reactions due to crowding have been predicted and observed but its effects on phase separation in biological mixtures typical of the cytoplasm have not been examined. Here, we describe a calculation based on the Flory-Huggins treatment of concentrated polymer solutions that sheds some light on this issue. We find that a background of 20 wt % of a high molecular weight species greatly reduces the concentrations needed to produce phase separation in a mixture of two incompatible macromolecules if one is more hydrophobic than the other. Given the high total concentration of macromolecules in cytoplasm, it is perhaps surprising that phases have not been observed. This issue is discussed and some explanations offered.

KEY WORDS: Flory-Huggins theory, Incompatibilty, Scaled particle theory, Phase diagrams.

I. Introduction: Phase Separation in Macromolecular Solutions

The tendency of aqueous solutions containing dilute macromolecular components to phase separate has been known since the pioneering studies of the Dutch microbiologist Beijerinck (1896), who discovered that two macroscopic liquid phases form upon mixing dilute aqueous solutions of gelatin and starch. Since then, the generality of demixing in aqueous solutions containing two or more dilute macromolecules has been established by the discovery of a large number of water-soluble polymer pairs which promote phase separation above a critical concentration (Albertsson, 1986; Zavslasky, 1984). Table I lists a number of macromolecule pairs which phase separate in water at 25°C (data from Albertsson, 1986; Zavslasky, 1984; Grinberg and Tolstoguzov, 1997; Polyakov *et al.,* 1997; and chapter by Tolstoguzov in this volume). Included in the table are the minimum total polymer concentration in mole-fraction (and weight-fraction) units required for phase separation and the average molecular weights of the polymer components. Two observations can be made. First, if the molecular weight of both macromolecular components is high, phase separation can occur at low total solute concentration (e.g., 0.6 and 0.2 weight percentage for the sodium dextran sulfate-methyl cellulose system). Second, phase separation has been observed for naturally occurring proteins and polysaccharides, for example, casein and gliadin. Extensive studies of protein-natural polysaccharide and protein–protein two-phase systems have been presented by Polyakov *et al.* (1997) and Grinberg and Tolstoguzov (1997), respectively. One important observation from these studies is that the

TABLE I
Critical Concentrations of Macromolecule Pairs that Form Two-Phase Systems at Room Temperature[a]

Polymer pairs	Critical concentrations		Molecular weight (KDa)
	Wt%	Mole fraction	
Dextran D68-Methyl cellulose[a]	0.76	6.3×10^{-8}	2200
	0.24	3.1×10^{-7}	140
Sodium Dextran Sulfate 68-Methyl cellulose[a]	0.6	5.0×10^{-8}	2200
	0.2	2.6×10^{-6}	140
Dextran-Poly(vinyl alcohol)[b]	2.6	8.6×10^{-6}	57.2
	2.0	6.9×10^{-6}	55
Dextran D48-PEG 6000[b]	4.1	1.6×10^{-6}	500
	3.6	1.2×10^{-6}	6
Casein-Amylopectin[c]	4.3	8.4×10^{-6}	100
	3.9	2.6×10^{-8}	30,000
Casein-Gliadin[d]	3.0	1.9×10^{-5}	30
	2.5	2.4×10^{-5}	20

[a] Data from polymer pairs taken from Albertsson (1986).
[b] Data from Zaslavsky (1984).
[c] Data from Grinberg and Tolstogusov (1997).
[d] Data from Polyakov *et al.* (1997).

critical concentrations of the macromolecules are, in general, higher for pairs of globular macromolecules than for pairs where one of the components is a random coil (e.g., soyabean globular fraction (8%)–casein (7%), and casein (5%)-sodium alginate (0.5%), respectively).

Table II lists the concentrations of the major soluble proteins localized in the cytosol of *Escherichia coli* (data from Neidhardt, 1987; Goodsell, 1991; Albe *et al.*, 1990). One is immediately struck by the very high concentrations of soluble macromolecular components, including oligonucletides, proteins, and glucans. These high-molecular-weight solutes make up roughly 30% of the cytosol on a weight/volume basis. The general tendency for mixtures of water-soluble polymers to phase separate suggests that demixing should occur in the cytosol if the total concentration of any two incompatible macromolecular components is sufficiently high. The critical concentration required for phase separation will depend on both the chemistry and the sizes of the two solutes. From the data in Table I, however, we expect demixing could occur if the total weight fraction of any two incompatible macromolecular solutes within the cytosol exceeded 0.2–4%. From Table II, it would appear that the concentrations of several soluble macromolecules within the cytosol are near or above this critical limit. Phase separation has been observed in calf lens cytosol (Clark, 1994; and chapter by Clark and Clark in this volume) and in mixtures of calf lens cytosolic proteins (Liu *et al.,* 1995, 1996). The cytosol has an upper critical solution temperature at 17°C, that is, above this temperature the cytosol exists as a single homogeneous phase. Below this temperature, the cytosol separates into two distinct phases with one protein-enriched phase (48%)

TABLE II
Macromolecular Composition of Cytosol of *E. coli* (g/l)

Total soluble cytosolic protein[a]	120 (g/l)
Glyceraldehyde 3-phosphate dehydrogenase[b]	9.4
Succinate dehydrogenase[b]	8.9
Glutamate decarboxylase[b]	7.7
Enolase[b]	6.6
Phosphoglycerate mutase[b]	6.5
Glycerol kinase[b]	6.3
Glycogen[a]	12
Ribosomal RNA and protein, tRNA and mRNA[a]	150
DNA[a]	15

[a] Cytostolic composition calculated from Neidhardt (1987) and Goodsell (1991).

[b] Cytostolic composition calculated from Albe *et al.* (1990).

and one protein-depleted (35%); below 30% and above 50%, the system is a one-phase system. More extreme enrichments are observed in mixtures of purified lens proteins (Liu *et al.,* 1995, 1996). This system is qualitatively different from the systems in Table I, however, which are only weakly temperature dependent. Furthermore, the critical macromolecular concentrations for the systems in Table I are 1–2 orders of magnitude lower than the calf lens cytosolic system. Polyakov *et al.* (1997) showed that incompatibility exists between native ovalbumin and thermo-denatured ovalbumin. This example suggests that protein–protein phase separation can occur by, for instance, temperature-induced conformational changes. Another type of macromolecular incompatibility which has biological relevance is the aqueous phase systems of schizophyllan, which is a rodlike polysaccharide. This single polymer can form up to three phases if two different molecular weight fractions of the polymer are present. In these systems, one of the phases contains ordered polymers while the other phase is isotropic (Itou and Teramoto, 1984).

Solid-liquid phase separation of macromolecules can exist in cytosol as, for example, the large granules of poly-β-hydroxybutyrate (PHB), in certain prokaryotes (FEMS Microbiology Reviews, 1992). However, this system is qualitatively different from the ordinary polymer incompatibility from Table I. The PHB polymer is strongly hydrophobic and is insoluble in water. The polymer incompatibility from Table I is not due to solvophobicity but rather to effective repulsion between the polymer pairs.

II. Macromolecular Crowding

Based on the above discussion, it may seem odd that phase separation has not been reported, to date, in cytoplasm. The high total concentration of macromolecules present in cytoplasm makes this even more surprising when consideration is given to what has come to be known as macromolecular crowding effects. It is now quite widely recognized that macromolecular reactions are strongly enhanced in the presence of high background concentrations of uninvolved macromolecules which volumetrically exclude other macromolecules, including putative reactants, from relatively large regions of solution. Such effects have been examined theoretically by applying a statistical mechanical method called Scaled Particle Theory (SPT) to calculate the effect of background concentrations of macromolecules on the activity coefficients of high molecular weight components of a solution (Minton, 1983; Berg, 1990). The predicted increases in activity coefficients can be enormous, implying that the effective concentration of two macromolecular reactants, i.e., their thermodynamic activities, can be much higher than their concentrations would indicate if high background concentrations of proteins, polysaccharides, or nucleic acids are present. While application of SPT has helped explain some anomalies observed *in vitro* compared to *in vivo* reactions, for instance, in the ionic strength dependence of association reactions between DNA and repressor proteins (Cayley *et al.,* 1991; Zimmerman and Trach, 1991) and rationalizes much other data in the literature (Zimmerman and Minton, 1993; Guttman *et al.,* 1995; Minton, 1983, 1995), this approach has not been applied explicitly to the question of phase separation in mixtures of biomolecules (Walter and Brooks, 1995).

In order to examine explicitly the effect of a high background concentration of an uninvolved macromolecule, we describe a calculation involving a lattice model of polymer solutions, rather than the hard sphere model invoked in SPT calculations. This approach carries the advantage that it is able to include the qualitative effects of the signs and magnitudes of the energetics of interaction among all the components, a feature which is not included overtly in hard sphere calculations. It has the disadvantage that the macromolecules are treated as random walk polymers, however, in which all parts of each molecule are free to sample any part of the lattice. This is not a good model for proteins, which typically exhibit strong secondary and tertiary structure. Nonetheless, this approach has proven useful in describing protein distributions in phase-separated polymer solutions (Walter *et al.,* 1991). As discussed below, the calculation strongly supports the possibility of phase separation occuring in cytoplasm and makes its lack of observation all the more surprising.

III. Flory-Huggins Theory

The simplest route to evaluating the potential for phase instability within the cytosol involves application of the mean-field theory of Flory and Huggins (Flory, 1942; Huggins, 1942) describing the equilibrium thermodynamic properties of mixtures of flexible polymers in the presence (or absence) of solvent. Flory-Huggins (FH) theory treats polymers in solution as an incompressible lattice, in which each lattice site is occupied by solvent or an appropriately scaled polymer segment. The theory is "mean field" in that it assumes that the composition of nearest neighbors around every lattice site is the same as the average composition of the phase. Under athermal conditions (i.e., where there is no net energy of attraction or repulsion between unlike components), FH theory gives a simple expression for the (combinatorial) entropy ΔS_{mix} of mixing m components in a lattice of N sites

$$\Delta S_{\text{mix}} = -RN\sum_{i=1}^{m} \frac{\phi_i}{M_i} \ln\phi_i \tag{1}$$

which captures the basic physics of mixing molecules of different size. In Eq. (1), ϕ_i and M_i are the volume fraction and molecular size parameter for component i, respectively. In this treatment, the size of the macromolecule as described by M_i is equal to the ratio of the volume of the macromolecule divided by the volume of a solvent molecule. Equation (1) predicts that the entropy of athermally mixing one or more high-molecular-weight polymers ($M_i > 1$) with solvent ($M_i = 1$) is positive, and thus that athermal mixing of macromolecules and solvent cannot result in liquid–liquid phase separation. However, for a given composition, the ΔS_{mix} predicted by Eq. (1) is significantly lower than that for mixing two equal-sized components:

$$\Delta S_{\text{mix}} = -RN \sum_{i=1}^{m} x_i \ln x_i \qquad \text{(molecules of equal size)} \tag{2}$$

where x_i is the mole fraction of component i.

In FH theory, the enthalpy of mixing ΔH_{mix} is described by the simple Bragg-Williams model derived from regular-solution theory (Hill, 1960):

$$\Delta H_{\text{mix}} = N\sum_{i=1}^{m-1} \sum_{j=1+1}^{m} \phi_i\phi_j W_{ij} = NRT \sum_{i=1}^{m-1} \sum_{j=i+1}^{m} \phi_i\phi_j X_{ij} \tag{3}$$

where w_{ij} is the effective pairwise nearest-neighbor interaction parameter (J mol^{-1}) and χ_{ij} is the more common Flory interaction parameter which expresses the energy of the pairwise interaction in thermal (RT) units. For polymeric components, χ_{ij} quantifies the energy of interaction between a segment (filling one lattice site) of polymer i and a segment of polymer j. A positive χ_{ij} signifies a net repulsive or unfavorable i–j interaction. Comparison of Eqs. (1) and (3) shows that while the entropy of mixing polymers in solvent decreases with increasing polymer size, ΔH_{mix} is independent of polymer size. As a result, weak polymer–polymer incompatibilities which result in a positive enthalpy of mixing can lead to phase separation. A striking example of this is seen with solvent-free mixtures of two infinitely long polymers, where any net repulsion between unlike polymer segments (i.e., any $\chi_{ij} > 0$) will lead to demixing. In contrast, an infinitely long polymer (1) will remain soluble in a solvent (2) of $M_2 = 1$ if χ_{12} does not exceed 0.5, demonstrating the strong dependence of ΔS_{mix} on the sizes of the components.

FH theory, originally derived for concentrated nonpolar mixtures of flexible unbranched polymers, is based on several simplifying assumptions. Nevertheless, it qualitatively captures much of the physics describing mixing of flexible macromolecules in water (Tompa, 1956). For example, FH theory has been successfully applied to the description of phase behavior (e.g., spinodal and coexistence curves) for a wide range of ternary polymer 1-polymer 2-water mixtures, including those used in aqueous two-phase extraction technology (Johansson *et al.,* 1998). For such systems, it predicts the dependence of phase behavior on temperature and on polymer molecular weight, chemistry, and concentration. Moreover, adaptations of FH theory which account for the presence of charge have been used to qualitatively model the partitioning of proteins and other biological macromolecules in aqueous two-phase systems (Johansson *et al.,* 1998). Again, although the arguments are not exact, the general dependence of partioning on key system variables is captured.

IV. The Potential for Liquid–Liquid Phase Separation in the Cytosol: A Flory-Huggins Model

A qualitative understanding of the effects of a high concentration of soluble macromolecules (e.g., proteins, polyglucans, etc.) on phase behavior can be gained by creating a model "cytosol" composed of water (component 1), two dilute macromolecular components 2 and 3 which are incompatible ($\chi_{23} > 0$) and which interact with the solvent differently ($\chi_{12} \neq \chi_{13}$), and a relatively high concentration of background (inert) polymer (component 4) which mixes athermally with all other components (all $\chi_{i4} = 0$). In this study, the coulombic interactions which are important for biopolymers are implicitly incorporated in the short range χ-parameter, i.e., the calculations are done assuming electrically neutral components. This is equivalent to assuming the ionic strength is sufficiently high to screen electrostatic interactions, such that they are short-range in character. FH theory then allows us to qualitatively predict the phase behavior of this system as a function of composition and strength of intermolecular interactions.

A. Phase Behavior of Ternary Mixtures

Let us first consider the case where the concentration of the inert macromolecule, component 4, is zero. Assume species 2 and 3 are both of high and equal molecular weight ($M_2 = M_3 = 200$) and fully soluble in water (χ_{12} and $\chi_{13} < 0.5$). Figure 1 shows the phase diagram for this system when the three Flory interaction parameters are given by $\chi_{12} = 0.38$, $\chi_{13} = 0$ and $\chi_{23} = 0.20$. Phase diagrams for ternary mixtures in which one component (water in this case) is in large excess are often presented in Cartesian coordinates, as shown. The ordinate in Fig. 1 plots the weight fraction of

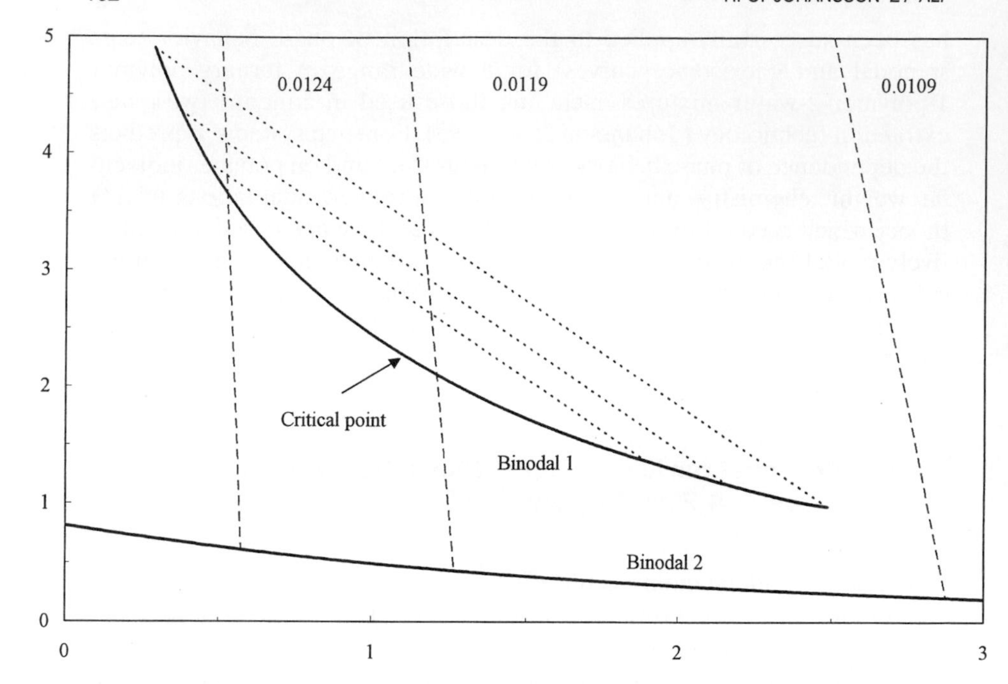

FIG. 1 Phase diagram for the two mixtures described in the text. The vertical axis is the concentration of the more hydrophobic material, component 2, in wt percentage; the horizontal axis gives the concentration of component 3 in wt percentage. In the absence of the background component 4, the binodal is the upper curve labeled Binodal 1, while in the presence of 20% of component 4, Binodal 2 is predicted. The three very steep dashed lines are tie lines associated with Binodal 2. The numbers near the top of the figure give the partition coefficient of component 4 in each of the systems described by the tie lines in the four-component system.

the more hydrophobic solute, macromolecule 2 in this case, since $\chi_{12} > \chi_{13}$. The abscissa plots the weight fraction of the hydrophilic polymer, which usually is enriched in the bottom phase of an aqueous two-phase system. The solid curve designated as Binodal 1 in Fig. 1, often called the coexistence curve (or binodal), separates the region in which the concentrations of the three components are such that they mix to form a single homogeneous phase (below the binodal) from a higher concentration region where the mixture phase separates into an aqueous top phase enriched in macromolecule 2 and an aqueous bottom phase enriched in component 3. The dashed-line segments shown within the two-phase region are examples of equilibrium tie lines, the intersections of which, with the binodal, give the compositions of the top and bottom phases for any total solution composition falling on the line. The position of the total composition on the tie line determines the volume ratio of the top and bottom phases. The volume ratio (top phase/bottom phase) goes from infinity to zero when moving from the left end point to the right end point of a tie line.

The critical point shown on Binodal 1 in Fig. 1 gives the minimum total macromolecule (2 and 3) concentration which will result in phase separation. Demixing occurs above a total macromolecule concentration of approximately 3.5% (w/w) in the illustrated example. However, the critical point can be lowered by increasing the molecular weight of either polymer or by increasing χ_{23}, the net-repulsive energy of interaction between segments of components 2 and 3. For instance, increasing χ_{23} from 0.19 to 0.4 and the molecular weight to 400 reduces the total concentration of macromolecule required for phase separation to 1.1% (not shown on the figure). Thus, even at relatively low concentrations, say 1.5% (w/w) total, two soluble macromolecular components within the cytosol could, in principle, promote liquid–liquid phase separation, provided they are incompatible and both of high molecular weight.

B. Crowding Effects: Phase Behavior in the Presence of an Athermal Macromolecule

Let us assume for the moment that no two soluble macromolecular components of the cytosol are sufficiently concentrated to promote demixing in pure water. This alone would not preclude liquid–liquid phase separation in the cytosol since the presence of the remaining macromolecular solutes, which occupy roughly 20 to 30 wt percentage of the fluid volume, could alter the conditions for phase instability. Returning to our model cytosol, we wish to determine the range of conditions, if any, where addition of a high-molecular-weight solute ($M_4 = 200$) which mixes athermally with all other components leads to a lowering of the coexistence curve for a mixture of two incompatible macromolecules in water. In accordance with the basic composition of the cytosol of *E. coli* (Table II), we set the total concentration of the athermal macromolecule to 20 wt percentage.

If the χ_{ij} parameters for the two dilute incompatible solutes 2 and 3 are those listed in Fig. 1, FH theory predicts a dramatic lowering of the coexistence curve to the location indicated as Binodal 2 in Fig. 1. For this example, the minimum total concentrations of polymers 2 and 3 required for phase separation are reduced from 3.5% (w/w) ($w_2 = 2.4$, $w_3 = 1.1$; w_i is the weight percentage of component i) to approximately 0.81% (w/w) ($w_2 = 0.8\%$, $w_3 = 0.01\%$).

Although our model is crude, it predicts that phase separation can occur at soluble macromolecule concentrations consistent with those listed in Table II. For example, the concentration of soluble glycogen ($M_i \gg 1000$)

in the cytosol of *E. coli* is 1.2% (w/w), which is eightfold higher than the minimum concentration of the more dilute polymer (M_i = 200) required to phase separate our model cytosol. Several other soluble components, particularly rRNA and tRNA, are present in similar high concentrations (Table II). Therefore, on a purely physicochemical basis, the possibility of compartmentalization of the cytosol into a set of two or more liquid phases cannot by discounted.

One may argue that the model cytosol proposed does not adequately reflect the true state, particularly the macromolecular diversity, of the cytosol. It is, therefore, important to note that we obtain the same basic result when the model is extended to include a mixture of athermal macromolecules which together comprise 20% (w/v) of the cytosol (data not shown). Thus, the result is general and not particularly sensitive to structure of the model system. It can also be argued that the FH lattice model is a poor approach to predicting phase separation conditions for proteins because they behave more like solid particles than flexible chains. However, as pointed out above, the entropy of mixing is small for any model of macromolecular solutions so the conditions for phase separation are determined primarily by the enthalpic interactions among the components. Since the values of the interaction parameters selected in our examples are in an acceptable range for biomolecules, the qualitative character of the behavior predicted should be correct.

Cell cytoplasm is a complex matrix of interconnected fibers and bilayers immersed within the cytosol. Thus, although there is no definitive evidence for the existence of liquid phase boundaries or multiple, distinct liquid phases within the cytosol, the complex network of surfaces within the cytoplasm may serve to mask the presence of a stable liquid phase whose volume is a small fraction of the total intracellular volume. It is, therefore, informative to explore the compositions and relative volumes of the two liquid phases predicted in our model calculations. In particular, we would like to know how the background macromolecule and the solvent partition between the two equilibrium phases.

The partition coefficient K_i for component i is defined as the concentration (given in FH theory by the volume fraction) ratio ϕ_i^t/ϕ_i^b, where superscripts t and b refer to the top and bottom phases, respectively. K_i is determined by the exponential of the difference in the excess chemical potential, μ_i^{ex}, of i in the top and bottom phases:

$$\ln K_i = \frac{1}{RT}\left[\left(\mu_i^{ex}\right)^b - \left(\mu_i^{ex}\right)^t\right] \tag{4}$$

where $\mu_i^{ex} = h_i^{ex} - T s_i^{ex}$ and is, therefore, a function of the excess partial molar enthalpy h_i^{ex} and entropy s_i^{ex} of i.

Consider a macromolecule i which interacts with the system such that h_i^{ex} is equal to zero in both phases. Under such conditions, the partitioning of i can still be uneven, provided the excess partial molar entropy of i in the two phases is different. FH theory predicts a nonzero value of $(s_i^{ex})^b - (s_i^{ex})^t$ when the number density (i.e., the total number of molecules per unit volume) of the two phases differs (Johansson *et al.,* 1998). The partition coefficient of i is then given by

$$\ln K_i^s = \frac{M_i}{\rho}\left[\frac{n^t}{V^t} - \frac{n^b}{V^b}\right] \tag{5}$$

where the superscript s indicates that Eq. (5) provides only the entropic contribution to partitioning; ρ is the number of lattice sites per unit volume, and n^t/V^t, for instance, is the number density of the top phase. The dependence of K_i on the difference in number density of each phase follows from the fact that the mixing entropy increases in proportion to the number of distinguishable ways of arranging the molecules in a homogeneous mixture, which, of course, is larger for larger numbers of molecules.

Enthalpy may also drive partitioning. Johansson *et al.* (1998) have shown that within FH theory, the enthalpic contribution to the partition coefficient of i (K^H) is given by

$$\ln K_i^H = -\frac{M_i}{RT}\left[(W_{ti} - E_t) - (W_{bi} - E_b)\right] \tag{6}$$

where W_{ti}, for instance, is the average energy of interaction between i and the top phase, and E_t is the average energy of the top phase in the absence of i due to all interactions between phase components. Partitioning of i to the top phase will, therefore, increase when the direct energy of interaction of i with the top phase becomes more favorable (w_{ti} becomes more negative), and when the energy of interaction between the remaining phase components, termed the phase self energy, becomes more unfavorable (E_t becomes more positive). In the latter case, partitioning is favored because insertion of i into the phase requires breaking unfavorable intermolecular contacts to form more energetically favorable contacts.

In our model cytosol, the background macromolecule 4 mixes athermally with all system components. Thus, both w_{ti} and w_{bi} are zero and the partition coefficient of 4, including both entropic and enthalpic components, is given by

$$\ln K_4 = \frac{M_4}{RT}(E_t - E_b) + \frac{M_4}{\rho}\left(\frac{n^t}{V^t} - \frac{n^b}{V^b}\right) \tag{7}$$

Equation (7) states that an inert (athermal) macromolecular solute with a degree of polymerization M_4 will partition unevenly if the self energies or the number densities of the two phases differ. It also shows that the magnitude of in K_4 scales linearly with the size of the solute, indicating that a large macromolecule may partition quite strongly to one of the phases.

In Fig. 1, the numbers adjacent to each tie line near the top of the chart give the calculated partition coefficient K_4 for the inert macromolecule. Under all conditions, K_4 is near zero, indicating that the background macromolecule partitions almost exclusively to the bottom phase. The preference of the athermal macromolecule for the phase enriched in the more hydrophilic polymer can be understood using Eq. (7). In the two-phase system containing inert polymer, the relatively hydrophobic top phase is essentially a binary solution, containing almost none of the hydrophilic polymer. It also contains less water due to the more favorable interaction of water with the hydrophilic polymer (i.e., $\chi_{12} > \chi_{13}$). As a result, the hydrophobic top phase has a significantly lower number density, causing the inert macromolecule to partition away from it.

The self energy of the top phase is higher than that of the bottom due to the unfavorable energy of interaction between the enriched hydrophobic polymer and water. Thus, enthalpy favors partitioning of the athermal macromolecule to the top phase, but not to the extent that it compensates the strong entropic driving force for partitioning. However, if the polymer–polymer interaction is much more repulsive than the polymer solvent interaction ($\chi_{23} \gg \chi_{12}$) in the system containing inert polymer, an essentially even partitioning of the inert polymer is obtained. In this case, phase separation in the model cytosol is *not* potentiated and the critical concentrations of components 2 and 3 are essentially unaffected by the presence of the inert species 4.

In the calculations on which Fig. 1 is based, χ_{14} is less than χ_{12}. Water, therefore, interacts more favorably with the athermal polymer than with the hydrophobic polymer and copartitions into the hydrophilic phase. As a result, the number density of the hydrophilic phase remains high despite the large concentration of inert polymer. The depletion of water from the hydrophobic phase results in further exclusion of the hydrophilic polymer, thereby shifting the phase boundary to the left as shown by Binodal 2. Moreover, the influx of inert polymer and water into the hydrophilic phase reduces the concentration of both the hydrophobic and the hydrophilic phase-forming polymers. Addition of the inert polymer, therefore, lowers the coexistence curve. Equation (7) states that the magnitude of the effect an inert macromolecule has on phase behavior depends strongly on the size of the macromolecule. Phase instability is enhanced by crowding effects because the partitioning of large athermal macromolecules is more extreme than that of low-molecular-weight components.

C. Comparison with Known Cytosol Structure

Figure 1 shows that in the presence of 20% inert polymer, phase separation can occur at a total hydrophobic polymer concentration of less than 0.5% (v/v), resulting in a top phase containing >30% (v/v) hydrophobic polymer and essentially no hydrophilic or inert polymer. The hydrophobic polymer is concentrated by a factor of approximately 60 through phase separation. This, combined with the low total concentration of hydrophobic polymer in the system (e.g., 1%), results in a top-phase volume which is less than 1% of the total volume of the two-phase system. Thus, natural incompatibilities between large solutes combined with macromolecular crowding effects can, in theory, promote segregation of certain macromolecular components, particularly more hydrophobic ones, into dense phases which occupy a very small fraction of the total fluid volume.

Based on these results, it is not wholly unreasonable to think that liquid-phase separation occurs in the cytosol, but remains undetected because of the small size of the hydrophobic phase. For example, the small hydrophobic phase may contact and wet the complex, relatively hydrophobic fibrous and bilayer structures throughout the cytoplasm, thereby distributing the phase into a thin film.

It could be that evolutionary pressures have selected against conditions that would lead to segregation of cytoplasm into two or more large liquid phases since partitioning could then impose on components which must interact during normal cellular function the necessity of diffusing over distances of order 0.1 μm. The time required for free diffusion scales with distance squared. Moreover, solute diffusivities in water scale with $MW^{-1/3}$, where MW is the molecular weight of the solute. Thus, large macromolecular solutes, such as proteins or oligonucleotides, diffuse slowly. Macromolecular crowding effects can slow diffusion much further (Zimmerman and Minton, 1993). Cellular functions which require rapid turnover, including protein synthesis and nucleic acids, presumably have evolved under conditions which allow the cellular components involved to be in close proximity. Formation of a thin liquid film which coats the fibrous matrix of the cytoplasm is one mechanism by which a relatively hydrophobic biomacromolecule required in basic cellular function could be distributed throughout the cytosol and thereby limit mass-transfer resistances. The fact that no such phases have been observed, however, may mean that cells overcome mass-transfer limitations (and avoid liquid-phase instabilities) in other ways.

Based on known solution thermodynamics, it is very likely that an aqueous solution containing a complex mixture of macromolecules at a total concentration of approximately 30% (v/v) will phase separate provided each macromolecule is solubilized in its monomeric form. The extraordinary macromolecule concentration requirements of the cytosol must, therefore, have applied considerable pressure on the evolutionary development of cells (Berg, 1990). One result is the set of well-defined macromolecular aggregates which are formed through highly evolved sets of specific interactions and are distributed throughout the cytoplasm. Examples include ribosomes, lysozomes, and glycogen granules.

As shown in Table II, soluble glycogen and oligonucleotides of the cytosol are very high in molecular weight ($MW > 1 \times 10^6$ g/mol). As discussed above, random coil macromolecules have, in general, larger mutual incompatibility than globular compact macromolecules. Thus, one would, for instance, expect that the neutral glycogen and the highly charged nucleic acids would be incompatible. However, the charges from nucleic acids are partially neutralized by the association of polyamines with the nucleic acids, reactions that are enhanced by macromolecular crowding effects. Numerous proteins are associated with DNA, which leads to a low probability for direct DNA-glycogen repulsion. Furthermore, the glycogen molecules are highly branched, which, in effect, reduces the probability of contact between the glucose monomers and other components in the system, compared to linear polysaccharides. Based on that for dextran, χ_{12} between glycogen and water is probably around 0.35 (Haynes, 1992). Considering this relatively high repulsive interaction between glycogen and water, one would expect phase separation when such a polymer is mixed with hydrophilic macromolecules, as the diagram in Fig. 1 shows. The fact that glycogen exists mainly in soluble "granules," composed of a single high-molecular-weight and highly branched glycogen chain, suggests that evolution has avoided the macromolecular incompatibility that would likely occur if glucose was stored into an unbranched random coil molecule. Berg has presented similar arguments that the high concentration of macromolecules in cytosol has evolutionarily forced macromolecules to adopt compact conformations.

Though the concentration of RNA is relatively high (Table II), much of the RNA is rRNA, which is concentrated in ribosomes. In these compact structures, many of the RNA molecules are effectively excluded from the other cytosolic macromolecules. For instance, if the individual components of ribosomes, which are roughly 30-nm sized aggregates formed from specific binding interactions among about 55 different protein molecules and 3 different rRNA molecules, are incubated *in vitro* under appropriate conditions, they spontaneously reform the ribosomal structure (Nomura). One purpose of the directed formation of ribosomes must be to bring enzymes required for protein synthesis in close contact with required ribosomal and messenger RNA, thereby minimizing mass-transfer limitations. However, the formation of ribosomes and other well-defined aggregate structures within the cytosol may also be driven by the need to avoid liquid phase separation in the cytosol.

V. Conclusion

Macromolecular incompatibility in highly concentrated solutions of high molecular weight materials is a very common phenomenon. However, no phase separation due to macromolecular incompatibility has been observed to date in natural cytosols, although the macromolecular concentrations of high molecular weight and chemically unlike components such as glycogen, nucleic acids, and proteins are very high. Calculations based on the Flory-Huggins model show that aqueous two-phase separation caused by two incompatible polymers can be strongly facilitated under crowded conditions (i.e., high background concentration of an inert athermal polymer). In these calculations, a small phase, highly enriched with a hydrophobic polymer, is obtained when a hydrophobic and a hydrophilic polymer are mixed with a concentrated athermal polymer. The apparent absence of phase separation in microscopic cytosol may be due to the compact conformations of the biological macromolecules, which is expected to reduce incompatibility or to the difficulty of observing small phases.

References

Albe, K. R., Butler, M. H., and Wright, B. E. (1990). Cellular concentrations of enzymes and their substrates. *J. Theor. Biol.* **143,** 163–195.

Albertsson, P.-Å. (1986). "Partition of Cell Particles and Macromolecules," 3rd ed. Wiley, New York.

Beijerinck, M. W. (1896). Ueber eine eigentümlichkeit der löslichen stärke. *Centralb. Bakteriol. Parasitenkd. Infektionskr.* **2,** 697–699.

Berg, O. G. (1990). The influence of macromolecular crowding on thermodynamic activity: Solubility and dimerization constants for spherical and dumbell-shaped molecules in a hard-sphere mixture. *Biopolymers* **30,** 1027–1037.

Cayley, S., Lewis, B. A., Guttman, H. J., and Record, M. T., Jr. (1991). Characterization of the cytoplasm of *Escherichia coli* K-12 as a function of external osmolarity; implications for protein-DNA interactions *in vivo*. *J. Mol. Biol.* **222,** 281–300.

Clark, J. I. (1994). Lens cytoplasmic protein solutions: Analysis of a biologically occurring aqueous phase separation. *In* "Methods in Enzymology" (H. Walter and G. Johansson, eds.), Vol. 228, pp. 525–537. Academic Press, San Diego, CA.

FEMS Microbiology Reviews (1992). Papers presented at the International Symposium on Bacterial Polyoxyalkanoates in Göttingen. *FEMS Microbiol. Rev.* **103,** 91–376.

Flory, P. J. (1942). Thermodynamics of high polymer solutions. *J. Chem. Phys.* **10,** 51.

Goodsell, D. S. (1991). Inside a living cell. *Trends Brochem. Sci.* **16,** 203–206.

Grinberg, V. Y., and Tolstoguzov, V. B. (1997). Thermodynamic incompatibility of proteins and polysaccharides. *Food Hydrocolloids* **11,** 145–158.

Guttman, H. J., Anderson, C. F., and Record, M. T., Jr. (1995). Analysis of thermodynamic data for concentrated hemoglobin solutins using scaled particle theory: Implications for a simple two-state model of water in thermodynamic analyses of crowding *in vitro* and *in vivo*. *Biophys. J.* **68,** 835–846.

Haynes, C. A. (1992). Separation of protein mixtures by extraction: Statistical-mechanical models of aqueous solutions containing polymers, salts, and globular proteins. Ph.D. Thesis, University of California, Berkeley.

Hill, T. L. (1960). "Introduction to Statistical Themodynamics." Addison-Wesley, London.

Huggins, M. L. (1942). Some properties of solutions of long-chain compounds. *J. Phys. Chem.* **46,** 151.

Itou, T., and Teramoto, A. (1984). Triphase equilibrium in aqueous solutions of the rodlike polysaccharide schizophyllan. *Macromolecules* **17,** 1419–1420.

Johansson, H.-O., Lundh, G., Karlström, G., Tjerneld, F. (1996). Effects of ions on partitioning of serum album and lysozyme in aqueous two-phase systems containing ethylene oxide-propylene oxide copolymers.

Johansson, H.-O., Karlström, G., Tjerneld, F., and Haynes, C. A. (1998). Driving forces for phase separation and partitioning in aqueous two-phase systems. *J. Chromatogr. B* **711,** 3–17.

Liu, C., Lomakin, A., Thurston, G. M., Hayden, D., Pande, A., Pande, J., Ogun, O., Asherie, N., and Benedek, G. B. (1995). Phase separation in multicomponent aqueous-protein solutions. *J. Phys. Chem.* **99,** 454–461.

Liu, C., Asherie, N., Lomarkin, A. Pande, J., Ogun, O., and Benedek, G. B. (1996). Phase separation in aqueous solutions of lens γ-crystallins: Special role of γ_s. *Proc. Natl. Acad. Sci. U.S.A.* **93,** 377–382.

Minton, A. P. (1983). The effect of volume occupancy upon the thermodynamic activity of proteins: Some biochemical consequences. *Mol. Cell. Biochem.* **55,** 119–140.

Minton, A. P., ed. (1995). Macromolecular crowding. *Biophys. Chem.* **57,** 1–110.

Neidhardt, F. C., ed. (1987). "*Escherichia coli* and *Salmonella typhimurium:* Cellular and Molecular Biology," Vol. 1. American Society for Microbiology, Washington, DC.

Nomura, M. (1973). Assembly of bacterial ribosomes. *Science* **179,** 864–873.

Polyakov, V. I., Grinberg, V. Y., and Tolstoguzov, V. B. (1997). Thermodynamic incompatibility of proteins. *Food Hydrocolloids* **11,** 171–180.

Tompa, H. (1956). "Polymer Solutions." Butterworth, London.

Walter, H., and Brooks, D. E. (1995). Phase separation in cytoplasm, due to macromolecular crowding, is the basis for microcompartmentation. *FEBS Lett.* **361,** 135–139.

Walter, H., Johansson, G., and Brooks, D. E. (1991). Partitioning in aqueous two phase systems: Recent results. *Anal. Biochem.* **155,** 215–242.

Zaslavsky, B. Y. (1984). "Aqueous two-phase Partitioning: Physical Chemistry and Bioanalytical Applications." Dekker, New York.

Zimmerman, S. B., and Minton, A. P. (1993). Macromolecular crowding: Biochemical, biophysical and physiological consequences. *Annu. Rev. Biophys. Biomol. Struct.* **22,** 27–65.

Zimmerman, S. B., and Trach, S. O. (1991). Estimation of macromolecule concentrations and excluded volume effects for the cytoplasm of *Escherichia coli. J. Mol. Biol.* **222,** 599–620.

Lens Cytoplasmic Phase Separation

John I. Clark and Judy M. Clark
University of Washington, School of Medicine, Seattle, Washington 98195

Cytoplasmic transparency is a unique feature of lens cells. The cytoplasm is a concentrated solution of crystallin proteins with minor constituents that include cytoskeletal proteins and lens specific intermediate filaments. Under normal physiological conditions, the proteins exist as a single transparent phase. With normal aging, progressive modification of the interactions between lens proteins occurs so that conditions within the lens become favorable for phase separation. These conditions produce intracellular inhomogeneities that approach or exceed the dimensions of the wavelength of visible light (400 to 700 nm) and light scattering from lens cells increases. The resulting opacification is the primary factor in the visual loss experienced in cataract, the leading cause of blindness in the world. We study biochemical factors that regulate the cytoplasmic phase separation and maintain normal cellular transparency.

KEY WORDS: Lens, Crystallins, Phase separation, Cellular transparency, Cataract, Cytoplasm.

I. Introduction

The only transparent cells in the human are found in the lens of the eye. In nonlens cells, differentiation includes the formation of intracellular compartments under conditions that resemble an aqueous phase separation in the cytoplasm (Walter *et al.,* 1985; Walter and Johansson, 1994). The cytoplasm is a complex polymer solution in which phase separation can produce filaments, particles, vesicles, and other structures that provide unique intracellular environments for specialized cellular functions. The formation of separate condensed and dilute cytoplasmic phases depends on cellular composition, ionic strength, pH, hydration, and other biophysical parameters that influence the collective interactions between the solute and solvent components of cells. The dimensions of these condensed phases can approach or exceed the wavelength of visible light (400 to 700 nm) and scatter light (Tanaka and Benedek, 1975). In lens cells, the cytoplasmic phase separation is carefully controlled to establish and maintain a single transparent cytoplasmic phase during the normal development of transparent biological cells (Clark, 1994a,b).

II. Development of Transparency

Lens development begins at approximately 40 to 50 days of gestational age, when the first transparent fiber cells appear in the human embryo (Kuwabara, 1975; Kuszak and Brown, 1994). The differentiation of transparent lens fiber cells is a synchronized process that results in a symmetric optical element. The lens epithelium is the source of all lens fiber cells. Undifferentiated epithelial cells contain organelles and cytoskeletal elements similar to those found in other epithelia throughout the body. As a result, the lens epithelium scatters light and appears as a thin bright line on the anterior surface of the lens in a slit lamp photograph (Fig. 1). Epithelial cells divide and migrate to the equator of the lens, where concen-

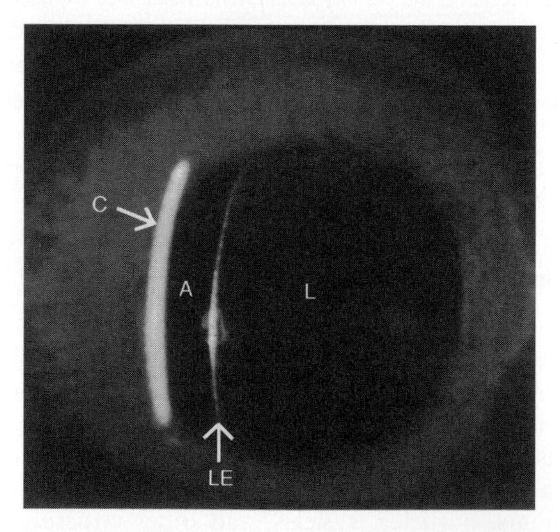

FIG. 1 Slit Lamp Ophthalmoscope view of an eye. The slit lamp provides an optical section of the eye. The broad bright curved band is the cornea (C). Behind the cornea (to the right) is a fluid-filled black space, the aqueous chamber (A). Behind the aqueous chamber, the scattering from the epithelial cells of the lens (LE) is seen as a thin bright line. The differentiated cells of the lens (L) appear dark because the cytoplasmic proteins are organized during differentiation to minimize light scattering.

tric layers of new differentiating cells are added. During differentiation, most large light-scattering structures disappear from the cell and the lens cytoplasm becomes homogeneous and uniformly transparent (Clark, 1994b). Cell nuclei, mitochondria, endoplasmic reticulum, Golgi, and many cytoskeletal structures are absent in the transparent lens fiber cell. The cellular mechanisms responsible for the disappearance of organelles from lens cells are not understood (Kuwabara, 1975).

Cell turnover does not occur in the human lens. The first cells to become transparent in the embryo remain in the center of the adult lens and are not replaced. The addition of newly differentiated cells at the periphery of the lens is responsible for growth of the tissue. In the absence of cell renewal, the transparent structure of human lens cells must be maintained for the lifetime of an individual. Without mitochondria and other organelles, the lens cells develop a specialized metabolism that utilizes soluble enzyme systems, including those of anaerobic glycolysis and the hexose monophosphate shunt (Kador, 1994). Only structures that are small relative to the wavelength of visible light or that contribute to the long-term stability of transparent cellular structure are retained in lens cells (Benedek, 1971; Clark, 1994b; Vaezy *et al.,* 1995) (Fig. 2).

III. Phase Separation and Lens Cytoplasmic Proteins

Once the organelles disappear from differentiating lens cells, the transparency of the concentrated cytoplasm depends largely on the interactions between lens proteins (Delaye and Tardieu, 1983; Veretout *et al.,* 1989; Tardieu *et al.,* 1992). Concentrated proteins in other cells, such as those found in epidermis and muscle, form large structural elements that scatter light. In contrast, lens proteins form a transparent and uniform single phase at high concentrations under physiological conditions. The predominant constituents of lens cytoplasm in humans are proteins of the alpha and beta/gamma crystallin families (Harding, 1991; Piatigorsky, 1992). The forces resulting from the net attractive interactions between individual crystallin proteins can result in the condensation of distinct coexisting phases. The conditions for phase separation in lens cells have been studied using the phase separation temperature, Tc, of the lens cytoplasm (Clark and Benedek, 1980a; Delaye *et al.,* 1981; Clark, 1994a). The Tc is a direct measure of the difference between the energies of interaction of lens proteins with the surrounding solvent constituents as described by the expression, $kTc = A[E_{ps} - (E_{ss} + E_{pp})/2]$, where k is the Boltzmann constant, A is a parameter determined by the detailed form of the phase diagram, and E_{ps}, E_{ss}, and E_{pp} are the protein–solvent, solvent–solvent, and protein–

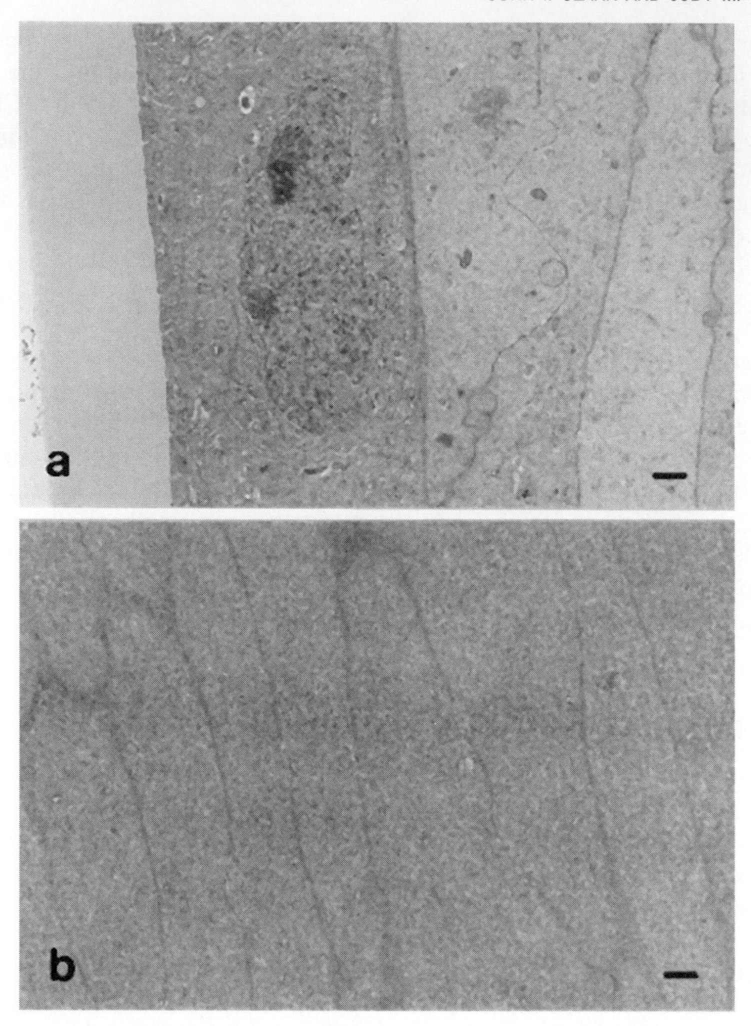

FIG. 2 Electron micrographs of undifferentiated and differentiated lens cells. Lens cells originate in an undifferentiated epithelium that contains organelles large enough to scatter light. Undifferentiated epithelial cells contain a nucleus, mitochondrion, Golgi apparatus, endoplasmic reticulum, cytoskeleton, and other organelles that are compartments for specialized metabolic processes (a). In lens cells, the large size of the intracellular compartments creates light-scattering centers, which are detrimental to the normal transparent function of lens cells. The epithelial layer is so thin that most light is easily transmitted to the retina. In transparent differentiated lens fiber cells, the organelles and cytoskeletal elements that could create large inhomogeneities are absent and are replaced by a smooth and uniform cytoplasm consisting of highly concentrated proteins (b). In differentiated transparent lens cells, the cytoplasm exists as a single phase that is homogeneous with respect to the dimensions of the wavelength of light (400–700 nm.). Bar = 1 micron.

protein interaction energies, respectively (Taratuta *et al.,* 1990; Pande *et al.,* 1991; Clark, 1994a; Hiraoka *et al.,* 1996). For aqueous solutions of lens proteins, the interaction energies are expected to be negative as a result of the attractive interactions between cellular constituents. Changes in the energy of interactions as small as kT or 0.6 kcal/mole can have a strong effect on the Tc and on the formation of cytoplasmic microstructure in lens cells.

IV. Phase Diagrams for Lens Cytoplasmic Proteins

The conditions for cytoplasmic transparency and opacity can be summarized in a phase diagram (Fig. 3). The phase diagram for lens cytoplasm is divided into two regions by the coexistence curve (Clark and Benedek, 1980a; Delaye *et al.,* 1981; Clark, 1994a). Outside the coexistence curve, lens cytoplasm exists as a single transparent phase. Inside the coexistence curve, the cytoplasm separates into spatially distinct phases of different dimensions and morphologies. The structures resulting from phase separation can act

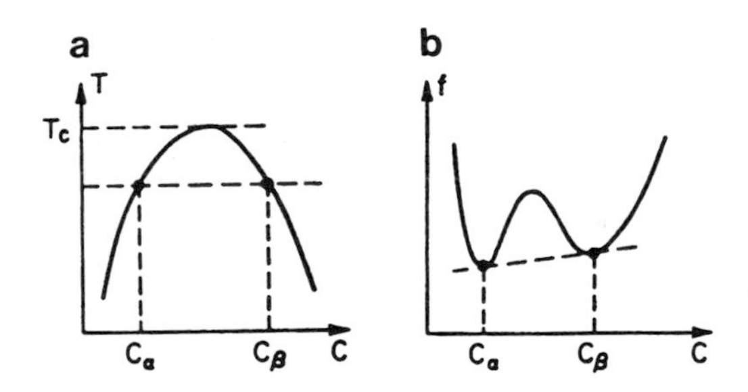

FIG. 3 Model phase diagram and free energy plots for lens cytoplasm. In the phase diagram for a two-phase system, the coexistence curve is the boundary between the conditions for opacity and the conditions for transparency (3a). Temperature is plotted on the Y-axis and concentration on the X-axis. At temperatures below Tc, the critical temperature or maximum temperature of the coexistence curve, the cytoplasmic proteins separate into distinct coexisting phases of different densities and concentrations given by the end points of the horizontal tie line for the given temperature. C_α is the concentration of the dilute phase and C_β, the concentration of the dense protein phase. Outside the coexistence curve, the cytoplasm exists as a single, uniform, transparent phase that is homogeneous at the scale of the wavelength of visible light (400–700 nm.). When the dimensions of the density fluctuations produced by the phase separation approach or exceed the wavelength of visible light, the cytoplasm scatters light. The theoretical analysis of the phase separation in multicomponent systems is dependent upon the parameters that characterize the energy of interaction of the protein solvent constituents (Liu *et al.,* 1995). In the model plot of the free energy, *f*, as a function of concentration it is assumed that two energy minima occur at concentrations corresponding to the concentrations of the separated phases, C_α and C_β (3b) (Langer, 1975).

as microcompartments for important metabolic or physiological functions (Walter and Brooks, 1995). In biological cells, the phase diagram can be valuable for defining conditions for the self-assembly of functional organelles from constituents of the cytoplasm. In lens cells, the phase diagram is important for identification of conditions which maintain a single homogeneous and transparent cytoplasmic phase without phase separation (Clark, 1994a).

The phase diagram for lens cytoplasm has been determined experimentally by measuring the phase separation temperature, Tc, as a function of protein concentration (Clark and Benedek, 1980a; Delaye *et al.,* 1981; Clark, 1994a). At the low protein concentrations found in newly differentiated lens cells, transparency is favored, but the index of refraction is inadequate to provide the normal optical function of the lens. As protein concentrations increase in lens cytoplasm, the interactions between components of the cytoplasm move the conditions to the right, approaching the coexistence curve and phase separation of cytoplasm. Continued concentration of lens proteins moves the conditions further to the right in the phase diagram, away from the coexistence curve. Mature differentiated cells at body temperature (37°C) and high protein concentration are outside the coexistence curve and well away from conditions for phase separation. Just as protein concentrations increase with normal lens cell differentiation, the variations in metabolism, ionic strength, pH, and composition that occur with aging and development of opacification can move the position of the coexistence curve in the phase diagram (Taratuta *et al.,* 1990; Berland *et al.,* 1992; Asherie *et al.,* 1996; Liu *et al.,* 1996). It is well documented that cellular hydration, which decreases the protein concentration of lens cytoplasm and shifts the position of the coexistence curve, favors opacification in the osmotic cataract, one of the most common forms of lens opacity (Chylack, 1981; Kador, 1994).

V. Effects of Composition on the Phase Diagram

The position of the coexistence curve in the phase diagram is sensitive to many parameters that influence the energies of interactions important for transparent lens cytoplasm (Siezen *et al.,* 1985; Broide *et al.,* 1991; Taratuta *et al.,* 1990; Clark, 1994a; Asherie *et al.,* 1996). In rabbits, mice, and rats, the coexistence curve and Tc of normal lens cytoplasm approaches physiological temperature at birth and the lens is opaque (Clark *et al.,* 1982; Clark and Carper, 1987; Clark and Steele, 1992). During normal development, the Tc and the coexistence curve and Tc move down, away from body temperature, to favor lens transparency at lower temperatures (Ishimoto *et al.*, 1979) (Fig. 4). By the time the newborn animal opens its eyes, several days after birth, the normal process of lens development has pushed the coexistence curve down to favor lens transparency. Apparently, normal lens

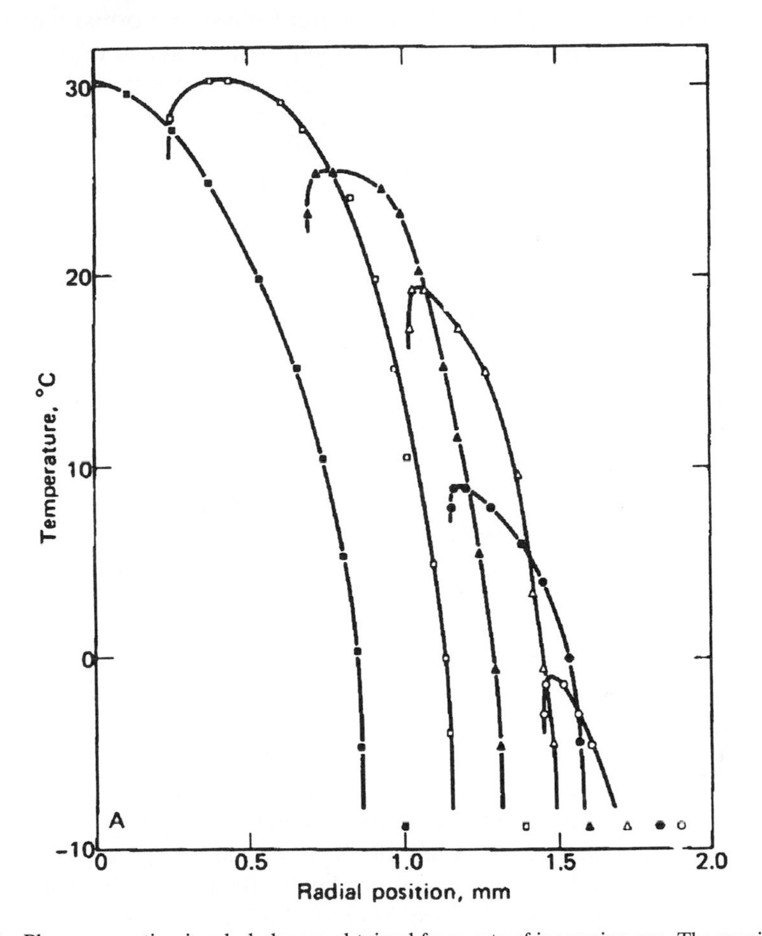

FIG. 4 Phase separation in whole lenses obtained from rats of increasing age. The maximum phase separation temperature in a 1-day-old rat occurs at 30°C in cells at the center of the lens, ("0" on the X-axis). In lenses from older animals, the maximum Tc decreases to lower temperature in more peripheral cells (to the right on the X-axis). The parabolic shape of the coexistence curve is similar at all ages. In a lens, the oldest cells are in the center and the age and protein concentration decrease with radial position from the center to the periphery. The changes observed with age are not accounted for solely on the basis on protein concentration. The relative positions of the coexistence curves in the phase diagram are the result of the collective interactions between the protein and solvent constituents of the lens cells. These interactions differ with the age of the animals and the stage of differentiation of the lens cells (Ishimoto *et al.*, 1979). ■ : 1-day-old; □ : 10-day-old; ▲ : 18-day-old; △ : 25-day-old; ● : 42-day-old; ○ : 56-day-old.

cells have endogenous mechanisms, genetic or metabolic, for controlling the protein and solvent composition and such mechanisms move the coexistence curve away from the conditions for phase separation and opacification. Tc is useful as a characteristic measure of mechanisms that regulate development and maintenance of a single transparent phase in normal lens cells. Abnormal changes in Tc can be used to identify natural and experimental modifications of protein and solvent interactions contributing to protein aggregation and opacification (Fig. 5). Similarly, chemical reagents that maintain normal Tc are potentially important for maintenance of the transparent single phase in lens cells (Clark and Benedek, 1980b; Hammer and Benedek, 1983; Clark *et al.,* 1987).

The effects of solvent conditions on Tc were confirmed in experimental studies (Clark, 1990). Selected solvents were diffused into lenses to change the composition of the cytoplasmic solvent. The effect of the solvents on Tc was directly proportional to the boiling point and heat of vaporization, empirical measures of hydrogen bonding between the solvents. While several factors contribute to the Tc, these empirical findings indicate the importance of hydrogen bonds between solvent and protein constituents of lens cells. Although contributions from hydrophobic, electrostatic, and van der Waals interactions to the Tc have not been measured experimentally, it can be expected that these weak noncovalent interactions are as important as hydrogen bonds for transparent or opaque lens cell structure.

In cells of various vertebrate species at various stages of differentiation, the precise composition of the cytoplasm can determine the energies of interactions between crystallins and their surroundings (Clark and Benedek, 1980a; Siezen *et al.,* 1985). As the composition varies with lens development, the phase diagram for lens cytoplasm moves downward away from body temperature, decreasing Tc and favoring transparency (Clark *et al.,* 1982). Subsequent studies determined that the expression of gamma crystallins having different Tc could contribute to the changes in the coexistence curve during development (Siezen *et al.,* 1985). Each gamma crystallin protein has a different Tc and the differences may be due to amino acid substitutions at only 3 or 4 out of 175 sites (Broide *et al.,* 1991). A more obvious effect of composition is demonstrated in birds, where high concentrations of delta crystallin substitute for much of the alpha and beta/gamma crystallins found in mammals (Harding, 1991; Piatigorsky, 1992; Tomarov and Piatigorsky, 1996). Delta crystallins have not been observed to undergo phase separation under physiological conditions, and avian lenses appear to resist protein aggregation and cataract formation. A very unusual example is found in the pigeon lens, which contains nearly 20% glycogen (Rabaey, 1963; Lo *et al.,* 1993). It is well documented that glycogen aggregates into large light-scattering particles. In the undifferentiated lens cells, the glycogen particles scatter light and the lens cells are opaque (Fig. 6). As lens development

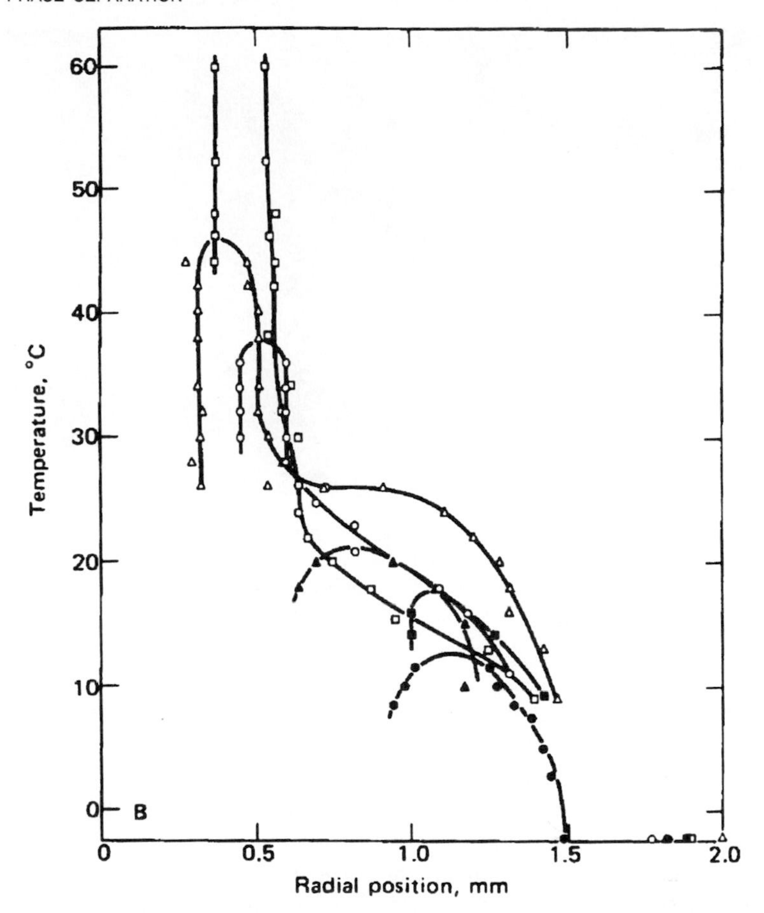

FIG. 5 Phase separation in lenses of animals developing cataracts as a result of a high galactose diet. The phase separation temperature, Tc, was plotted as a function of radial position for lenses of different ages. As predicted, the phase diagrams for lenses developing cataracts became abnormal and the phase separation temperature, Tc, occurred at abnormally high temperatures, indicating abnormal interactions in lens cytoplasm. The galactose diet resulted in abnormal lens hydration, ionic strength, ATP levels, and posttranslational modification of proteins, all of which contribute to abnormal concentration and composition of the lens cytoplasm, phase separation temperature, protein aggregation, and opacification (Ishimoto *et al.*, 1979). ■ : normal lens at 31 days; ● : normal lens at 38 days; △,□ : galactose lens at 34 days; ○ : galactose lens at 35 days.

progresses, the glycogen particles are dispersed throughout the lens cytoplasm to create a uniform and homogeneous cytoplasm of high refractive index. By the time the newborn fledgling leaves the nest, the lens cells are transparent. It will be interesting when the phase diagram for cytoplasm

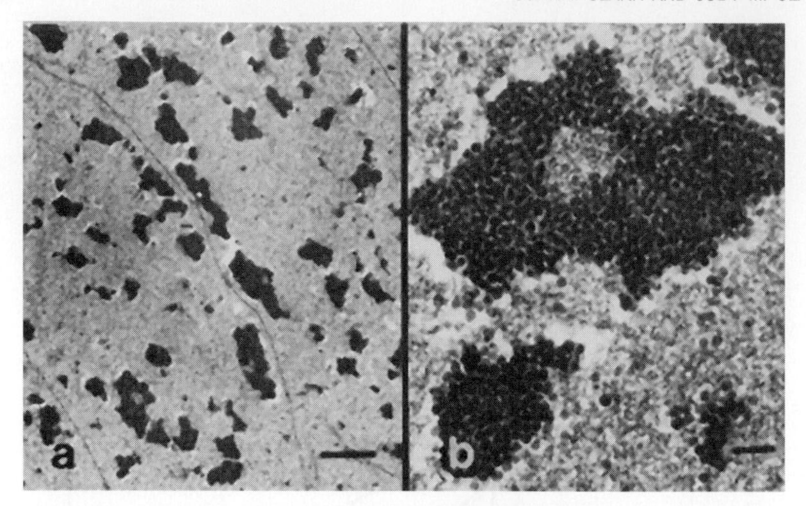

FIG. 6 Electron micrographs of pigeon lenses. In a newborn pigeon, the lens cells contain large glycogen particles that produce opacity. With lens development, the glycogen particles are dispersed uniformly throughout the cells to produce a single homogeneous transparent phase (figures courtesy of K. W. Lo). Bar = 1 micron in (a) and 0.3 micron in (b).

from bird lenses is determined so that the effects of the delta crystallins and glycogen on the coexistence curve can be understood. While there is diversity in the composition of lens cells across species (Piatigorsky, 1992; Tomarov and Piatigorsky, 1996), natural modifications to lens proteins may have a strong effect on the position of the coexistence curve in the phase diagram. The effects of synthetic posttranslational modification of gamma crystallin using N-bromoacetyl ethanolamine phosphate and N-ethylmaleimide were found to alter the Tc of gamma crystallins (Pande *et al.*, 1993). Both noncovalent and covalent interactions can modify interactions between lens proteins and protect against phase separation and opacification (Clark and Benedek, 1980b; Hammer and Benedek, 1983; Friberg *et al.*, 1996). Taken together, these results emphasize the importance of the cellular environment on interactions responsible for phase separation in lens cells.

VI. Endogenous Mechanisms for Regulation of Phase Separation

Studies using purified lens proteins and intact lenses suggest that the biological lens has endogenous mechanisms for regulation of phase separation. In the absence of organelles, expression of crystallins and dehydration increase the normal protein concentrations to nearly 80% in some mammalian species. The high protein concentrations are responsible for the high index of refraction necessary for the refractive power of the lens. The diversity of the proteins in the lens cytoplasm suggests that transparency is not the result of interactions between members of a single protein family. Even minor constituents may have an important function in maintenance of lens cell transparency. As in other cells, differentiating lens cells have a well developed cytoskeleton that participates in the organization of transparent cell structure (Maisel, 1985). Cytoskeletal proteins comprise less than 10% of the total protein the human lens but their function as an intracellular lattice may be very important for the early stages of differentiation when the concentration of crystallin proteins increases most rapidly (Matsushima *et al.,* 1997). It will be of interest to determine the effects of cytoskeletal proteins on the phase separation behavior of lens proteins. It should be noted that cytoskeletal elements are a potential source of light scattering in lens. Apparently, the cytoskeletal proteins are involved with interactions required for establishment of transparent lens cell structure and then disappear when their structural function has been completed. Once this function is completed, cytoskeletal proteins may be processed systematically to form fragments, which decrease light scattering during normal development (Sandilands *et al.,* 1995; Prescott *et al.,* 1996; Matsushima *et al.,* 1997). Abnormal processing of cytoskeletal proteins might be expected to result in a disruption of the organization of the cytoskeletal lattice and lead to opacity, a situation which has been reported in the selenite model for cataract rats (Matsushima *et al.,* 1997). The importance of the cytoskeleton for interactions responsible for the cytoplasmic phase separation in lens cells needs to be investigated in detail.

The role of metabolism in phase separation has been suggested in many studies of cataract formation (Chylack, 1981; Kador, 1994). For example, the high concentration of cytoplasmic proteins increases the osmotic activity of lens cells (Duncan and Jacob, 1984; Veretout *et al.,* 1989; Tardieu *et al.,* 1992). The maintenance of normal cellular hydration is an important function of transport systems in lens cell membranes. Disruption of membrane transport can produce abnormal hydration of lens cells which is the basis for "osmotic cataracts" (Chylack, 1981). In lens fibers, phase separation is sensitive to the regulation of ions and water which involves ATP dependent metabolism. In some animals, the levels of ATP in lens cells are as high as in muscle cells (Greiner *et al.,* 1985). Large reservoirs of ATP appear to be necessary to support membrane transport of ions, water, and nutrients and to protect against abnormal phase separation and osmotic cataract. While ATP may be involved with metabolism that maintains normal phase separation in lens cells, ATP has been shown to be involved directly in interactions of alpha crystallins with other proteins (Muchowski and Clark, 1998). Alpha crystallins form complexes that may assist or chaperone the normal folding of lens crystallins during protein synthesis (Horwitz, 1992; Muchowski *et al.,* 1997). Protection against unfolded conformations that alter normal interactions between lens proteins to favor phase separation and protein aggregation is especially important in lens cytoplasm. The action of alpha crystallin as a molecular chaperone may inhibit abnormal interactions that favor phase separation and light scattering (Horwitz, 1992; Muchowski *et al.,* 1997). The effects of cellular metabolism, either direct or indirect, on cytoplasmic phase separation is an important area of investigation.

It should be emphasized that the liquid–liquid phase separation (coacervation) in lens cells is the result of collective interactions between solvent and solute constituents of lens cells. Phase separation does not result from a specific chemical action or biochemical mechanism. A variety of interactions in lens cells is involved to protect against abnormal phase separation and protein aggregation. Recent studies of selected endogenous biochemicals and synthetic chemical reagents conducted on Tc and on inhibition of opacification *in vivo* (Hiraoka *et al.,* 1996) indicated that a number of chemical functional groups have anticataract activity. Various metabolic and physiological challenges, including oxidation resulting from radiation exposure or glycation in diabetes, are associated with posttranslational modifications to lens proteins. The modified proteins have abnormal interactions that are represented by an abnormal phase diagram. It can be assumed that a broad range of chemical functionalities may be required to inhibit protein aggregation and protect against opacification *in vivo*. In a complex macromolecular solution of lens cytoplasmic proteins, the diversity in the chemical nature of the interactions between lens constituents may protect against protein condensation resulting from modification of interactions between lens cell constituents.

VII. Phase Separation and Cellular Structures

Phase separation can result in formation of a variety of cytoplasmic structures: networks, sheets, cylinders, spheres, tubules, filaments, and micelles (Vaezy *et al.,* 1990; Radlick and Koretz, 1992). It is not difficult to imagine how some of the structures may serve as barriers, microcompartments, enzyme complexes, or lattice elements that have functional and structural significance. The phase diagram describes conditions for formation of coexisting phases in lens cytoplasm and not the specific morphology of the phase. This point is readily appreciated by looking at the diversity of structures found in lens cells of different types of opacities (see Fig. 7). In some lens cells, the separate phases appear as dense spherical droplets surrounded by dilute cytoplasm. In some instances, the droplets appear to be continuous with a dispersed network of aggregated proteins. In other cells, the droplets appear associated with the cell membranes, as if nucleation sites on the membranes initiated their formation. In some cells, the condensed proteins appear as dense networks that partition the cytoplasm into microvolumes that are large enough to scatter light (Fig. 7). Multiple forms of the condensed phase are often observed. Formation of large structures is not necessary for opacity, however. A modest change in texture can result in a sharp increase in light scattering (Clark *et al.,* 1980; Vaezy *et al.,* 1995; Giblin *et al.,* 1995; Al-Ghoul and Costello, 1996). Often the spatial fluctuations in opaque cytoplasm are difficult to observe and quantitative methods are needed to characterize the dimensions of the condensed form of the cytoplasmic aggregates (Clark *et al.,* 1980; Vaezy and Clark, 1995; Vaezy *et al.,* 1995). While the phase diagram describes the conditions for formation

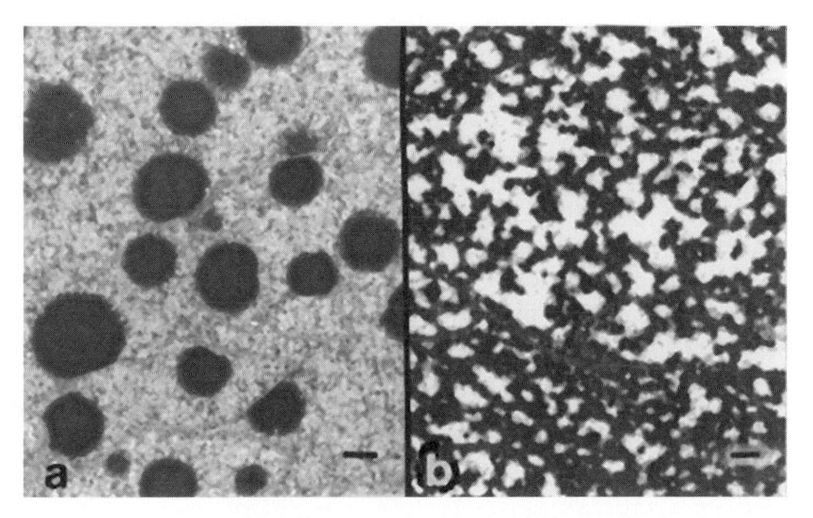

FIG. 7 Electron micrographs of phase separation in mouse lens cytoplasm. Phase separation in cytoplasmic proteins can produce diverse structures having a different sizes and shapes. In (a), the cytoplasmic proteins condense into droplets which differ in size in cells of different ages and stages of differentiation. In (b), the dense phase appears as a coarse network of cytoplasmic proteins. The pattern of the smaller structures in the micrograph resembles the larger structures, suggesting similarity across the various dimensions of the phases. The form and dimensions of the condensed phase are determined by the differences in cytoplasmic composition and the attractive interactions described by the equation for Tc. The size and the shape of the condensed phase may vary with the specific mechanism of phase separation and opacification in lens cytoplasm. This possibility remains to be demonstrated experimentally. Bar = 1 micron.

of condensed and dilute cytoplasmic phases and the assembly of well-defined complexes or aggregates, the mechanism and the exact size or form of the separated phases cannot be determined from a phase diagram. The relationships between the diversity in structure and the diversity of the cellular regulation of phase separation needs to be considered in much more detail.

VIII. Concluding Remarks

At the molecular level, it is known that the surfaces of crystallin protein are capable of a variety of interactions that can be modified by posttranslational alteration of the side chains of surface peptides (Tardieu *et al.,* 1992). In the past, protein modification in lens was thought to be associated with abnormal protein interactions, aggregation, and lens opacification (Takemoto and Gopalakrishnan, 1994; Takemoto, 1997). More recently, it was observed that the modification of crystallin proteins begins in normal lenses at a very early age in humans and continues throughout our lives (Smith *et al.,* 1997; Bloemendahl *et al.,* 1997). Rather than being unfavorable for transparency, modifications occurring during the earliest stages of differentiation and development may be protective and favor the single transparent phase in lens cytoplasm. While modification to lens proteins is not represented specifically in the expression for Tc, the balance between the net attractive forces between all lens cell constituents is the basis for the use of Tc as an indicator of conditions that favor opacity or transparency in lens cells. When the boundary between the single transparent phase and the two separated phases in the phase diagram involves a few kcal/mole, even modest alterations in interactions between proteins and solvent constituents can mean the difference between transparency or opacity of a lens cell. The phase diagram for lens cytoplasm provides a useful approach to understanding the transparency of lens cells *in vivo.*

Acknowledgments

We appreciate the support of grants EY04542 and EY 01730 from the National Eye Institute. We thank Ms. C. Ganders for technical support and P. Muchowski for reading the manuscript.

References

Al-Ghoul, K. J., and Costello, M. J. (1996). Fiber cell morphology and cytoplasmic texture in cataractous and normal human lens nuclei. *Curr. Eye Res.* **15,** 533–542.

Asherie, N., Lomakin, A., and Benedek, G. B. (1996). Phase diagram for colloidal solutions. *Phys. Rev. Lett.* **77,** 4832–4835.

Berland, C. R., Thurston, G. M., Kondo, M., Broide, M. L., Pande, J., Ogun, O., and Benedek, G. B. (1992). Solid-liquid phase boundaries of lens protein solutions. *Proc. Natl. Acad. Sci. U.S.A.* **89,** 1214–1218.

Benedek, G. B. (1971). Theory of transparency of the eye. *Appl. Opt.* **10,** 459–473.

Bloemendal, H., Van de Gaer, K., Benedetti, E. L., Dunia, I., and Steely, H. T. (1997). Towards a human crystallin map. *Ophthalmic Res.* **29,** 177–190.

Broide, M. L., Berland, C. R., Pande, J., Ogun, O., and Benedek, G. B. (1991). Binary-liquid phase separation of lens protein solutions. *Proc. Natl. Acad. Sci. U.S.A.* **88,** 5660–5664.

Chylack, L. T., Jr. (1981). Sugar cataracts—possibly the beginning of medical anti-cataract therapy. *In* "Mechanisms of Cataract Formation in the Human Lens" (G. Duncan, ed.), pp. 237–252. Academic Press. New York.

Clark, J. I. (1990). Phase separation and H-bonding in cells of the ocular lens. *Biopolymers* **30,** 995–1000.

Clark, J. I. (1994a). Lens cytoplasmic protein solutions: Analysis of a biologically occurring aqueous phase separation. *In* "Methods in Enzymology" (H. Walter and G. Johansson, eds.), Vol. 228, pp. 525–537. Academic Press, San Diego, CA.

Clark, J. I. (1994b). Development and maintenance of lens transparency. *In* "Principles and Practice of Ophthalmology" (D. M. Albert and F. A. Jakobiec, eds.), pp. 114–123. Saunders, Philadelphia.

Clark, J. I., and Benedek, G. B. (1980a). Phase diagram for cell cytoplasm for the calf lens. *Biochem. Biophys. Res. Commun.* **95,** 482–489.

Clark, J. I., and Benedek, G.B. (1980b). Effects of glycols, aldehydes and acrylamide on phase separation and opacification in the calf lens. *Invest. Ophthalmol. Vis. Sci.* **19,** 771–776.

Clark, J. I., and Carper, D. L. (1987). Phase separation in lens cytoplasm is genetically linked to cataract formation in the Philly mouse. *Proc. Natl. Acad. Sci. U.S.A.* **84,** 122–125.

Clark, J. I., and Steele, J. E. (1992). Phase separation inhibitors, P.S.I. and prevention of selenite cataract. *Proc. Natl. Acad. Sci. U.S.A.* **89,** 1720–1724.

Clark, J. I., Mengel, L., Bagg, A., and Benedek, G. B. (1980). Cortical opacity, calcium concentration and fiber membrane structure in the calf lens. *Exp. Eye Res.* **31,** 399–410.

Clark, J. I., Giblin, F. J., Reddy, V. N., and Benedek, G. B. (1982). Phase separation in X-irradiated lenses of the rabbit. *Invest. Ophthalmol. Vis. Sci.* **22,** 186–190.

Clark, J. I., Osgood, T. B., and Trask, S. J. (1987). Inhibition of phase separation by reagents that prevent x-irradiation cataract in vivo. *Exp. Eye Res.* **45,** 961–967.

Delaye, M. A., and Tardieu, A. (1983). Short range order of crystallin proteins accounts for eye lens transparency. *Nature,* (*London*) **302,** 415–417.

Delaye, M. A., Clark, J. I., and Benedek, G. B. (1981). Coexistence curves for phase separation in the calf lens cytoplasm. *Biochem. Biophys. Res. Commun.* **100,** 908–914.

Duncan, G., and Jacob, T. J. C. (1984). The lens as a physiochemical system. *In* "The Eye" (H. Davson, ed.), Vol. 1B, pp. 159–206. Academic Press, New York.

Friberg, G., Pande, J., Ogun, O., and Benedek, G. B. (1996). Pantethine inhibits the formation of high Tc protein aggregates in γB crystallin solutions. *Curr. Eye Res.* **15,** 1182–1190.

Giblin, F. J., Padgaonkar, V. A., Leverenz, V. R., Lin, L. R., Lou, M. F., Unakar, N. J., Dang, L., Dickerson, J. E., and Reddy, V. N. (1995). Nuclear light scattering, disulfide formation and membrane damage in lenses of older guinea pigs treated with hyperbaric oxygen. *Exp. Eye Res.* **60,** 219–235.

Greiner, J. V., Kopp, S. J., and Glonek, T. (1985). Distribution of phosphatic metabolites in crystalline lens. *Invest. Ophthalmol. Vis. Sci.* **26,** 537–544.

Hammer, P., and Benedek, G. B. (1983). Effect of naturally occurring cellular constituents on phase separation and opacification in calf lens nuclear homogenates. *Curr. Eye Res.* **2,** 809–814.

Harding, J. (1991). "Cataract, Biochemistry, Epidemiology and Pharmacology." Chapman & Hall, London.

Hiraoka, T., Clark, J. I., Li, X. Y., and Thurston, G. M. (1996). Effect of selected anti-cataract agents on opacification in the selenite cataract model. *Exp. Eye Res.* **62,** 11–21.

Horwitz, J. (1992). Alpha crystallin can function as a molecular chaperone. *Proc. Natl. Acad. Sci. U.S.A.* **89,** 10449–10453.

Ishimoto, C., Goalwin, P. W., Sun, S., Nishio, I., and Tanaka, T. (1979). Cytoplasmic phase separation in formation of galactosemic cataract in lenses of young rats. *Proc. Natl. Acad. Sci. U.S.A.* **76,** 4414–4416.

Kador, P. F. (1994). Biochemistry of the lens: Intermediary metabolism and sugar cataract formation. *In* "Principles and Practice of Ophthalmology" (D. M. Albert and F. A. Jakobiec, eds.), pp. 146–167. Saunders, Philadelphia.

Kuszak, J. R., and Brown, H. G. (1994). Embryology and anatomy of the lens. *In* "Principles and Practice of Ophthalmology" (D. M. Albert and F. A. Jakobiec, eds.), pp. 82–96. Saunders, Philadelphia.

Kuwabara, T. (1975). Maturation of the lens cell: A morphological study. *Exp. Eye Res.* **20,** 427–443.

Langer, J. S. (1975). Spinodal decomposition. *In* "Fluctuations, Instabilities and Phase Transitions." (T. Riste, ed.), Nato Advanced Study Institute, pp. 19–42. Plenum, New York.

Liu, C., Lomakin, A., Thurston, G. M., Hayden, D., Pande, A., Pande, J., Ogun, O., Asherie, N., and Benedek, G. B. (1995). Phase separation in multicomponent aqueous-protein solutions. *J. Phys. Chem.* **99,** 454–461.

Liu, C., Asherie, N., Lomakin, A., Pande, J., Ogun, O., and Benedek, G. B. (1996). Phase separation in aqueous solutions of lens γ-crystallins: Special role of γs. *Proc. Natl. Acad. Sci. U.S.A.* **93,** 377–382.

Lo, W. K., Kuck, J. F., Shaw, A. P., and Yu, N. T. (1993). The altricial pigeon is born blind with a transient glycogen cataract. *Exp. Eye Res.* **56,** 121–126.

Maisel, H. (1985). "The Ocular Lens: Structure, Function and Pathology." Dekker, New York.

Matsushima, H., David, L. L., Hiraoka, T., and Clark, J. I. (1997). Loss of cytoskeletal proteins and lens cell opacification in the selenite cataract model. *Exp. Eye Res.* **64,** 387–395.

Muchowski, P. J., and Clark, J. I. (1998). ATP enhanced molecular chaperone functions of the small heat shock protein αB crystallin. *Proc. Natl. Acad. Sci. USA* **95,** 1004–1009.

Muchowski, P. J., Bassuk, J. A., Lubsen, N. H., and Clark, J. I. (1997). Human alpha B crystallin: Small heat shock protein and molecular chaperone. *J. Biol. Chem.* **272,** 2578–2582.

Pande, J., Berland, C., Ogun, O., Melhuish, J., and Benedek, G. B. (1991). Suppression of phase separation in solutions of bovine gamma IV crystallin by polar modification of the sulfur containing amino acids. *Proc. Natl. Acad. Sci. U.S.A.* **88,** 4916–4920.

Pande, J., Ogun, O., Nath, C., and Benedek, G. B. (1993). Suppression of phase separation in bovine γIV crystallin solutions: Effect of modification by charged versus uncharged polar groups. *Exp. Eye Res.* **57,** 257–264.

Piatigorsky, J. (1992). Lens crystallins—innovation associated with changes in gene regulation. *J. Biol. Chem.* **267,** 4277–4280.

Prescott, A. R., Sandilands, A., Hutcheson, A. M., Carter, J. M., and Quinlan, R. A. (1996). The intermediate filament cytoskeleton of the lens: An ever changing network through development and differentiation. A minireview. *Ophthalmic Res.* **28,** 58–61.

Rabaey, M. (1963). Glycogen in the lens of birds' eyes. *Nature,* (*London*) **198,** 206–207.

Radlick, L. W., and Koretz, J. F. (1992). Biophysical characterization of alpha crystallin aggregates: validation of the micelle hypothesis. *Biochim. Biophys. Acta* **1120,** 193–200.

Sandilands, A., Prescott, A. R., Carter, J. M., Hutcheson, A. M., Quinlan, R. A., Richards, J., and Fitzgerald, P. G. (1995). Vimentin and CP49/filensin from distinct networks in the lens which are independently modulated during lens fiber cell differentiation. *J. Cell Sci.* **108,** 1397–1406.

Siezen, R. J., Fisch, M. R., Slingsby, C., and Benedek, G. B. (1985). Opacification of γ–crystallin solutions from calf lens in relation to cold cataract formation. *Proc. Natl. Acad. Sci. U.S.A.* **82,** 1701–1705.

Smith, J. B., Ma, Z., Hanson, S. A., Smith, D. L., Lampi, K., and David, L. L. (1997). Age related changes in young human lens crystallins. *Invest Ophthalmol. Vis. Sci.* **38,** S205a.

Takemoto, L. J. (1997). Changes in the C-terminal region of alpha A crystallin during human cataractogenesis. *Int. J. Biochem. Cell Biol.* **29,** 311–315.

Takemoto, L. J., and Gopalakrishnan, S. (1994). Alpha A crystallin: Quantitation of C-terminal modification during lens aging. *Curr. Eye Res.* **13,** 879–883.

Tanaka, T., and Benedek, G. B. (1975). Observation of protein diffusivity in intact human and bovine lenses with application to cataract. Invest. *Ophthalmol. Vis. Sci.* **14,** 449–456.

Taratuta, V. G., Holschbach, A., Thurston, G. M., Blankschtein, D., and Benedek, G. B. (1990). Liquid liquid phase separation of aqueous lysozyme solutions: Effects of pH and salt. *J. Phys. Chem.* **94,** 2140–2144.

Tardieu, A., Veretout, F., Krop, B., and Slingsby, C. (1992). Protein interactions in the calf eye lens. *Eur. Biophys. J.* **21,** 1–12.

Tomarev, S. I., and Piatigorsky, J. (1996). Lens crystallins of invertebrtes—diversity and recruitment from detoxification enzymes and novel proteins. *Eur. J. Biochem.* **235,** 449–465.

Vaezy, S., and Clark, J. I. (1995). Characterization of the cellular microstructure of ocular lens using 2-D power law analysis. *Ann. Biomed. Eng.* **23,** 482–490.

Vaezy, S., Clark, J. I., and Clark, J. M. (1990). Fourier analysis of two structural organizations in opaque mouse lenses. *Proc. Int. Congr. Electron Microsc., 12th,* pp. 554–556.

Vaezy, S., Clark, J. I., and Clark, J. M. (1995). Quantitative analysis of lens cell microstructure in selenite cataract using 2-D Fourier analysis. *Exp. Eye Res.* **60,** 245–255.

Veretout, F., Delaye, M., and Tardieu, A. (1989). Molecular basis of eye lens transparency. *J. Mol. Biol.* **205,** 713–728.

Walter, H., and Brooks, D. E. (1995). Phase separation in cytoplasm, due to macromolecular crowding, is the basis for microcompartmentation. *FEBS Lett.* **361,** 135–139.

Walter, H., and Johansson, G., eds. (1994). "Methods in Enzymology," Vol. 228. Academic Press, San Diego, CA.

Walter, H., Brooks, D. E., and Fisher, D., eds. (1985). "Partitioning in Aqueous Two-phase Systems: Theory, Methods, Uses and Application to Biotechnology." Academic Press, New York.

Cytoarchitecture and Physical Properties of Cytoplasm: Volume, Viscosity, Diffusion, Intracellular Surface Area

Katherine Luby-Phelps
Department of Physiology, The University of Texas Southwestern Medical Center, Dallas, Texas 75235-9040

Classical biochemistry is founded on several assumptions valid in dilute aqueous solutions that are often extended without question to the interior milieu of intact cells. In the first section of this chapter, we present these assumptions and briefly examine the ways in which the cell interior may depart from the conditions of an ideal solution. In the second section, we summarize experimental evidence regarding the physical properties of the cell cytoplasm and their effect on the diffusion and binding of macromolecules and vesicles. While many details remain to be worked out, it is clear that the aqueous phase of the cytoplasm is crowded rather than dilute, and that the diffusion and partitioning of macromolecules and vesicles in cytoplasm is highly restricted by steric hindrance as well as by unexpected binding interactions. Furthermore, the enzymes of several metabolic pathways are now known to be organized into structural and functional units with specific localizations in the solid phase, and as much as half the cellular protein content may also be in the solid phase.

KEY WORDS: Cytoplasm, Viscosity, Diffusion, Partitioning, Particle Transport.

I. Introduction

Much of the current paradigm for cellular biochemistry has been extrapolated from studies of dilute solutions containing a single enzyme and a single substrate whose interaction is diffusion-limited. While this reductionist approach has led to many valuable insights over the past several decades, measurements of the physical properties of cells indicate that the interior of a cell departs from these ideal conditions in several important ways. Long considered a minority view, there now is a dawning awareness on the part of investigators that the nonideality of the cell interior may necessitate an alternate view of cellular biochemistry. In this chapter, we will address five assumptions implicit in the prevailing paradigm, review the extent to which they are valid for cells, and point out areas in need of further study. These assumptions are that: (1) the reaction volume is infinite, (2) the solution is dilute, (3) the concentrations of substrates are much higher than the concentrations of their enzymes, (4) the solution is well-defined, and (5) the solution is homogeneous. Each of these points has been raised by previous reviewers (for example, Srere, 1967; Fulton, 1982; Agutter *et al.,* 1995). This chapter will provide an overview of the field and focus on the most recent contributions. For the sake of brevity, the reader is referred to recent review articles for more detailed discussion of some areas, and I apologize to any author whose work is not cited individually.

II. Examining the Assumptions

A. Assumption 1: Infinite Volume

An underlying assumption of classical physical biochemistry is that the reaction volume is infinite. The familiar concept of concentration depends on this assumption. In reality, because a cell is surrounded by a limiting membrane, which is only selectively permeable, the assumption of infinite volume is patently false. The question is whether the finite volume of cells can be neglected, as it frequently is. Single cells range in size from the smallest mycoplasma (0.3 μm in diameter) to the eggs of the African clawed toad, *Xenopus laevis* (1 mm in diameter). The relevant volumes thus range from 1×10^{-17} liters to about one-half microliter. Bacteria typically have a volume of the order of 2×10^{-16} liter (Halling, 1989). The volume of an average mammalian tissue culture cell is approximately 4 picoliter (Alberts *et al.,* 1994). For eukaryotic cells, in which many biochemical pathways are sequestered within membrane-bounded compartments (organelles) of submicrometer dimensions, the reaction volume may be very small indeed (6×10^{-20} liters for a spherical vesicle 50 nm in diameter). In addition, the internal membrane-bounded compartments together occupy up to half the volume of the cell, further reducing the free volume available to reactants diffusing in the cytoplasm that bathes the organelles.

The small, finite volume of cells and intracellular compartments means that, at physiologically relevant molar concentrations, the number concentration of a particular molecular species may be surprisingly low. In the example given by Halling (1989), a species present in a single copy per bacterium will have a nominal concentration of nearly 10^{-8} M. In an average eukaryotic cell, a protein present at a whole cell concentration of 1 nM would have number concentration on the order of 1000 copies. At these low copy numbers, number fluctuations may limit the availability of a species at a given subcellular location and time. In addition, a large surface with even low affinity for that species could adsorb the entire population. Even for ions, number fluctuations can be significant if the volume is small enough. For example, the pH of the endosomal compartment in eukaryotic cells is maintained at around 6. For a typical, spherical endosome 250 nm in diameter, this amounts to one free proton. As the endosome matures, the interior becomes acidified to pH 5 by the action of a proton pump in the surrounding membrane. Assuming constant volume, this requires an increase of only 48 free protons. Thus, it may be more accurate to understand the interior of this compartment in terms of the protonation of specific residues rather than a global pH. In general, for small cells or intracellular compartments, the number concentration of a particular molecular species may be more informative than its molar concentration.

B. Assumption 2: The Cell Interior Is Dilute

A fundamental assumption of the physical chemistry of dilute solutions is that interactions between solute molecules can be neglected. However, the total concentration of macromolecules inside cells is very high, with proteins by far the most abundant species. As reviewed by Fulton (1982), the protein content of cells is in the range 17 to 35% by weight. Estimation of the protein concentration in the cytoplasm of mammalian tissue culture cells from refractive index measurements indicates a value of 20 to 30 g/100 ml (Lanni *et al.*, 1985). The low spatial resolution of these meaurements doesn't allow a distinction between structural and soluble proteins, and the proportion of the total that is actually soluble continues to be the subject of some debate, as discussed in more detail below.

A 20% solution of protein is very crowded and significant interactions between solute molecules can be expected: For a typical 50 kDa protein, this is well above the theoretical "overlap" concentration of 13 g/100 ml (Chang *et al.*, 1987). Although the physical volume fraction of protein at a concentration of 20% is well below the threshold for close packing of hard spheres (Kertesz, 1981), some proteins have been crystallized at concentrations in this range. How crowded a milieu the cytoplasm is can be appreciated visually from the drawings of Goodsell (1991, 1993), who has depicted the cell interior with macromolecular components at the correct concentration and drawn to scale (Fig. 1).

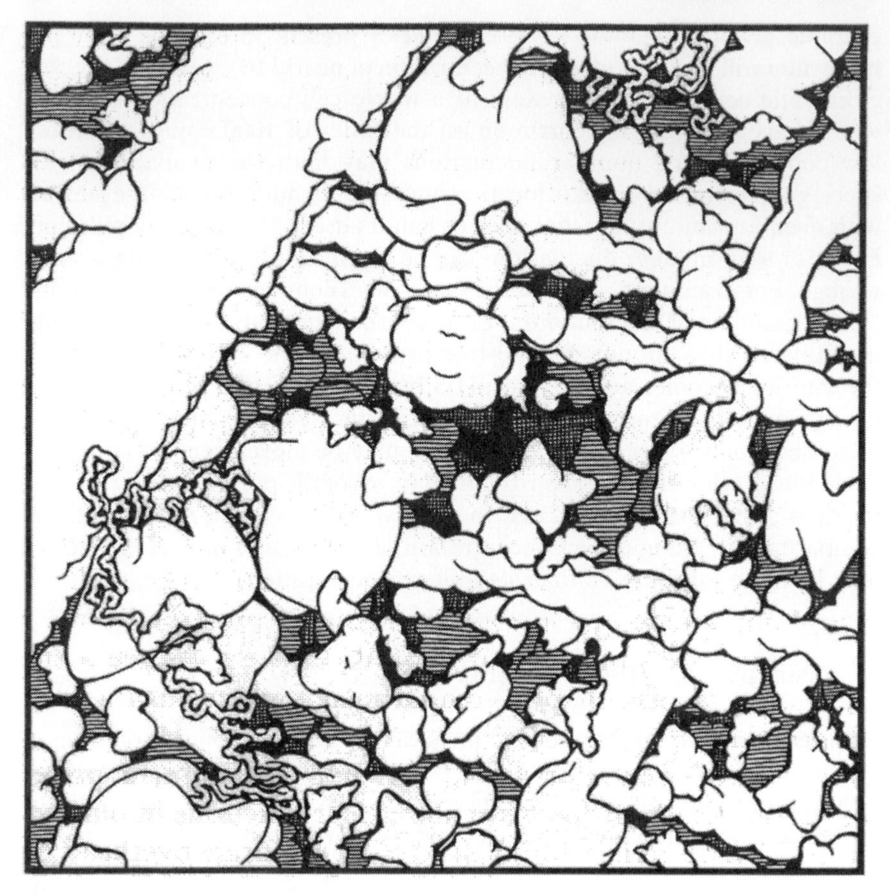

FIG. 1 A view of the cytoplasm of the yeast, *S. cerevisiae,* with components drawn to scale and at the correct concentrations. Soluble proteins, ribosomes with mRNA, microtubules, actin filaments, and intermediate filaments are all represented. Small solutes are omitted from the drawing. Magnification 1,000,000 ×. Reprinted with permission (Goodsell, 1993).

The biochemistry of crowded solutions can be shown both theoretically and experimentally to deviate from what occurs in dilute solution (Minton, 1997; see also chapter by Johansson *et al.* in this volume). The increased pathlength required to circumvent obstacles can retard significantly the translational diffusion of particles comparable in size to the crowding solute, slowing the rates of diffusion-limited reactions accordingly. The entropy benefit to the system arising from a reduction in particle number may drive association of components that associate only weakly or not at all in dilute solution. Excluded volume effects may drive conformation changes in a particular species. Because the threshold for crowding effects is steep, small changes in cellular volume may thus lead to large changes in the chemical activities of the solutes. Many examples of altered kinetics, specific activity or substrate specificity of enzyme-catalyzed reactions in crowded solutions can be found in the literature (Minton, 1997). Crowding has also been shown to drive self-assembly reactions and phase separations of biomolecules *in vitro* (Minton, 1997). For a more detailed discussion of the effects of macromolecular crowding, the reader is directed to the chapter by Johansson *et al.* in this volume.

C. Assumption 3: Substrate Concentrations are Greater than Enzyme Concentrations

Not only are macromolecules generally present in high concentration, but the concentration of a specific enzyme or protein is generally much higher in cells than in conventional *in vitro* assays, and is often much in excess of the K_d as determined *in vitro.* The Ca^{2+}-dependent regulatory protein, calmodulin (CaM), which binds to and activates a variety of target enzymes in response to intracellular Ca^{2+} signals, serves as a good illustration. The K_{act} of CaM for myosin light chain kinase, its major known target in smooth muscle, is around 1 nM, but the concentration of CaM in smooth muscle is on the order of 38 μM (Tansey *et al.,* 1994; Zimmermann *et al.,* 1995) The concentration of enzymes confined within organelles such as mitochondria or peroxisomes can be even higher. As pointed out some time ago by Srere (1967), if the K_ds *in vivo* are the same as measured *in vitro,* a substantial proportion of substrates may be bound to enzymes, making the concentration of available substrate rate limiting. (Srere, 1967). In order to proceed efficiently under these conditions, channeling of substrates from enzyme to enzyme in a particular metabolic pathway may be necessary. See the chapter to by Ovádi and Srere in this volume for further discussion.

D. Assumption 4: The Cytoplasm is a Simple, Well-defined Solution

An inspection of any metabolic pathway chart will show that the interior of the cell contains a complex web of enzyme pathways, many of which compete for common substrates. It appears from the number of different mRNAs that a typical eukaryotic cell synthesizes 10,000 to 20,000 different proteins (Alberts *et al.,* 1994). Because most of these have not been characterized biochemically, their effect on any given reaction pathway is unknowable. In addition, most regulatory enzymes and proteins are promiscuous, interacting with a large number of target proteins, at least *in vitro.* To continue with the example of calmodulin introduced above, it has more than 50 known targets, including kinases, phosphatases, ion channels, structural proteins, and transcription factors (Rhoads and Friedberg, 1997). *In vitro,* the affinities of most of these targets for binding by Ca^{2+}–CaM are within an order of magnitude. It is difficult to see how discrimination among competing interactions can occur in the intact cell without preexisting colocalization of enzyme and substrate, or regulatory protein and target. For some pathways, such as the tricarboxylic acid cycle, this has been achieved by sequestering them within an organelle or immobilizing the enzyme on a membrane. However, the majority of "soluble" proteins have long been assumed to be distributed homogeneously in the cell interior.

E. Assumption 5: The Cytoplasm is a Homogeneous Solution

Traditional chemistry depends on the assumption that the solution is homogeneous and can be treated as a continuum. This assumption is not valid for intact cells. To begin with, the presence of membrane-bounded subcompartments confers an intrinsic inhomogeneity on the interior of an eukaryotic cell. Additionally, the organelles themselves are not randomly distributed (see chapter by Aw in this volume). The localization of mitochondria near sites of high ATP consumption, for example, in striated muscle, is well known. The nonrandom disposition of organelles during differentiation can confer both structural and functional polarity on cells, as in the case of columnar epithelial cells or neurons. Cells expend a large proportion of their energy supply in establishing and maintaining this nonrandom array of organelles.

The cytoplasm that bathes the organelles also cannot be considered a homogeneous solution. It is permeated by a network of self-assembling polymeric protein fibers known collectively as the cytoskeleton. The three best studied of these are the microfilaments, the microtubules, and the intermediate filaments. The microfilaments are homopolymers 8 nm in diameter formed from the globular protein, actin. Actin is very abundant in most cells, with a concentration of about 4 mg/ml in a typical eukaryotic cell (Bray and Thomas, 1975). Roughly half this amount is polymeric, giving a value of 3×10^5 μm of filament per cell (based on 370 monomers per μm). Microtubules are heteropolymers of the related globular proteins, α- and β-tubulin. The subunits self-assemble to form a rigid, hollow tube 25 nm in diameter. There are typically about 150 microtubules in a tissue culture cell and each is 50 to 100 μm in length (Hiller and Weber, 1978). Intermediate filaments are highly stable polymers formed by lateral, coiled-coil associations between the polypeptides of a broad family that includes vimentin, the cytokeratins, and the neurofilaments (Chou *et al.,* 1997). The concentration of vimentin in a mammalian tissue culture cell of mesenchymal origin is on the order of 150 μg/ml, all of which is polymerized (Lai *et al.,* 1993), resulting in an estimated 10,000 μm filament per cell (based on 32 monomers per 10 nm diameter $\times$ 48 nm long association domain).

Although the three types of cytoskeletal filaments are often treated as independent systems, a mounting body of evidence supports the idea that they are physically interconnected. Whole-mount electron microscopy of tissue culture cells extracted with nonionic detergents revealed a three-dimensional network in which all three filament types could be identified with frequent connections among them (Webster *et al.,* 1978; Schliwa and van Blerkom, 1981). Similar views have been obtained by electron microscopy of rapidly frozen tissue from which the ice has been sublimed (Heuser and Kirschner, 1980). In living cells, treatment specifically targeting one of these three systems frequently leads to rearrangement of one or both of the other two. In a recent example, a peptide that specifically disassembles intermediate filaments *in vitro* leads to collapse of the microtubules and microfilament systems when microinjected into intact cells (Eriksson *et al.,* 1992). In addition, specific proteins, such as plectin and map2C, can be shown to crosslink one filament type to another both *in vitro* and *in vivo* (Chou *et al.,* 1997; Cunningham *et al.,* 1997). Thus, the cytoplasmic volume may be better described as an aqueous gel than as a homogeneous solution (Fig. 2).

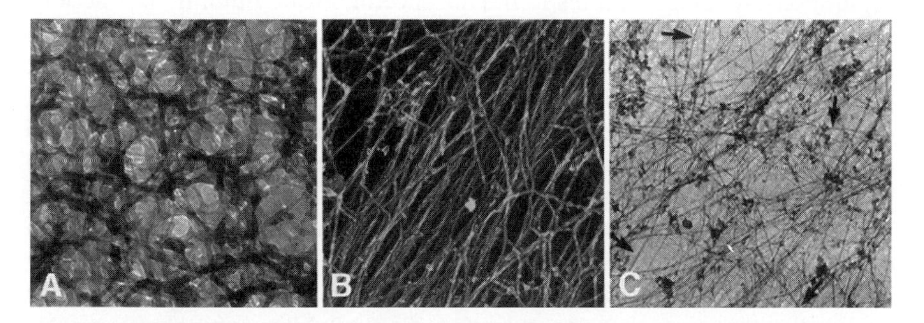

FIG. 2 Three views of the cytoplasm by electron microscopy. (A) The microtrabecular lattice in a freeze-dried, whole-mounted fibroblast. Magnification 80,000. Reproduced from *The Journal of Cell Biology,* 1984, vol. 99, pp. 3s–12s by copyright permission of The Rockefeller University Press. (B) The appearance of the cytoplasm in a platinum replica of a fibroblast that was extracted for 30 min with 0.5% Triton before freeze-drying. Magnification 70,000. Reproduced from *The Journal of Cell Biology,* 1980, vol. 86, pp. 212–234 by copyright permission of The Rockefeller University Press. (C) The appearance of the cytoplasm in a fibroblast lightly extracted with Triton, then critical point dried and whole mounted. Magnification 17,500. Arrows indicate microtubules. Reproduced from Schliwa and van Blerkom (1981) by copyright permission of The Rockefeller University Press.

The individual cytoskeletal filaments themselves are not isotropically distributed within the network. The microfilaments form bundles known as stress fibers, and the ends of the bundles are often fixed at specific sites, such as adhesion plaques in the ventral membrane of a fibroblast or adherens junctions in the lateral membranes of adjoining epithelial cells. By whole-mount electron microscopy, stress fibers and smaller microfilament bundles appear as local inhomogeneities embedded in the larger cytoskeletal network (Small *et al.,* 1978; Provance *et al.,* 1993). In most cells, the majority of microtubules radiate from an organizing center, or centrosome, usually located near the cell nucleus. Intermediate filaments are often colinear with these microtubules, especially in the more peripheral regions of the cell.

The inhomogeneities contributed by the cytoskeleton vary temporally as well as spatially. The assembly state and subcellular distribution of all three types of cytoskeletal filament are physiologically regulated. Well-known examples of this include the disassembly of the radial array of microtubules and subsequent assembly of the microtubule spindle at mitosis, and the reorganization of the actin cytoskeleton in response to growth factor stimulation or tumor viruses. Recently, progress has been made in defining the signaling and control pathways that lead to these changes, and accessory protein candidates for mediating these changes have been identified (for reviews, see Maccioni and Cambiazo, 1995; Johansen, 1996; Lim *et al.,* 1996; Pereira and Schiebel, 1997; Tapon and Hall, 1997). In pure solutions, microfilaments and microtubules undergo spontaneous phase separations, forming bundles in the absence of strong binding interactions between filaments. Three types of phase separation are recognized: filament length- and concentration-dependent, entropy-driven formation of nematic phases embedded in the isotropic phase; excluded volume-driven demixing of filaments with different flexibilities or of filaments and spheres; and bundling of filaments in the presence of polycations analogous to the counterion condensation of DNA (Herzfeld, 1996; Tang and Janmey, 1996). The composition and type of phases can be modified by changing the length or concentration of the filaments; manipulating the free volume by crowding the solution with inert macromolecules (such as polyethylene glycol (PEG) or Ficoll); changing the concentration or availability of polycations; or stabilizing bundles with cross-linking proteins. Similar phase separations of cytoskeletal components *in vivo* may contribute to the inhomogeneous distribution described above. Specific proteins are known that regulate the length or total content of filaments, as well as proteins that stabilize bundles by cross-linking the filaments (Carlier and Pantaloni, 1997; Pereira and Schiebel, 1997). These proteins are presumed to be the immediate effectors of signal-initiated reorganization of the cytoskeleton. In addition to the regulated formation and dissolution of more or less ordered phases, individual microtubules in living cells have been shown to be dynamically unstable, alternately growing and shrinking (Waterman-Storer and Salmon, 1997). This means that local inhomogeneities fluctuate on a time scale of a few minutes.

The inhomogeneities described above present large intracellular surface areas upon which phase separations might occur. The surface area of specific membrane compartments has been measured in many cell types. The total outer surface area of mitochondria is on the order of a few hundred to a few thousand μm^2/cell (Mori *et al.,* 1982; Hoppeler *et al.,* 1987). At the other end of the spectrum, the total surface area of the abundant endoplasmic reticulum in secretory cells, such as the Leydig cells of the testis, may be as high as 30,000 μm^2/cell (Mori *et al.,* 1982). Altogether, for eukaryotic cells, the internal membrane surface area far exceeds the area of the plasma membrane. For hepatocytes, it is estimated to be approximately 100,000 μm^2/cell, which is 50 times the area of the plasma membrane (Alberts *et al.,* 1994). It is interesting to note that many cell types modify the surface area of their internal organelles in response to growth conditions. While this is often interpreted as a consequence of modifications of organelle volume, the corresponding changes in intracellular surface area are significant and could potentially play a regulatory role in metabolism.

The cytoskeleton also presents a significant surface area. Calculating from the known size, shape, and concentration of actin microfilaments, microtubules, and intermediate filaments, the naked cytoskeleton would contribute about 3000 μm^2 of surface area in a typical mammalian tissue culture cell. This may be an underestimate if other types of filaments exist or when one takes into account other components immobilized on the cytoskeleton. Whole-mount electron microscopy of unextracted cells reveals not the well-defined cytoskeleton that is observed after detergent extraction, but a three-dimensional network of irregularly shaped strands that was designated the microtrabecular lattice (MTL) by Porter and colleagues (Wolosewick and Porter, 1979) (Fig. 2C). Great care was taken to minimize fixation artifacts in the preparation of these specimens, suggesting that the MTL may accurately reflect the appearance of the cytoskeletal network in the living state. The surface area of the MTL estimated by morphometry of stereo electron micrographs is 69,000–91,000 μm^2 per cell (Gershon *et al.,* 1985). Thus, the surface area of the MTL and internal membranes together could be as high as 200,000 μm^2 per cell in some cell types under certain conditions. It has been proposed that many enzyme pathways are immobilized and localized on the surfaces of internal membranes and that substrates are transported to these enzymes by bulk streaming of the cytoplasm (Wheatley and Malone, 1993; Wheatley, 1998).

III. Physical Properties of Cytoplasm (Measured)

It is obvious from the foregoing discussion that the nonideality of the cell interior could have profound effects on its physical and chemical properties. In this section, we will survey experimental data aimed at studying these properties in the cytoplasm of intact cells. Three areas of research that have seen the most activity are the behavior of intracellular water, the constraints on diffusion and partitioning of inert tracer particles, and the diffusibility of endogenous proteins and vesicles.

A. Water in Intact Cells

The idea that macromolecules in solution contain bound water molecules is a concept of long standing in physical biochemistry. As the crystal structures of more and more proteins have become available, it has become clear that not only is hydration necessary for the function of particular enzymes, but insertion or removal of water molecules can drive conformational changes and macromolecular associations. What is less commonly considered is the possibility that in addition to water that is an integral part of a protein's folded conformation, the mobility of water near hydrophobic surfaces may be reduced by the formation of hydrogen-bonded clathrate structures. Thermodynamically, reduced mobility of water translates into a reduced chemical activity compared with bulk water. The solvent properties of this perturbed water may also differ from bulk water in a solute-dependent manner. Conversely, dissolved macromolecules can also have water destructuring effects due to the dielectric field around charged or polar groups that attracts the polar water molecules and prevents hydrogen bonding (Cameron *et al.,* 1997). A different way of conceptualizing the same effects is to think in terms of water density (Wiggins, 1996). The local concentration of water molecules (and, hence, water activity) will be higher around charged (hydrophilic) surfaces and lower around hydrophobic surfaces.

The high concentration of macromolecules inside a living cell, and the large intracellular surface area presented by proteins and membranes, raises the question of whether a significant proportion of total cellular water differs from the bulk. This has been the subject of both experimentation and debate for many years (see chapter by Garlid in this volume). Opinions range from the point of view that only a low percentage of cellular water is bound (Cooke and Kuntz, 1974), to the claim that most cellular water is ordered to some degree (Ling, 1988; Ling *et al.,* 1993).

Attempts to address this issue experimentally fall into two broad methodological classes: measurement of the properties of cellular water itself and measurement of the rotational mobility of solutes, either endogenous molecules or probes that have been introduced into living cells. Approaches to measuring the properties of intracellular water can be further divided into measurement of water mobility and measurement of water activity by studying the colligative properties of cells.

1. Measurements of Water Mobility

Two methods have been used to measure the mobility of cellular water: proton NMR and quasielastic neutron scattering (QENS). Both methods have the advantage of being noninvasive. Since a typical cell contains 70% water, the ^1H -NMR signal from cells is dominated by the water protons. In pure water, the transverse relaxation time (T_2) reflects the randomization of the proton dipoles subsequent to orientation in a magnetic field and is, thus, a measure of rotational diffusion. In more complex milieuxs, T_2 may be shortened due to local fields in the sample or exchange of protons among populations with different properties. T_2 measured from cells generally is about 1/50 that of pure water protons (Foster *et al.,* 1976). Assuming that the mechanism of accelerated relaxation is fast exchange of protons between two or more populations of water with different mobilities, T_2 is the weighted average of the relaxation times from all water fractions in the cell. When T_2 is corrected to 100% water content, it appears that the average rotational mobility of cellular water is reduced about twofold (Clegg, 1984; Rorschach *et al.,* 1991; Cameron *et al.,* 1997). However, because water molecules can diffuse over distances comparable to the dimensions of a cell in the relatively long time required for these measurements, the reduction in T_2 could reflect encounters with physical barriers in the cell rather than altered rotational mobility. QENS measurements can potentially resolve this ambiguity because of the short distance scale over which the measurements are made. Unfortunately, because the samples must be held under nonphysiological conditions for several days, the use of this technique has been limited to a few hardy cell and tissue types, such as the brine shrimp *Artemia* and frog sartorius muscle (Rorschach *et al.,* 1973; Trantham *et al.,* 1984). As found by NMR, the apparent self-diffusion coefficient of intracellular water measured by QENS is significantly reduced compared with pure water. In the cases where QENS data and NMR data are both available for the same cell or tissue, there is remarkable quantitative agreement between the two parameters (Cameron *et al.,* 1997). However, quantitative interpretation of these data is ambiguous unless the relaxation times and proportions of each water fraction can be measured by independent methods. Furthermore, it may not be correct to assume that exchange between water fractions is rapid. For example, in a study of water mobility in barnacle muscle, the tightly bound fraction was assayed independently by its inability to exchange with deuterium. A second, less tightly bound fraction was identified that exchanged with deuterium at an intermediate rate, but remained unfrozen at −34°C. Together these two fractions accounted for only 3% of the water protons in barnacle muscle, with the rest having the properties of bulk water (Foster *et al.,* 1976). It should also be kept in mind that alternate mechanisms of relaxation such as dipole interactions, exchange of water protons with the protons of proteins, and slow exchange of water across the plasma membrane can also reduce T_2 and cannot always be ruled out experimentally.

2. Measurement of Colligative Properties of Cells and Biomolecules

The colligative properties of a solution are directly related to water activity. These include osmotic pressure, freezing point depression, and vapor pressure. While all of these have been applied to study the properties of protein solutions, only the first two are readily applicable to living cells. The appropriate use of osmotic pressure measurement to study the activity of water in biological specimens recently was the subject of an excellent review (Parsegian *et al.,* 1995). We have already seen that a small fraction of water in barnacle muscle had a depressed freezing point, suggesting that its activity is significantly lower than that of bulk water (Foster *et al.,* 1976). Reports of the nonideal osmotic response of some cells indicate a much larger fraction of water with perturbed activity. The water content of lens fiber cells, *Xenopus* oocytes, sea urchin eggs, and frog sartorius muscle, was found to scale linearly as a function of inverse osmotic pressure (Π) over a limited range as expected for an osmometer (for review, see Cameron *et al.,* 1997). However, linear extrapolation of the data to infinite Π gave a nonzero intercept that was interpreted as "osmotically unresponsive" water. This fraction ranged from 30 to 90% of total water content of the specimens. A recent paper addressed water activity in neurons by combining ^{1}H-NMR imaging of single cells with osmotic perturbation (Hsu *et al.,* 1996). These investigators found that T_2 varied with changes in tonicity over the very limited range tested, as expected for a perfect osmometer. Under the same conditions, the apparent diffusion coefficient of the protons was unaffected.

Several ambiguities attend the use of osmotic approaches to study the activity of cellular water that are not a factor for the study of water activity in solution. For example, it is rarely possible to rule out active responses of the cell to changes in extracellular tonicity. In addition, volume changes may not be directly coupled to osmotic pressure, because the ratio of plasma membrane surface area to cell volume often is very large due to convolution of the membrane (Clegg, 1986).

Measurements of the colligative properties of specific proteins *in vitro* have also been used to predict the amount of perturbed water in cells (for review, see Cameron *et al.,* 1997). For example, the nonzero intercept of plots of water content vs inverse osmotic pressure for BSA indicate that this protein contains 4 g of osmotically unresponsive water per gram of protein. Assuming that BSA is representative of other cytoplasmic proteins, extrapolation of this result to the estimated protein concentration in cytoplasm suggests that the osmotic response of all cellular water molecules would be perturbed. Studies of ion partitioning and water flow through protein gels *in vitro* suggest that the activity of water in such gels is greatly reduced (Van Steveninck *et al.,* 1991, Wiggins *et al.,* 1991; Ito *et al.,* 1992). If the cytoskeleton renders the interior of an intact cell more like a gel than a solution, significant changes in water activity and mobility are predicted on the basis of these experiments.

3. Measurements of Solute Rotational Mobility

The second approach to examining the properties of cellular water has been to study the rotational mobility of solutes, which primarily reflects the viscosity of the solvent (water) with a negligible contribution from any barriers to long-range translational motion. The rotational mobility of endogenous glycolytic enzymes was measured in the yeast *Saccharomyces cerevisae* by biosynthetically labeling them with 5-fluorotryptophan (Williams *et al.,* 1997). NMR gave relaxation times for phosphoglycerate kinase and hexokinase that were consistent with a solvent viscosity between 1 and 2 cP. Similarly, the rotational mobility of deoxymyoglobin was measured by ^{1}H-NMR in bovine heart muscle by taking advantage of its unique proton signature. In this case, a value consistent with a solvent viscosity of 2.3 cP was measured (Livingston *et al.,* 1983). These values are in general agreement with the average reduction in water mobility measured by proton NMR.

An alternative approach taken by several laboratories is to measure the rotational mobility of small spin probes or fluorescent molecules that have been introduced into living cells. Values of cytoplasmic viscosity as high as 11 cP were estimated from the behavior of two small spin probes in tissue culture cells (Lepock *et al.,* 1983; Mastro *et al.,* 1984). Similarly high values were suggested based on the steady-state fluorescence polarization exhibited by small fluorescent probes (Lindmo and Steen, 1977; Hashimoto and Shinozaki, 1988). However, the binding of these probes to intracellular components and the subcellular location of the probes were not assessed in these studies, so it is not clear that the rotational behavior of the probes reflects cytoplasmic viscosity. In addition, the high steady-state polarization of the fluorescent probes could not be directly interpreted in terms of the rotational mobility of the probe since the fluorescence lifetime of the probe in the cells was not measured. More recently, a series of papers from the Verkman laboratory have reported the use of time resolved fluorescence anisotropy imaging of the small fluorescent probe BCECF to measure solvent viscosity (Fushimi and Verkman, 1991; Periasamy *et al.,* 1991, 1992). Although the probe showed significant binding to intracellular components, these investigators developed an algorithm for data analysis by which free probe could be distinguished from bound. Measurements of lifetime showed that the lifetime of the probe was unaffected by the intracellular environment. In several tissue culture cell lines, these studies indicated that cytoplasmic solvent viscosity is not significantly different from bulk water (Fushimi and Verkman, 1991). However, cells such as sea urchin eggs and kidney tubule cells, that adapt to anisoosmotic conditions by making small osmolytes, did exhibit a two- or threefold increased solvent viscosity (Periasamy *et al.,* 1991, 1992). No significant spatial variation in viscosity was detected and nuclear viscosity was not different from the cytoplasm. More recently, time-resolved fluorescence anisotropy imaging of sulforhodamine B was used to measure the solvent viscosity of a range of cell types (Srivastava and Krishnamoorthy, 1997). The free and bound fractions of the probe could be distinguished unequivocally by a marked difference in fluorescence lifetime. In all cell types studied, the apparent solvent viscosity was very similar to bulk water, with modest spatial variations on a distance scale of 1 μm. It was found that plant cells generally have a higher viscosity and more spatial variability than do animal cells. A somewhat different approach has been to utilize steady-state fluorescence ratio imaging to map solvent viscosity, using a viscosity-sensitive fluorophore attached to dextran (Luby-Phelps *et al.,* 1993). Again, solvent viscosity in the cytoplasm of two mammalian tissue culture cell lines did not appear significantly different from bulk water and no spatial variation was observed beyond that accounted for by systematic error.

These results would appear, at first glance, to be in conflict with data on water mobility from NMR and QENS. However, an inherent limitation of using solute rotation to probe water viscosity is that steric hindrance or insolubility may limit access of the probes to compartments where water is perturbed. In addition, the limited spatial resolution of standard imaging apparatus would allow surface associated layers of ordered water up to 200 nm thick to escape detection. The spatial resolution can be improved by using total internal reflection fluorescence (TIRF) imaging to measure the mobility of only those fluorescent probes within 50 nm of the ventral membrane of a cell. TIRF combined with time-resolved anisotropy indicated that the rotation of BCECF adjacent to the membrane was unimpeded relative to bulk water (Bicknese *et al.,* 1993). In summary, the average mobility of water in living cells does not appear to be greatly altered compared with pure water, but significant layers of more or less ordered water cannot be ruled out.

B. Hindrance of Long-Range Translational Diffusion

1. Diffusion of Protein-sized Tracer Macromolecules

In addition to possible effects of water structure on the rotational mobility of solutes, the long-range translational diffusion of macromolecules in the cytoplasm could be hindered by the excluded volume effects of crowding and by steric constraints imposed by obstacles such as the cytoskeleton. Studies of the translational diffusion of hydrophilic, electroneutral tracer particles suggest that such long-range effects are important in intact cells. Two different fluorescence microscopic methods have been used in these studies to obtain the cytoplasmic diffusion coefficients of fluorescent, size-fractionated dextrans and Ficolls that have been microinjected into intact cells. The first approach, fluorescence recovery after photobleaching (FRAP), uses a laser beam focused at the specimen plane of a fluorescence microscope to introduce a gradient of fluorescence in the sample by irreversible photolysis (bleaching) of fluorophores in the path of the laser beam. The subsequent exchange of bleached tracers with unbleached tracers diffusing in from outside the bleached region leads to relaxation of the gradient and recovery of fluorescence in the bleached region. The time constant (τ) that characterizes the recovery is inversely proportional to the effective diffusion coefficient (D) of the fluorescent tracer (Elson and Qian, 1989). In an alternative approach that avoids the possible artifacts associated with high-power laser illumination, low light level, time-lapse video recording was used to monitor the spread of fluorescence down the axons of neurons that had been loaded with tracers by whole cell patch at the soma. D was calculated from the one-dimensional equation for diffusion (Popov and Poo, 1992).

Although dextrans and Ficolls do not appear to bind to intracellular components, the results of these studies show that they diffuse significantly more slowly in intact cells than in dilute solution (Luby-Phelps *et al.,* 1987; Hou *et al.,* 1990; Popov and Poo, 1992; Arrio-Dupont *et al.,* 1996; Seksek *et al.,* 1997). Comparison of the diffusion of the same tracers in concentrated solutions of unlabeled dextran or Ficoll suggests that the crowdedness of cytoplasm in fibroblasts resembles 12–13% dextran or Ficoll (Hou *et al.,* 1990). A similar result was obtained from FRAP studies of the small fluorescent dye, BCECF (Kao *et al.,* 1993). Although a significant fraction of

BCECF was bound inside cells, this was corrected for in the analysis of the FRAP data. A crowdedness similar to 12–13% dextran or Ficoll would mean that roughly half of total cytoplasmic protein, or about 110 mg/ml, is in solution and the remainder is structural. An estimate of 135 mg/ml soluble protein was obtained by FRAP of fluorescent dextrans in cultured myotubes (Arrio-Dupont *et al.*, 1996). The hindrance on translational diffusion due to macromolecular crowding may be more severe in the vicinity of membrane surfaces. By combining TIRF with FRAP, the diffusion coefficient of BCECF within 50 nm of the ventral surface of epithelial cells in culture was found to be as much as tenfold lower than in water (Swaminathan *et al.*, 1996).

Superimposed on this apparent crowding effect, several groups have reported a significant size-dependence of the two-dimensional diffusion coefficient of tracers in cytoplasm relative to aqueous buffer (D/Do) (Fig. 3). In Swiss 3T3 fibroblasts, D/Do declined from 0.3 to 0.04 for particles ranging from 3 to 25 nm in hydrodynamic radius (Luby-Phelps *et al.*, 1987).

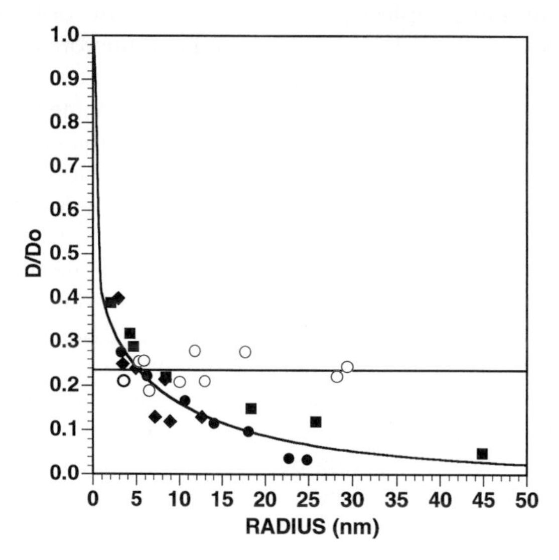

FIG. 3 Relative diffusion coefficients of inert fluorescent tracers in fibroblasts (filled circles, open circles), neurons (squares), and myotubes (diamonds). Data taken from Luby-Phelps *et al.* (1987), Popov and Poo (1992), Arrio-Dupont *et al.* (1996), and Seksek *et al.* (1997). Curve is nonlinear least-squares best fit of data from Luby-Phelps *et al.* (1987) (closed circles) to a model in which a network of filaments and a crowded background solution contribute independently to the hindrance on diffusion. The model predicts a network volume fraction of 0.11 and a protein concentration in the background of 110 mg/ml (Hou *et al.*, 1990). Line is linear least-squares fit to data from Seksek *et al.*, (1997) (open circles) with slope constrained to zero. The reason for the differing size-dependence of D/Do in the two studies of fibroblasts remains unexplained at present.

In cultured *Xenopus* neurons D/Do declined from 0.39 to 0.05 for particles ranging from 2 to 44.8 nm in hydrodynamic radius (Popov and Poo, 1992). In skeletal muscle myotubes, D/Do declined from 0.4 to 0.13 for particles ranging from 2.9 to 12.6 nm in radius (Arrio-Dupont *et al.,* 1996). Such size-dependence is consistent with sieving by a network of obstructions. The most likely candidate for such a network is the cytoskeleton. In support of this idea, in neurons the size-dependent constraint on diffusion was partially relieved by treating the cells with cytochalasin D, a drug that depolymerizes actin filaments (Popov and Poo, 1992).

The existence of sieving has been challenged in a recent paper which repeated earlier studies of the diffusion of dextrans and Ficolls in intact cells using a FRAP apparatus with greatly improved time resolution (Seksek *et al.,* 1997). In the more recent study, no size dependence of D/Do was found and these investigators concluded that the diffusion of macromolecules in cytoplasm is slowed only by crowding. It is not immediately obvious why an increase in time resolution should lead to such different results, and several differences in methodology need to be addressed before deciding which view of the cytoplasm is more accurate. For example, the interpretation of high time-resolution FRAP data potentially is complicated by photophysical effects, such as reversible photobleaching, that occur on the same timescale as the measurements. Anomalous temperature dependence of the rate of fluorescence recovery was observed in this study and interpreted as reversible photobleaching. The authors argued from circumstantial evidence that this was avoided by carrying out all measurements at room temperature rather than the physiological temperature of 37°C. Another potentially significant difference is the method of analyzing the FRAP recovery curves. Correct determination of τ dependent critically on the accurate estimation of the fluorescence intensity at infinite time, $F(\infty)$. Most algorithms for obtaining τ involve fitting the curve either to a truncated approximation of the nonconverging series that describes diffusive recovery (Axelrod *et al.,* 1976) or to an empirical relation that has been demonstrated to produce equivalent results (Yguerabide *et al.,* 1982), and then extrapolating to $F(\infty)$. In the more recent study, $F(\infty)$ was taken to be the apparent plateau value of the actual recovery curve at an arbitrarily long time after the bleach. If this plateau value was determined prematurely, τ will be underestimated and D overestimated. In addition, the fluorescence recovery will appear incomplete and the estimated percentage recovery will decline with increasing tracer size. In the previous studies, the fluorescence recovery of all but the largest tracers was nearly 100%. In the more recent paper, recoveries for tracers of all sizes were significantly less than 100% and declined with increasing tracer size. This was attributed to anomalous subdiffusion of tracer in subcompartments of cytoplasm whose viscosity differs from the bulk, perhaps due to some sort of phase separation.

2. Size-dependent Partitioning of Tracers

Fluorescence ratio imaging of large and small tracers comicroinjected into well spread fibroblasts reveals the existence of subcompartments of the cytoplasm into which the large tracers do not partition (Fig. 4) (Luby-Phelps and Taylor, 1988). Whole-mount electron microscopy of these compartments shows them to be very flat and to contain numerous filament bundles embedded in a meshwork of thin filaments (Provance *et al.,* 1993). Immunofluorescence microscopy suggests that actin and actin-binding proteins are components of the bundles and meshwork, while microtubules and intermediate filaments are absent from the excluding compartment (Provance *et al.,* 1993). Mitochondria, endoplasmic reticulum, and small vesicles also are absent from the excluding compartments and are restricted to the perinuclear region or to channels defined by microtubules radiating from the perinuclear region to the peripheral membrane of the cell. The boundary between excluding and nonexcluding compartments was discernible as an abrupt change in ultrastructure and cell height, but no membrane or other physical barrier was evident. Overall, the excluding compartments appeared ultrastructurally similar to previously described "peripheral cytoplasmic domains" (Bridgman *et al.,* 1986) and to the leading lamellipodium of migrating cells (Abercrombie *et al.,* 1971; Small *et al.,* 1978).

The presence of a finely divided filament network in the excluding compartments suggested a molecular sieving mechanism for exclusion. However, the reduced cell height in these regions could not be ruled out as a mechanism. Using quantitative fluorescence ratio imaging to measure the partition coefficient for tracers into these regions as a function of tracer size, it was determined that sieving is the more likely mechanism (Janson *et al.,* 1996). While the estimated mean pore size of the sieve is 50 nm in radius, the partition coefficient decays as the cube of tracer radius, leading to significant exclusion of particles well below the nominal percolation cutoff (Janson *et al.,* 1996).

3. Diffusion of Vesicle-sized Particles

The restriction of the machinery for vesicle transport to nonexcluding compartments, and the size-dependent exclusion of large particles from excluding compartments, suggest that vesicle trafficking is restricted to nonexcluding compartments. In working out a complete model of vesicle trafficking, it is important to know whether and how far a vesicle can diffuse in this compartment. Inspection of the data for diffusion of tracers up to 30 nm in radius in nonexcluding regions suggests that diffusion of the much larger vesicles might be highly restricted. In fact, it has long been noted that Brownian motion of endogenous intracellular particles is highly

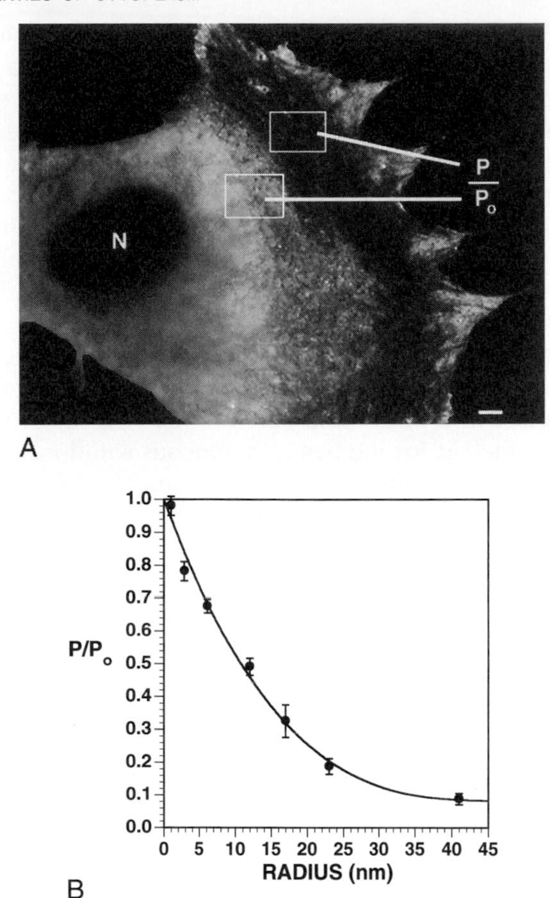

FIG. 4 Size-dependent partitioning of inert tracers in cultured fibroblasts. (A) Fluorescence ratio map of the distribution of 17 nm radius particles relative to 1 nm radius particles following injection into the cytoplasm of a single cell. Variations in intensity are directly proportional to the relative concentration of large to small particles in different regions of the cell. Regions of lower intensity are domains that exclude large particles. P is the fluorescence ratio value in excluding regions, and Po is the ratio value in nonexcluding regions. The nucleus (N) excludes large particles due to the finite size of pores in the nuclear envelope. Bar is 5 μm. (B) Partition coefficient (P/Po) plotted as a function of particle radius. Curve is nonlinear least-squares fit to a model in which exclusion is due to molecular sieving by a network of obstructions. Although the percolation size cutoff for partitioning into the excluding domains is 50 nm, the size-dependence is steep, resulting in significant exclusion of particles at sizes well below the percolation cutoff. (B) is reprinted with permission (Janson *et al.,* 1996).

constrained in living cells and that increased Brownian motion is an indicator of cellular distress (for example, Green, 1968; Wheatley and Malone, 1993). We have recently addressed this question directly by two-dimensional, single-particle tracking of 80 nm radius green fluorescent microspheres in living fibroblasts (Jones *et al.,* 1998). In order to minimize binding to intracellular components, the microspheres were made hydrophilic and electroneutral by derivatizing the surface with hydroxyls and preincubating them with rhodamine dextran to block exposed polystyrene surfaces. No adsorption of protein to the beads prepared in this way could be detected after incubation with concentrated cell lysates. By averaging the initial slopes of plots of mean squared displacement vs time, we obtained a mean D_{cyto} of 2.6×10^{-11} cm^2/sec for beads in SW3T3 fibroblasts and 4.0×10^{-11} cm^2/sec in CV1 fibroblasts. This is 500- to 1000-fold lower than the diffusion coefficient for the beads in aqueous solution. At this rate, in 1 min the bead would exhibit a mean displacement of <0.5 μm. D_{cyto} for the beads is very similar to values reported in the literature for the diffusion coefficient of endogenous vesicles in a variety of cell types. Using quasielectric light scattering, Felder and Kam (1994) obtained a mean value of 2.5×10^{-10} cm^2/sec for endogenous vesicles of unknown size in neutrophils. A value of 3×10^{-11} cm^2/sec has been reported for fluorescently stained chromaffin granules near the plasma membrane (Steyer *et al.,* 1997). Values ranging from 3.9×10^{-12} to 7.4×10^{-11} cm^2/sec were measured for secretory vesicles tagged with green fluorescent protein (Burke *et al.,* 1997). The very restricted diffusive mobility of vesicles necessitates their active transport on microtubules for efficient delivery to distant sites in the cell. On the other hand, it also ensures that beads detaching from a microtubule remain nearby, increasing the probability of reattachment.

In addition to slow diffusion due to percolation through a network of obstructions, it is possible that beads and vesicles might be confined ("caged") for some portion of their trajectory in the pores of the network. Tests for caging involve comparison of the actual area covered by an individual bead or vesicle over a long observation time with the mean squared displacement predicted from the diffusion coefficient measured over short observation times (Saxton 1995; Simson *et al.,* 1995). Applying these criteria to the trajectories of beads in CV1 and 3T3 cells indicates that some have a >90% probability of arising from caged diffusion (Jones *et al.,* 1998). The mean cage size estimated from this type of analysis is 1.5 μm in radius. The very slow diffusion coefficients measured over short times suggest that the cages are not simply fluid-filled pores, but contain additional constraints on diffusion.

To examine the role of the cytoskeleton in restricting the diffusion of vesicle-sized beads, we selectively disassembled each of the three main filament systems as independently as possible and examined the effects on bead diffusion (Jones *et al.,* 1998). Treatment of cells with nocodazole for times short enough to depolymerize most microtubules without noticeably affecting intermediate filaments did not remove the barrier to bead diffusion. This result is not surprising in view of the relatively low number concentration and wide interfilament spacing of the microtubules in these cell types. Decreasing the amount of polymerized actin by transient overexpression of the actin monomer-sequestering protein, thymosin, or by treatment with the actin-depolymerizing drug, latrunculin B, also had no effect on bead diffusion. Similarly, another actin-depolymerizing drug, cytochalasin D, did not alter the diffusivity of endogenous granules in neutrophils (Felder and Kam, 1994). However, treatment of cells with acrylamide, which has been shown to depolymerize intermediate filaments (Eckert, 1986) caused a nearly 10-fold increase in the mean diffusion coefficient of the beads. In qualitative agreement with this, bead diffusion decreased about 4-fold when vimentin intermediate filament protein was expressed in a vimentin-null background (Holwell *et al.,* 1997). However, even in the absence of intermediate filaments, the beads diffuse very slowly, suggesting that there are additional constraints on diffusion that do not involve the three major components of the cytoskeleton. In addition, preliminary analysis of the data suggests that caging of the beads is not abolished by any of these treatments.

4. Diffusion of Proteins and Other Macromolecules within Cytoplasm

As mentioned above, the intracellular diffusion of proteins that have hydrophobic domains and ionizable surface groups might be even slower than for inert tracer particles, due to binding interactions with intracellular components. In fact, fluorescent analogs of almost every protein ever studied by FRAP, including many that have no known binding sites inside cells, diffuse very slowly in intact cells (Table I). In general, the cytoplasmic diffusion coefficient is uncorrelated with the radius of the protein and a significant immobile component is evident in the FRAP recovery curves (Wojcieszyn *et al.,* 1981; Kreis *et al.,* 1982; Wang *et al.,* 1982; Jacobson and Wojcieszyn, 1984; Salmon *et al.,* 1984; Luby-Phelps *et al.,* 1986). Notable exceptions appear to be non-neuronal enolase (Pagliaro *et al.,* 1989; Arrio-Dupont *et al.,* 1997) and certain genetically engineered mutants of green fluorescent protein (Yokoe and Meyer, 1996; Swaminathan *et al.,* 1997). Both of these proteins recover at about the rate predicted for an inert spherical tracer of the same size—two to four times slower than in dilute aqueous solution. The crystal structure of GFP was recently reported to be a dimer "beta can" structure with very few ionizable or hydrophobic groups on its surface, which might explain its apparent lack of binding

TABLE I

Diffusion Coefficients of Some Proteins in Living Cells[a]

Protein	Aqueous D[b] (μm^2/sec)	Radius (nm)	Cytoplasmic D(μm^2/sec)	D/Do	% mobile	Reference
Insulin	nd	1.6	0.9	(0.007)	87	Jacobson and Wojcieszyn (1984)
CaM	102	2.1	<4	0.039	81	Luby-Phelps *et al.*, (1985)
Lactalbumin	102	2.1	6.9	0.07	42	Luby-Phelps *et al.*, (1985)
Green fluorescent protein	nd		43	(0.49)	nd	Yokoe and Meyer (1996)
	87	2.5	27	0.31	82	Swaminathan *et al.*, (1997)
Ovalbumin	69	3.1	5.9	0.086	78	Luby-Phelps *et al.*, (1985)
BSA	nd	3.5	1.7	0.027	97	Jacobson and Wojcieszyn (1984)
	67	3.2	6.8	0.1	77	Luby-Phelps *et al.*, (1985)
Creatine kinase	65	3.3	<4.5	0.07	50–80	Arrio-Dupont *et al.*, (1997)
Enolase (non-neuronal)	60	3.6	7–11	0.11–0.18	100	Pagliaro *et al.*, (1989)
	56	3.8	13.5	0.24	100	Arrio-Dupont *et al.*, (1997)
Aldolase	47	4.6	6–11	0.12–0.23	>77	Pagliaro and Taylor (1988)
IgG	46	4.7	6.7	0.146	54	Luby-Phelps *et al.*, (1985)
Apoferritin	nd	6.1	1.6	(0.045)	96	Jacobson and Wojcieszyn (1984)

[a] Numbers in parentheses are calculated using the Stokes-Einstein relation to obtain the aqueous diffusion coefficient expected for a globular protein of the given radius.

[b] nd, not determined.

(Ormo *et al.,* 1996; Yang *et al.,* 1996). A more detailed examination of the binding of specific proteins is contained in the chapter by D. E. Brooks in this volume.

An interesting example of the binding of proteins in cytoplasm is calmodulin, which is commonly assumed to be freely diffusible at resting intracellular Ca^{2+} levels because, in simple, dilute solutions, its binding to most target proteins requires Ca^{2+}. However, FRAP of fluorescent analogs of CaM in fibroblasts and in unstimulated smooth muscle cells shows that at most 5 to 10% is freely diffusing even under conditions where Ca^{2+} is low and the fluorescent analog is in 50% excess of the endogenous concentration of CaM (Luby-Phelps *et al.,* 1985, 1995). Both slowly exchanging and transient binding interactions can be inferred from the complex photobleaching recovery curves for obtained fluorescent CaM in living cells. The low diffusibility of CaM at resting Ca^{2+} levels is also indicated by a study of the rapid kinetics of myosin light chain phosphorylation and force development in permeabilized portal vein smooth muscle cells upon activation by photolysis of caged ATP or Ca^{2+}. Based on the strong dependence of these parameters on exogenous calmodulin, it was calculated that only about 0.25 μM of endogenous calmodulin is freely diffusible under resting conditions (Zimmermann *et al.,* 1995).

It is not clear how much of the binding observed for proteins in intact cells reflects specific interactions of these proteins with targets defined *in vivo.* Several of the proteins tested are serum proteins that do not have known intracellular targets. However, given the modular organization of proteins, these proteins may contain protein–protein interaction domains that allow them to bind to heterologous targets inside the cell. Preliminary studies of the diffusion of site-directed mutants of CaM in living cells show that binding is not strongly dependent on the domains of CaM that have been implicated in binding to specific targets *in vitro.* This suggests that binding is mediated by alternative domains or is nonspecific. Regardless of whether the unexpected binding interactions of proteins in cytoplasm are specific, they are indicative of real effects that must be taken into account in understanding the behavior of proteins in the cytoplasm of intact living cells.

In favor of the specificity of such interactions, many proteins previously assumed to be homogeneously distributed in the aqueous phase of cytoplasm are now known to be resident on membranes or on the cytoskeleton, at least under certain conditions. Several examples are known of so-called scaffold or adaptor proteins that serve as docking sites for the assembly of multikinase signaling complexes at the plasma membrane near receptors (Pawson and Scott, 1997). In addition, kinases that exist as multiple isozymes with similar substrate specificity *in vitro* may utilize differential localization to discriminate between substrates *in vivo* (for example, Jaken, 1996; Mochly-Rosen and Gordon, 1998).

Another example of the localization of a pathway in the solid phase of the cytoplasm is glycolysis. Several key enzymes in this pathway have been shown to interact with actin filaments and microtubules *in vitro* (Masters, 1992; Knull and Minton, 1996; Ovádi and Srere, 1996; Reitz *et al.,* 1997). Binding domains on actin for aldolase, phosphofructokinase, and GAPDH have been identified antigenically (Mejean *et al.,* 1989) and an actin-binding domain on aldolase has been mapped by deletion and site-directed mutagenesis (O'Reilly and Clarke, 1993; Wang *et al.,* 1996). In some cases, enzyme activity is modified by binding to cytoskeletal filaments and the binding is modified by substrate, suggesting that glycolysis might be regulated by association with the cytoskeleton (Knull and Minton, 1996; Ovádi and Srere, 1996). There is evidence for similar interactions *in vivo.* Aldolase and GAPDH have been localized to actin-containing stress fibers in cultured fibroblasts (Pagliaro and Taylor, 1988; Minaschek *et al.,* 1992), and cell cycle changes in glycolytic activity are correlated with changes in actin filament content in some cell types (Bereiter-Hahn *et al.,* 1995). GAPDH, aldolase, creatine kinase, and glycerol-3-phosphate dehydrogenase (GPDH) are colocalized at the M-lines in skeletal muscle (Arnold and Pette, 1968; Arrio-Dupont *et al.,* 1997; Wojtas *et al.,* 1997). Mutation or deletion of the flight muscle-specific isoform of GPDH in *Drosophila* leads to flightlessness and is accompanied by mislocalization of GAPDH and aldolase (Wojtas *et al.,* 1997). Taken together, these data suggest the existence of a glycolytic complex localized at specific sites in the cytoplasm and perhaps regulated by interaction with the cytoskeleton.

Many RNA messages and polysomes also are localized to specific regions or structures of cells (for reviews, see Fulton and L'Ecuyer, 1993; Suprenant, 1993; Bassell and Singer, 1997). This has been studied most extensively in developing cardiac muscle, in migrating fibroblasts, in developing embryos, and in neurons. The mechanisms responsible for mRNA localization are just beginning to be understood. For several messages, the 3′ UTR has been shown to be both necessary and sufficient to confer specific localization of mRNA (Kislauskis and Singer, 1991; Wilson *et al.,* 1995; Ainger *et al.,* 1997). In many systems, localization has been shown to depend on an intact microtubule cytoskeleton (Suprenant, 1993) and, in neurons, microtubule-dependent transport of RNPs containing specific mRNAs has been directly observed (Ainger *et al.,* 1993; Knowles *et al.,* 1996).

Messenger RNAs are often colocalized with the proteins they encode, suggesting that translation of these messages occurs at the site of localization (Singer *et al.,* 1989; Fulton and L'Ecuyer, 1993; Cohen, 1996; Cohen and Kuda, 1996). In fact, mislocalization of messages has been shown to result in altered phenotypes (Kislauskis *et al.,* 1994; L'Ecuyer *et al.,* 1995). Consistent with this idea, a component of the translational machinery, elongation factor 1α, binds actin and colocalizes on the cytoskeleton with RNPs (Condeelis, 1995; Knowles *et al.,* 1996). Specific mRNAs and polysomes are found to copurify with microtubules from sea urchin egg extracts (Hamill *et al.,* 1994).

Such organization may be the rule rather than the exception. Early experiments involving the centrifugation of intact cells, in which stratification order reflects specific gravity, revealed very little cellular protein in the aqueous layer where soluble proteins should be found (Zalokar, 1960; Kempner and Miller, 1968a,b). More recently, the release of cellular contents following permeabilization of the plasma membrane of some cells has been reported to be incomplete or slower than expected for freely diffusing molecules. For example, mouse L929 cells released only 10–15% of cellular protein during 30 min of permeabilization with dextran sulfate despite the fact that molecules as large as 400,000 kDa were able to diffuse into the permeabilized cells from outside (Clegg and Jackson, 1988). When exogenous glucose and ATP were supplied, the permeabilized cells were found to carry out efficient glycolysis at linear rates. However, exogenously added glycolytic intermediates were inefficient in entering the glycolytic pathway (Clegg and Jackson, 1990). Intact cells dehydrated to 15% of their normal water content still carried out glycolysis at significant rates, suggesting that this pathway does not rely on a dilute aqueous phase (Clegg *et al.,* 1990). In experiments with bovine eye lenses or red blood cells permeabilized with nonionic detergents, as little as 1 to 7% of total protein was found to leak out over an 8 hr period (Kellermayer *et al.,* 1994; Cameron *et al.,* 1996). However, both these cell types are unusual in that they contain a single major protein at very high concentration. It is not clear whether proteins in most cells are similarly immobile. These findings have led some investigators to propose that nearly all cellular protein is bound to membrane and cytoskeletal surfaces within the cell and that the aqueous phase is dilute rather than crowded.

A different approach to this problem has been to implant a reference phase into intact cells as a means of assessing the diffusivity of proteins *in vivo* (Paine, 1984). In these experiments, a drop of molten gelatin was injected into *Xenopus* oocytes and allowed to gel. After waiting 20 h to allow equilibration of cellular proteins with the aqueous pores of the implanted gel, the cells were rapidly frozen at liquid nitrogen temperature, and the implanted gel and the cytoplasm were sampled by cryomicrodissection. The proteins in each sample were analyzed by two-dimensional polyacrylamide gel electrophoresis and silver staining. Based on the optical density of specific spots, it was estimated that 73% of the polypeptides analyzed exhibited some degree of binding that led to asymmetry of their distribution between the cytoplasm and the reference phase. For individual spots, the nondiffusive fraction ranged from 5 to 80%. At first glance, this does not seem as dramatic as the results of the permeabilization experiments. However, the long time allowed for equilibration in the reference phase experiment would favor the detection of only the highest affinity interactions, while the slow release of protein from permeabilized cells would include more transient interactions that retard diffusion. Such transient interactions have been detected for specific proteins by FRAP as described above.

IV. Concluding Remarks

There appears to be ample evidence that the interior of a living cell is not well described by the dilute solution paradigm. Although the average mobility of cellular water is no more than twofold reduced compared to bulk water, existing data do not rule out a small but significant population of immobilized water molecules immediately adjacent to membrane and cytoskeletal surfaces. The evidence suggests that the aqueous phase is crowded with macromolecules, yet at least 50% of cytoplasmic proteins appear to be resident in the solid phase. Brownian motion of particles in the aqueous phase is hindered by crowding and perhaps obstructed by the cytoskeletal network. Transient binding interactions further slow the long-range translational diffusion of proteins. For vesicle-sized particles, this hindrance renders long-range diffusion negligible and necessitates active transport. Several phenomena have been observed that might indicate the existence of microphases in the cytoplasmic volume, including the caged diffusion of microinjected beads and the size-dependent partitioning of inert tracers between excluding and nonexcluding domains of the cytoplasm. The potential impact of actual intracellular conditions on the kinetics, mechanisms, and regulation of metabolism make it imperative to reexamine continuum descriptions of cellular biochemistry that have been extrapolated from reductionist experiments carried out in dilute solution.

Acknowledgments

This chapter is dedicated to the memory of Keith R. Porter (1912–1997), who taught me to question accepted paradigms. I also gratefully acknowledge Dick McIntosh for first pointing out the finite volume problem to me several years ago, Ivan Cameron for helpful discussion of his data, and The National Science Foundation for their generous support of my research (MCB-9604594).

References

Abercrombie, M., Heaysman, J. E. M., and Pegrum, S. M. (1971). The locomotion of fibroblasts in culture. IV. Electron microscopy of the leading lamella. *Exp. Cell Res.* **67,** 359–367.

Agutter, P. S., Malone, P. C., and Wheatley, D. N. (1995). Intracellular transport mechanisms: a critique of diffusion theory. *J. Theor. Biol.* **176,** 261–272.

Ainger, K., Avossa, D., Morgan, F., Hill, S. J., Barry, C., Barbarese, E., and Carson, J. H. (1993). Transport and localization of exogenous myelin basic protein mRNA microinjected into oligodendrocytes. *J. Cell Biol.* **123,** 431–441.

Ainger, K., Avossa, D., Diana, A. S., Barry, C., Barbarese, E., and Carson, J. H. (1997). Transport and localization elements in myelin basic protein mRNA. *J. Cell Biol.* **138,** 1077–1087.

Alberts, B., Bray, D., Lewis, J., Raff, M., Roberts, K., and Watson, J. D. (1994). "Molecular Biology of the Cell," 3rd ed. Garland Publishing, New York.

Arnold, H., and Pette, D. (1968). Binding of glycolytic enzymes to structure proteins of the muscle. *Eur. J. Biochem.* **6,** 163–171.

Arrio-Dupont, M., Cribier, S., Foucault, G., Devaux, P. F., and Dalbis, A. (1996). Diffusion of fluorescently labeled macromolecules in cultured muscle cells. *Biophys. J.* **70,** 2327–2332.

Arrio-Dupont, M., Foucault, G., Vacher, M., Douhou, A., and Cribier, S. (1997). Mobility of creatine phosphokinase and beta-enolase in cultured muscle cells. *Biophys. J.* **73,** 2667–2673.

Axelrod, D., Koppel, D. E., Schlessinger, J., Elson, E., and Webb, W. W. (1976). Mobility measurement by analysis of fluorescence photobleaching recovery kinetics. *Biophys. J.* **16,** 1055–1069.

Bassell, G., and Singer, R. H. (1997). mRNA and cytoskeletal filaments. *Curr. Opin. Cell Biol.* **9,** 109–115.

Bereiter-Hahn, J., Stubig, C., and Heymann, V. (1995). Cell cycle-related changes in F-actin distribution are correlated with glycolytic activity. *Exp. Cell Res.* **218,** 551–560.

Bicknese, S., Periasamy, N., Shohet, S. B., and Verkman, A. S. (1993). Cytoplasmic viscosity near the cell plasma membrane: Measurement by evanescent field frequency-domain microfluorimetry. *Biophys. J.* **65,** 1272–1282.

Bray, D., and Thomas, C. (1975). The actin content of fibroblasts. *Biochem. J.* **147,** 221–228.

Bridgman, P. C., Kachar, B., and Reese, T. S. (1986). The structure of cytoplasm in directly frozen cultured cells. II. Cytoplasmic domains associated with organelle movements. *J. Cell Biol.* **102,** 1510–1521.

Burke, N. V., Han, W., Li, D., Takimoto, K., Watkins, S. C., and Levitan, E. S. (1997). Neuronal peptide release is limited by secretory granule mobility. *Neuron.* **19,** 1095–1102.

Cameron, I. L., Hardman, W. E., Fullerton, G. D., Miseta, A., Koszegi, T., Ludany, A., and Kellermayer, M. (1996). Maintenance of ions, proteins and water in lens fiber cells before and after treatment with non-ionic detergents. *Cell Biol. Int.* **20,** 127–137.

Cameron, I. L., Kanal, K. M., Keener, C. R., and Fullerton, G. D. (1997). A mechanistic view of the non-ideal osmotic and motional behavior of intracellular water. *Cell Biol. Int.* **21,** 99–113.

Carlier, M. F., and Pantaloni, D. (1997). Control of actin dynamics in cell motility. *J. Mol. Biol.* **269,** 459–467.

Chang, T., Kim, H., and Yu, H. (1987). Diffusion through coarse meshes. *Macromolecules* **20,** 2629–2631.

Chou, Y. H., Skalli, O., and Goldman, R. D. (1997). Intermediate filaments and cytoplasmic networking: New connections and more functions. *Curr. Opin. Cell Biol.* **9,** 49–53.

Clegg, J. S. (1984). Properties and metabolism of the aqueous cytoplasm and its boundaries. *Am. J. Physiol.* **246,** R133–R151.

Clegg, J. S. (1986). L-929 cells under hyperosmotic conditions: Volume changes. *Cell. Physiol.* **129,** 367–374.

Clegg, J. S., and Jackson, S. A. (1988). Glycolysis in permeabilized L-929 cells. *Biochem. J.* **255,** 335–344.

Clegg, J. S., and Jackson, S. A. (1990). Glucose metabolism and the channeling of glycolytic intermediates in permeabilized L-929 cells. *Arch. Biochem. Biophys.* **278,** 452–460.

Clegg, J. S., Jackson, S. A., and Fendl, K. (1990). Effects of reduced cell volume and water content on glycolysis in L-929 cells. *J. Cell. Physiol.* **142,** 386–391.

Cohen, N. S. (1996). Intracellular localization of the mRNAs of argininosuccinate synthetase and argininosuccinate lyase around liver mitochondria, visualized by high-resolution in situ reverse transcription-polymerase chain reaction. *J. Cell Biochem.* **61,** 81–96.

Cohen, N. S., and Kuda, A. (1996). Argininosuccinate synthetase and argininosuccinate lyase are localized around mitochondria: an immunocytochemical study. *J. Cell Biochem.* **60,** 334–340.

Condeelis, J. (1995). Elongation factor 1 alpha, translation and the cytoskeleton. *Trends Biochem. Sci.* **20,** 169–170.

Cooke, R., and Kuntz, I. D. (1974). The properties of water in biological systems. *Annu. Rev. Biophys.* **3,** 95–126.

Cunningham, C. C., Leclerc, N., Flanagan, L. A., Lu, M., Janmey, P. A., and Kosik, K. S. (1997). Microtubule-associated protein 2c reorganizes both microtubules and microfilaments into distinct cytological structures in an actin-binding protein-280-deficient melanoma cell line. *J. Cell Biol.* **136,** 845–857.

Eckert, B. S. (1986). Alteration of the distribution of intermediate filaments in PtK1 cells by acrylamide. II: Effect on the organization of cytoplasmic organelles. *Cell Motil. Cytoskel.* **6,** 15–24.

Elson, E. L., and Qian, H. (1989). Interpretation of fluorescence correlation spectroscopy and photobleaching recovery in terms of molecular interactions. *Methods Cell Biol.* **30,** 307–332.

Eriksson, J. E., Brautigan, D. L., Vallee, R., Olmsted, J., Fujiki, H., and Goldman, R. D. (1992). Cytoskeletal integrity in interphase cells requires protein phosphatase activity. *Proc. Natl. Acad. Sci. U.S.A.* **89,** 11093–11097.

Felder, S., and Kam, Z. (1994). Human neutrophil motility: Time-dependent three-dimensional shape and granule diffusion. *Cell Motil. Cytoskel.* **28,** 285–302.

Foster, K. R., Resing, H. A., and Garroway, A. N. (1976). Bounds on "bound water": Transverse nuclear magnetic resonance relaxation in barnacle muscle. *Science* **194,** 324–326.

Fulton, A. B. (1982). How crowded is the cytoplasm? *Cell* (*Cambridge, Mass.*) **30,** 345–347.

Fulton, A. B., and L'Ecuyer, T. (1993). Cotranslational assembly of some cytoskeletal proteins: Implications and prospects. *J. Cell Sci.* **105,** 867–871.

Fushimi, K., and Verkman, A. S. (1991). Low viscosity in the aqueous domain of cell cytoplasm measured by picosecond polarization microfluorimetry. *J. Cell Biol.* **112,** 719–725.

Gershon, N. D., Porter, K. R., and Trus, B. L. (1985). The cytoplasmic matrix: Its volume and surface area and the diffusion of molecules through it. *Proc. Natl. Acad. Sci. U.S.A.* **82,** 5030–5034.

Goodsell, D. S. (1991). Inside a living cell. *Trends Biochem. Sci.* **16,** 203–206.

Goodsell, D. S. (1993). "The Machinery of Life" Springer-Verlag, New York.

Green, L. (1968). Mechanism of movements of granules in melanocytes of fundulus heteroclitus. *Proc. Natl. Acad. Sci. U.S.A.* **59,** 1179–1186.

Halling, P. J. (1989). Do the laws of chemistry apply to living cells? *Trends Biochem. Sci.* **14,** 317–318.

Hamill, D., Davis, J., Drawbridge, J., and Suprenant, K. A. (1994). Polyribosome targeting to microtubules: Enrichment of specific mRNAs in a reconstituted microtubule preparation from sea urchin embryos. *J. Cell Biol.* **127,** 973–984.

Hashimoto, Y., and Shinozaki, N. (1988). Measurement of cytoplasmic viscosity by fluorescence polarization in phytohemagglutinin-stimulated and unstimulated human peripheral lymphocytes. *J. Histochem. Cytochem.* **36,** 609–613.

Herzfeld, J. (1996). Entropically driven order in crowded solutions: From liquid crystals to cell biology. *Acc. Chem. Res.* **29,** 31–37.

Heuser, J. E., and Kirschner, M. W. (1980). Filament organization revealed in platinum replicas of freeze-dried cytoskeletons. *J. Cell Biol.* **86,** 212–234.

Hiller, G., and Weber, K. (1978). Radioimmunoassay for tubulin: A quantitative comparison of the tubulin content of different established tissue culture cells and tissues. *Cell (Cambridge, Mass.)* **14,** 795–804.

Holwell, T. A., Schweitzer, S. C., and Evans, R. M. (1997). Tetracycline regulated expression of vimentin in fibroblasts derived from vimentin null mice. *J. Cell Sci.* **110,** 1947–1956.

Hoppeler, H., Hudlicka, O., and Uhlmann, E. (1987). Relationship between mitochondria and oxygen consumption in isolated cat muscles. *J. Physiol. (London)* **385,** 661–675.

Hou, L., Lanni, F., and Luby-Phelps, K. (1990). Tracer diffusion in F-actin and Ficoll mixtures. Toward a model for cytoplasm. *Biophys. J.* **58,** 31–43.

Hsu, E. W., Aiken, N. R., and Blackband, S. J. (1996). Nuclear magnetic resonance microscopy of single neurons under hypotonic perturbation. *Am. J. Physiol.* **271,** C1895–C1900.

Ito, T., Suzuki, A., and Stossel, T. P. (1992). Regulation of water flow by actin-binding protein-induced actin gelation. *Biophys. J.* **61,** 1301–1305.

Jacobson, K., and Wojcieszyn, J. (1984). The translational mobility of substances within the cytoplasmic matrix. *Proc. Natl. Acad. Sci. U.S.A.* **81,** 6747–6751.

Jaken, S. (1996). Protein kinase c isozymes and substrates. *Curr. Opin. Cell Biol.* **8,** 168–173.

Janson, L. W., Ragsdale, G. K., and Luby-Phelps, K. (1996). Mechanism and size cutoff for steric exclusion from actin-rich cytoplasmic domains. *Biophys. J.* **71,** 1228–1234.

Johansen, K. M. (1996). Dynamic remodeling of nuclear architecture during the cell cycle. *J. Cell. Biochem.* **60,** 289–296.

Jones, J. D., Ragsdale, G. K., Rozelle, A., Yin, H. L., and Luby-Phelps, K. (1998). Diffusion of vesicle-sized particles in living cells is restricted by intermediate filaments. *Mol. Biol. Cell* **8,** 174a.

Kao, H. P., Abney, J. R., and Verkman, A. S. (1993). Determinants of the translational mobility of a small solute in cell cytoplasm. *J. Cell Biol.* **120,** 175–184.

Kellermayer, M., Ludany, A., Miseta, A., Koszegi, T., Berta, G., Bogner, P., Hazlewood, C. F., Cameron, I. L., and Wheatley, D. N. (1994). Release of potassium, lipids, and proteins from nonionic detergent treated chicken red blood cells. *J. Cell. Physiol.* **159,** 197–204.

Kempner, E. S., and Miller, J. H. (1968a). The molecular biology of *Euglena gracilis.* IV. Cellular stratification by centrifuging. *Exp. Cell Res.* **51,** 141–149.

Kempner, E. S., and Miller, J. H. (1968b). The molecular biology of *Euglena gracilis.* V. Enzyme localization. *Exp. Cell Res.* **51,** 150–156.

Kertesz, J. (1981). Percolation of holes between overlapping spheres: Monte Carlo calculation of the critical volume fraction. *J. Phys. (Paris)* **42,** L393–L395.

Kislauskis, E. H., and Singer, R. H. (1991). Characterization of sequences responsible for peripheral localization of β-actin mRNA. *J. Cell Biol.* **115,** 158a.

Kislauskis, E. H., Zhu, X., and Singer, R. H. (1994). Sequences responsible for intracellular localization of β-actin messenger RNA also affect cell phenotype. *J. Cell Biol.* **127,**(2) 441–451.

Knowles, R. B., Sabry, J. H., Martone, M. E., Deerinck, T. J., Ellisman, M. H., Bassell, G. J., and Kosik, K. S. (1996). Translocation of RNA granules in living neurons. *J. Neurosci.* **16,** 7812–7820.

Knull, H., and Minton, A. P. (1996). Structure within eukaryotic cytoplasm and its relationship to glycolytic metabolism. *Cell Biochem. Funct.* **14,** 237–248.

Kreis, T. E., Geiger, B., and Schlessinger, J. (1982). Mobility of microinjected rhodamine actin within living chicken gizzard cells determined by fluorescence photobleaching recovery. *Cell* (*Cambridge, Mass.*) **29,** 835–845.

Lai, Y. K., Lee, W. C., and Chen, K. D. (1993). Vimentin serves as a phosphate sink during the apparent activation of protein kinases by okadaic acid in mammalian cells. *J. Cell. Biochem.* **53,** 161–168.

Lanni, F., Waggoner, A. S., and Taylor, D. L. (1985). Structural organization of interphase 3T3 fibroblasts studied by total internal reflection fluorescence microscopy. *J. Cell Biol.* **100,** 1091–1102.

L'Ecuyer, T. J., Tompach, P. C., Morris, E., and Fulton, A. B. (1995). Transdifferentiation of chicken embryonic cells into muscle cells by the 3′ untranslated region of muscle tropomyosin. *Proc. Natl. Acad. Sci. U.S.A.* **92,** 7520–7524.

Lepock, J. R., Cheng, K.-H., Campbell, S. D., and Kruuv, J. (1983). Rotational diffusion of tempone in the cytoplasm of chinese hamster lung cells. *Biophys. J.* **44,** 405–412.

Lim, L., Manser, E., Leung, T., and Hall, C. (1996). Regulation of phosphorylation pathways by p21 GTPases. The p21 Ras-related Rho subfamily and its role in phosphorylation signalling pathways. *Eur. J. Biochem.* **242,** 171–185.

Lindmo, T., and Steen, H. B. (1977). Flow cytometric measurement of the polarization of fluorescence from intracellular fluorescein in mammalian cells. *Biophys. J.* **18,** 173–187.

Ling, G. N. (1988). A physical theory of the living state: Application to water and solute distribution. *Scanning Microsc.* **2,** 899–913.

Ling, G. N., Niu, Z., and Ochsenfeld, M. (1993). Predictions of polarized multilayer theory of solute distribution confirmed from a study of the equilibrium distribution in frog muscle of twenty-one nonelectrolytes including five cryoprotectants. *Physiol. Chem. Phys. Med. NMR* **25,** 177–208.

Livingston, D. J., LaMar, G. N., and Brown, W. D. (1983). Myoglobin diffusion in bovine heart muscle. *Science* **220,** 71–73.

Luby-Phelps, K., and Taylor, D. L. (1988). Subcellular compartmentalization by local differentiation of cytoplasmic structure. *Cell Motil. Cytoskel.* **10,** 28–37.

Luby-Phelps, K., Lanni, F., and Taylor, D. L. (1985). Behavior of a fluorescent analogue of calmodulin in living 3T3 cells. *J. Cell Biol.* **101,** 1245–1256.

Luby-Phelps, K., Taylor, D. L., and Lanni, F. (1986). Probing the structure of cytoplasm. *J. Cell Biol.* **102,** 2015–2022.

Luby-Phelps, K., Castle, P. E., Taylor, D. L., and Lanni, F. (1987). Hindered diffusion of inert tracer particles in the cytoplasm of mouse 3T3 cells. *Proc. Natl. Acad. Sci. U.S.A.* **84,** 4910–4913.

Luby-Phelps, K., Mujumdar, S., Mujumdar, R., Ernst, L., Galbraith, W., and Waggoner, A. (1993). A novel fluorescence ratiometric method confirms the low solvent viscosity of the cytoplasm. *Biophys. J.* **65,** 236–242.

Luby-Phelps, K., Hori, M., Phelps, J., and Won, D. (1995). Ca^{2+}-regulated dynamic compartmentalization of calmodulin in living smooth muscle cells. *J. Biol. Chem.* **270,** 21532–21538.

Maccioni, R. B., and Cambiazo, V. (1995). Role of microtubule-associated proteins in the control of microtubule assembly. *Physiol. Rev.* **75,** 835–864.

Masters, C. (1992). Microenvironmental factors and the binding of glycolytic enzymes to contractile filaments. *Int. J. Biochem.* **24,** 405–410.

Mastro, A. M., Babich, M. A., Taylor, W. D., and Keith, A. D. (1984). Diffusion of a small molecule in the cytoplasm of mammalian cells. *Proc. Natl. Acad. Sci. U.S.A.* **81,** 3414–3418.

Mejean, C., Pons, F., Benyamin, Y., and Roustan, C. (1989). Antigenic probes locate binding sites for the glycolytic enzymes glyceraldehyde-3-phosphate dehydrogenase, aldolase and phosphofructokinase on the actin monomer in microfilaments. *Biochem. J.* **264,** 671–677.

Minaschek, G., Groschel-Stewart, U., Blum, S., and Bereiter-Hahn, J. (1992). Microcompartmentation of glycolytic enzymes in cultured cells. *Eur. J. Cell Biol.* **58,** 481–428.

Minton, A. P. (1997). Influence of excluded volume upon macromolecular structure and associations in 'crowded' media. *Curr. Opin. Biotechnol.* **8,** 65–69.

Mochly-Rosen, D., and Gordon, A. S. (1998). Anchoring proteins for protein kinase C—A means for isozyme selectivity. *FASEB J.* **12,** 35–42.

Mori, H., Hiromoto, N., Nakahara, M., and Shiraishi, T. (1982). Stereological analysis of Leydig cell ultrastructure in aged humans. *J. Clin. Endocrinol. Metab.* **55,** 634–641.

O'Reilly, G., and Clarke, F. (1993). Identification of an actin binding region in aldolase. *FEBS Lett.* **321,** 69–72.

Ormo, M., Cubitt, A. B., Kallio, K., Gross, L. A., Tsien, R. Y., and Remington, S. J. (1996). Crystal structure of the aequorea victoria green fluorescent protein. *Science* **273,** 1392–1395.

Ovádi, J., and Srere, P. A. (1996). Metabolic consequences of enzyme interactions. *Cell Biochem. Funct.* **14,** 249–258.

Pagliaro, L., and Taylor, D. L. (1988). Aldolase exists in both the fluid and solid phases of cytoplasm. *J. Cell Biol.* **107,** 981–991.

Pagliaro, L., Kerr, K., and Taylor, D. L. (1989). Enolase exists in the fluid phase of cytoplasm in 3T3 cells. *J. Cell Sci.* **94,** 333–342.

Paine, P. L. (1984). Diffusive and nondiffusive proteins in vivo. *J. Cell Biol.* **99,** 188s–195s.

Parsegian, V. A., Rand, R. P., and Rau, D. C. (1995). Macromolecules and water: Probing with osmotic stress. *In* "Methods in Enzymology," (M. L. Johnson and G. K. Ackers, eds.), Vol. 259, pp. 43–94. Academic Press, San Diego, CA.

Pawson, T., and Scott, J. D. (1997). Signaling through scaffold, anchoring, and adaptor proteins. *Science* **278,** 2075–2080.

Pereira, G., and Schiebel, E. (1997). Centrosome-microtubule nucleation. *J. Cell Sci.* **110,** 295–300.

Periasamy, N., Armijo, M., and Verkman, A. S. (1991). Picosecond rotation of small polar fluorophores in the cytosol of sea urchin eggs. *Biochemistry* **30,** 11836–11841.

Periasamy, N., Kao, H. P., Fushimi, K., and Verkman, A. S. (1992). Organic osmolytes increase cytoplasmic viscosity in kidney cells. *Am. J. Physiol.* **263,** C901–C907.

Popov, S., and Poo, M. M. (1992). Diffusional transport of macromolecules in developing nerve processes. *J. Neurosci.* **12,** 77–85.

Porter, K. R. (1984). The cytomatrix: a short history of its study. *J. Cell Biol.* **99,** 3s–12s.

Provance, D. W., MacDowall, A., Marko, M., and Luby-Phelps, K. (1993). Cytoarchitecture of size-excluding compartments in living cells. *J. Cell Sci.* **106,** 565–578.

Reitz, F. B., and Pagliaro, L. (1997). Does regulatory protein play a role in glucokinase localization? *Hormone & Metabolic Research.* **29,** 317–321.

Rhoads, A. R., and Friedberg, F. (1997). Sequence motifs for calmodulin recognition. *FASEB J.* **11,** 331–340.

Rorschach, H. E., Chang, D. C., Hazlewood, C. F., and Nichols, B. L. (1973). The diffusion of water in striated muscle. *Ann. N. Y. Acad. Sci.* **204,** 445–452.

Rorschach, H. E., Lin, C., and Hazlewood, C. F. (1991). Diffusion of water in biological tissues. *Scanning Microsc., Suppl.* **5,** S9–S10.

Salmon, E. D., Saxton, W. M., Leslie, R. J., Karow, M. L., and McIntosh, J. R. (1984). Diffusion coefficient of fluorescein-labeled tubulin in the cytoplasm of embryonic cells of a sea urchin: Video image analysis of fluorescence redistribution after photobleaching. *J. Cell Biol.* **99,** 2157–2164.

Saxton, M. J. (1995). Single-particle tracking: effects of corrals. *Biophys. J.* **69,** 389–398.

Schliwa, M., and van Blerkom, J. (1981). Structural Interaction of cytoskeletal components. *J. Cell Biol.* **90,** 222–235.

Seksek, O., Biwersi, J., and Verkman, A. S. (1997). Translational diffusion of macromolecule-sized solutes in cytoplasm and nucleus. *J. Cell Biol.* **138,** 131–142.

Simson, R., Sheets, E. D., and Jacobson, K. (1995). Detection of temporary lateral confinement of membrane proteins using single-particle tracking analysis. *Biophys. J.* **69,** 989–993.

Singer, R. H., Langevin, G. L., and Lawrence, J. B. (1989). Ultrastructural visualization of cytoskeletal mRNAs and their associated proteins using double-label in situ hybridization. *J. Cell Biol.* **108,** 2343–2353.

Small, J. V., Isenberg, G., and Celis, J. E. (1978). Polarity of actin at the leading edge of cultured cells. *Nature (London)* **272,** 638–639.

Srere, P. A. (1967). Enzyme concentrations in tissues. *Science* **158,** 936–937.

Srivastava, A., and Krishnamoorthy, G. (1997). Cell type and spatial location dependence of cytoplasmic viscosity measured by time-resolved fluorescence microscopy. *Arch. Biochem. Biophys.* **340,** 159–167.

Steyer, J. A., Horstmann, H., and Almers, W. (1997). Transport, docking and exocytosis of single secretory granules in live chromaffin cells. *Nature* (*London*) **388,** 474–478.

Suprenant, K. A. (1993). Microtubules, ribosomes, and RNA: Evidence for cytoplasmic localization and translational regulation. *Cell Motil. Cytoskel.* **25,** 1–9.

Swaminathan, R., Bicknese, S., Periasamy, N., and Verkman, A. S. (1996). Cytoplasmic viscosity near the cell plasma membrane—translational diffusion of a small fluorescent solute measured by total internal reflection-fluorescence photobleaching recovery. *Biophys. J.* **71,** 1140–1151.

Swaminathan, R., Hoang, C. P., and Verkman, A. S. (1997). Photobleaching recovery and anisotropy decay of green fluorescent protein GFP-S65T in solution and cells: Cytoplasmic viscosity probed by green fluorescent protein translational and rotational diffusion. *Biophys. J.* **72,** 1900–1907.

Tang, J. X., and Janmey, P. A. (1996). The polyelectrolyte nature of F-actin and the mechanism of actin bundle formation. *J. Biol. Chem.* **271,** 8556–8563.

Tansey, M., Luby-Phelps, K., Kamm, K. E., and Stull, J. T. (1994). Ca^{2+}-dependent phosphorylation of myosin light chain kinase decreases the Ca^{2+} sensitivity of light chain phosphorylation within smooth muscle cells. *J. Biol. Chem.* **269,** 9912–9920.

Tapon, N., and Hall, A. (1997). Rho, Rac and Cdc42 GTPases regulate the organization of the actin cytoskeleton. *Curr. Opin. Cell Biol.* **9,** 86–92.

Trantham, E. C., Rorschach, H. E., Clegg, J. S., Hazlewood, C. F., Nicklow, R. M., and Wakabayashi, N. (1984). Diffusive properties of water in artemia cysts as determined from quasi-elastic neutron scattering spectra. *Biophys. J.* **45,** 927–938.

Van Steveninck, J., Paardekooper, M., Dubbelman, T. M. A. R., and Ben-Hur, E. (1991). Anomalous properties of water in macromolecular gels. *Biochim. Biophys. Acta* **1115,** 96–100.

Wang, J., Morris, A. J., Tolan, D. R., and Pagliaro, L. (1996). The molecular nature of the F-actin binding activity of aldolase revealed with site-directed mutants. *J. Biol. Chem.* **271,** 6861–6865.

Wang, Y., Lanni, F., McNeil, P. L., Ware, B. R., and Taylor, D. L. (1982). Mobility of cytoplasmic and membrane-associated actin in living cells. *Proc. Natl. Acad. Sci. U.S.A.* **79,** 4660–4664.

Waterman-Storer, C. M., and Salmon, E. D. (1997). Microtubule dynamics: Treadmilling comes around again. *Curr. Biol.* **7,** R369–R372.

Webster, R. E., Henderson, D., Osborn, M., and Weber, K. (1978). Three-dimensional electron microscopical visualization of the cytoskeleton of animal cells: Immunoferritin identification of actin- and tubulin-containing structures. *Proc. Natl. Acad. Sci. U.S.A.* **75,** 5511–5515.

Wheatley, D. N. (1998). Diffusion theory, the cell and the synapse. *BioSystems* **45,** 151–163.

Wheatley, D. N., and Malone, P. C. (1993). Heat conductance, diffusion theory and intracellular metabolic regulation. *Biol. Cell* **79,** 1–5.

Wiggins, P. M. (1996). High and low density water and resting, active and transformed cells. *Cell Biol. Int.* **20,** 429–435.

Wiggins, P. M., van Ryn, R. T., and Ormrod, D. G. C. (1991). Donnan membrane equilibrium is not directly applicable to distributions of ions and water in gels or cells. *Biophys. J.* **60,** 8–14.

Williams, S. P., Haggle, P. M., and Brindle, K. M. (1997). F-19 nmr measurements of the rotational mobility of proteins in vivo. *Biophys. J.* **72,** 490–498.

Wilson, I. A., Brindle, K. M., and Fulton, A. M. (1995). Differential localization of the mRNA of the M and B isoforms of creatine kinase in myoblasts. *Biochem. J.* **308,** 599–605.

Wojcieszyn, J. W., Schlegel, R. A., Wu, E., and Jacobson, K. A. (1981). Diffusion of injected macromolecules within the cytoplasm of living cells. *Proc. Natl. Acad. Sci. U.S.A.* **78,** 4407–4410.

Wojtas, K., Slepecky, N., von Kalm, L., and Sullivan, D. (1997). Flight muscle function in Drosophila requires colocalization of glycolytic enzymes. *Mol. Biol. Cell.* **8,** 1665–1675.

Wolosewick, J. J., and Porter, K. R. (1979). Microtrabecular lattice of the cytoplasmic ground substance. *J. Cell Biol.* **82,** 114–139.

Yang, F., Moss, L. G., and Phillips, G. N. (1996). The molecular structure of green fluorescent protein. *Nat. Biotechnol.* **14,** 1246–1251.

Yguerabide, J., Schmidt, J. A., and Yguerabide, E. E. (1982). Lateral mobility in membranes as detected by fluorescence recovery after photobleaching. *Biophys. J.* **39,** 69–75.

Yokoe, E., and Meyer, T. (1996). Spatial dynamics of gfp-tagged proteins investigated by local fluorescence enhancement. *Nat. Biotechnol.* **14,** 1252–1256.

Zalokar, M. (1960). Cytochemistry of centrifuged hyphae of *Neurospora. Exp. Cell Res.* **19,** 114–132.

Zimmermann, B., Somlyo, A. V., Ellis-Davies, G. C., Kaplan, J. H., and Somlyo, A. P. (1995). Kinetics of prephosphorylation reactions and myosin light chain phosphorylation in smooth muscle. Flash photolysis studies with caged calcium and caged ATP. *J. Biol. Chem.* **270,** 23966–23974.

Intracellular Compartmentation of Organelles and Gradients of Low Molecular Weight Species

Tak Yee Aw
Department of Molecular and Cellular Physiology, Louisiana State University Medical Center, Shreveport, Louisiana 71130

Intracellular compartmentation of metabolites without intervening membranes is an important concept that has emerged from consideration of the metabolic inhomogeneities associated with a highly organized and structured cytoplasm within mammalian cells. This recognition is primarily due to the development of experimental approaches to measure metabolite or ion concentrations at specific subcellular sites, thereby providing a means to study concentration gradients within the aqueous cytoplasm in intact cells. The presence of mitochondrial clusters has been shown to create gradients of low molecular weight species, such as O_2, ATP, and pH, with important implications for substrate supply for function and regulation of cellular processes. Moreover, the existence of kinetically distinct precursor pools has been shown to result in functional compartmentation of biochemical pathways, such as DNA replication and carbohydrate metabolism. The creation of these specialized microzones of metabolism in accordance with their association with cellular organelles or membranal structures may be integral to normal function and regulation of adult mammalian cells.

KEY WORDS: Mitochondrial distribution, Mitochondrial O_2 dependence, Intracellular oxygen gradients, Heterogeneity in ATP supply, Intracellular pH gradients, Microzonation of metabolites, Microcompartmentation.

I. Introduction

The concept that asymmetry occurs within mammalian cells has long been recognized. The intrinsic asymmetry within the cytoplasm of differentiated cells is reflected in the polarized distribution of cell ultrastructures and the associated heterogeneous chemistry and zonation of metabolism. The initial studies that elegantly demonstrated an association of asymmetry of structures with asymmetry of function within cells were described over 30 years ago by Garfinkel (1963) and Garfinkel and Lajtha (1963). They found that exogenously added radiolabeled glycine was preferentially utilized for hippurate biosynthesis in the mitochondria and did not equilibrate with the cellular glycine pool in kidney (Garfinkel, 1963; Garfinkel and Lajtha, 1963). This selective partitioning of glycine corresponded to the localization of high density of mitochondria adjacent to the plasma membrane of the cells. Thus, despite the absence of membranal barriers, metabolic inhomogeneity occurs within cell cytoplasm due to functional compartmentation of metabolites and substrates with cellular organelles.

In more recent years, our laboratory has revisited the concept of metabolic inhomogeneity and its significance for cell function. We found that the spatial distribution of mitochondria within liver and kidney cells can create metabolite gradients and intracellular compartmentation of high flux systems (Aw and Jones, 1985; Jones, 1986; Jones *et al.,* 1987), thus resulting in specific microcompartmentation of metabolism within the cytoplasmic milieu of these cells. Depending on metabolite and substrate fluxes and consumption, as well as on diffusional distances, the intracellular compartmentation of low molecular weight species may be a crucial and generalized mechanism for optimizing specialized cell functions within mammalian cells. We have termed this metabolic inhomogeneity "microzonation," to designate specific zones of heterogeneous metabolism within the cytoplasm. Conceptually, microzonation of function within cells is analogous to the zonation of metabolism in different regions of an organ, such as the liver, as described by Jungermann (1988).

Despite the obvious fundamental importance of intracellular metabolic heterogeneity in regulation of cell function, research progress in this field has been slow. It is noteworthy that the major concepts of intracellular compartmentation, diffusion-mediated metabolite gradients, inhomogeneity of structure and function, that were proposed years earlier are still viable today. This review summarizes our current knowledge on this subject. The chapter will focus on the specific concepts pertaining to intracellular compartmentation and gradients of low molecular weight species in association with heterogeneous distribution of cell organelles, such as the mitochondria. This discussion will be pertinent to understanding metabolic inhomogeneity within the cytoplasmic milieu of cells in the absence of delineating membranal barriers and the influence of such heterogeneous chemistry on control of cell function.

II. Compartmentation of Mitochondria within Mammalian Cells

Mitochondrial oxidative phosphorylation essentially provides the energy input that is needed for the highly specialized structures and functions of differentiated cells. The volume occupied by mitochondria in cells is highly variable and can range from 15 to 50% of the total cell volume. The distance between the site of ATP production in the mitochondria and the site of ATP utilization, such as at the plasma membrane, can impose a limitation to optimal ATP supply and consumption, thus necessitating the juxtaposition of mitochondria to sites of high ATP utilization. However, there would be no evolutionary driving force for creating and maintaining a defined mitochondrial distribution juxtapose to regions of high ATP utilization if diffusion does not pose a barrier for ATP supply, and if ATP-consuming systems are uniform throughout the cell. In reality, mitochondrial distribution within cells is often highly structured and parallels the distribution of ATP-consuming systems.

A. Tissue Specific Mitochondrial Distribution

1. Heterogeneous Distribution of Mitochondria in Differentiated Cells

Detailed morphometric studies of the subcellular distribution of mitochondria show that mitochondria have defined distributions in fully differentiated adult mammalian cells. For example, in transport epithelial cells, high densities of mitochondria are often adjacent to the basal region of the cell. In the small intestine (Aw *et al.,* 1992) and the proximal and distal kidney tubules (Bulger and Dobyan, 1982), mitochondrial clusters are juxtaposed to the basal membrane where Na^+, K^+ ATPase is located, while in gastric parietal cells, mitochondria are clustered next to the membranal H^+-ATPase. Clusters of mitochondria several microns across occur within the retina epithelial cell, while clustered mitochondria congregate at synaptic terminals of neuronal cells (Fawcett, 1981). In skeletal muscle, mitochondria may actually form large syncytia (Kirkwood *et al.,* 1986; Jones and Aw, 1988). The existence of these regional discrete zones or clusters of mitochondria will have critical implications for tissue oxygenation.

2. Mitochondrial Distribution in Neonatal Cells and Its Metabolic Implications

Studies on mitochondrial distribution in fetal and neonatal cells are sketchy, and the available data from electron micrographs of hepatocytes from newborn rats reveal fewer mitochondria and a more uniform mitochondrial distribution throughout the cell cytoplasm as compared to adult cells (Aw and Jones, 1987, Aw and Jones, 1988a,b; Aw, 1991). The lack of clustering of mitochondria in neonatal cells results in marked differences in oxygenation characteristics from that of adult hepatocytes. For instance, the O_2 requirement for maximal oxidation of mitochondrial cytochromes in fetal and neonatal cells is an order of magnitude lower than in adult cells and more closely resembles the value for isolated mitochondria (Aw and Jones, 1987, 1988a,b). Thus, the immature and less differentiated fetal and neonatal cells possess a significantly lower O_2 requirement for maximal mitochondrial function, a characteristic that is well suited to a low arterial pO_2 environment *in utero.*

Several important metabolic implications are associated with the homogeneous mitochondrial distribution and low O_2 dependence in perinatal tissues. The low half-maximal value (P_{50}) for O_2 dependence of mitochondrial function ensures that the plasma O_2 concentrations will support normal cellular respiration despite the hyperbolic shape of the fetal hemoglobin dissociation curve. Because mitochondrial clustering results in steep intracellular O_2 gradients and poor mitochondrial oxygenation (see Section III below), perinatal tissues are less susceptible to O_2 deficiency and anoxic failure, consistent with hypoxic tolerance (Dawes, *et al.,* 1959). Furthermore, cellular biochemical processes that are dependent secondarily on function of mitochondrial cytochrome oxidase (which has a low P_{50} for mitochondrial function) will be relatively insensitive to O_2 lack. Finally, the absence of large mitochondrial aggregates allows for greater metabolic homogeneity in perinatal tissues.

B. Physiological Modulation of Mitochondrial Distribution

While it is appreciated that differentiated adult tissues exhibit characteristic intracellular mitochondrial distribution that is heterogeneous, tissue specific, and commensurate with functional and metabolic needs, an important issue that is poorly understood, and has hitherto not been addressed, is whether this spatial arrangement of mitochondria is dynamic. Earlier studies from Jones' laboratory have shown that altered physiological states can cause mitochondrial redistribution (Sillau *et al.,* 1988; Jones *et al.,* 1990, 1991). They found that exposure of rats to chronic hypoxia (normobaric hypoxia at 10% O_2) induces a uniform distribution of mitochondria in liver cells and an associated change in cellular respiration that is characteristic with altered mitochondrial spatial arrangement, namely, a decreased P_{50} for O_2 dependence of mitochondria function. Organ development represents another physiological cue for spatial structuring of mitochondria. In our studies, we found that an increase in O_2 dependence of oxidation of mitochondrial cytochromes and pyridine nucleotides in rat hepatocytes occurs in concert with postnatal development of the animal and the associated increase in mitochondrial number and mitochondrial cluster size (Aw, 1991). Collectively, these findings suggest that the localization of mitochondria in cells is a dynamic process and that distribution or redistribution of mitochondria can occur in response to physiological signals or metabolic demands.

1. Nutrients and Differential Polarization of Mitochondria in Intestinal Cells

Our recent studies in the actively absorbing intestinal epithelium provide a clear illustration of mitochondrial polarization in response to different metabolic demands, such as that associated with the fed and fasted states. An association between nutrient feeding and mitochondrial distribution in intestinal cells was first reported in suckling rats in the early 1970s (Gonzalez-Licea, 1970, 1971). Gonzalez-Licea (1971) found that feeding of fat, but not glucose, resulted in basal orientation of intestinal mitochondria. Given that the adult intestine is frequently subjected to varied nutrient availability associated with different dietary states, nutrient control of mitochondrial distribution is expected to occur. As shown in Fig. 1, we found that the distribution of mitochondria in intestinal cells is asymmetric and differs under fed and fasted conditions with preferential polarization of the mitochondria to the basal pole during feeding and to the apical pole during fasting. Within a single cell, an average of 75% of the mitochondria was predominantly localized to the infra- and perinuclear regions under fed conditions while the reverse was observed under fasted conditions. Examination of the frequency of distribution reveals that most enterocytes from fed animals exhibit an infra- to supranuclear (I-to-S) distribution of 70 to 30% (Fig. 2), whereas a majority of cells exhibited I-to-S distribution greatly in favor of an apical localization in the fasted state. Moreover, differential shifts in mitochondrial profiles in cells occur in conjunction with the specific nutrients of carbohydrate, protein, or lipid (Fig. 2). Notably, lipid feeding induced a basal polarization of mitochondria in contrast to glucose-induced apical migration of mitochondria. This dramatic apical-to-basal translocations of mitochondria in response to specific nutrient supply may reflect differential energy demands of the enterocyte during feeding and fasting. Collectively, these results show that nutritional status is a key determinant of mitochondrial movement and underscores the importance of intracellular mitochondrial localization and function in absorptive cells. Spatial restructuring of mitochondria may, in fact, represent a general adaptative mechanism for optimal organelle-function relationship in all mammalian cells.

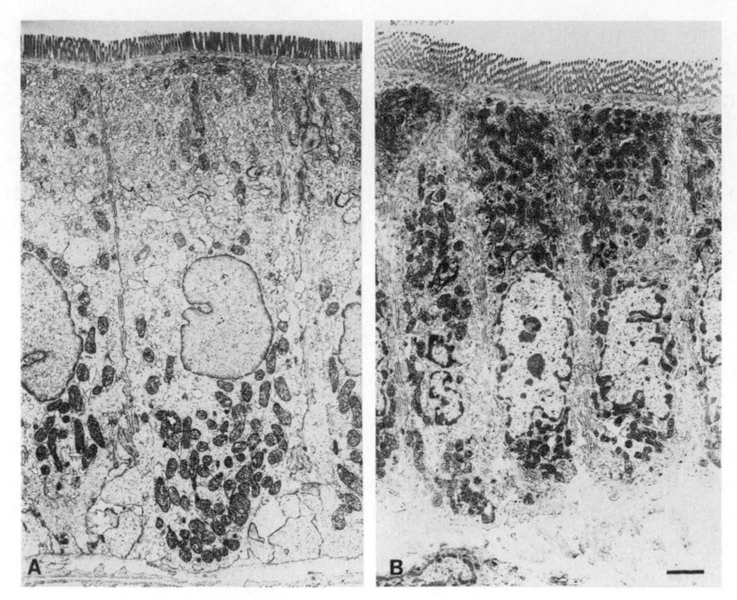

FIG. 1 Electron micrographs showing mitochondrial distribution profiles in jejunal enterocytes from adult rat small intestine. The views shown are of whole cells with visible apical and basal regions from a villus cut longitudinally. (A) fed *ad libitum* with chow; (B) 24h fasted. Scale bar equals 2 μm, $\times$ 5000.

At present, little is known of how mitochondria achieve their cell specific intracellular arrangement and of the molecular determinants of mitochondrial distribution. During axonal transport (Papasozomemos *et al.,* 1982) and establishment of cells in culture (Heggeness *et al.,* 1978), mitochondria have been shown to associate with microtubules, perhaps by specific microtubule-associated proteins. Deep-etch studies of nerve axon by Hirokawa (1982) have demonstrated crosslinking between mitochondria and microtubules and between mitochondria and intermediate filaments. Thus, it appears that specific proteins provide attachment of mitochondria to the cellular cytoskeletal elements along which mitochondria can move and redistribute in response to physiological signals.

III. Intracellular Gradients of O_2

The initial studies on O_2 diffusion in tissues and tissue oxygenation were pioneered by Krogh (1919). These earlier studies provided indications for imposed O_2 gradients, but the factors that affect intracellular O_2 supply have not been rigorously defined.

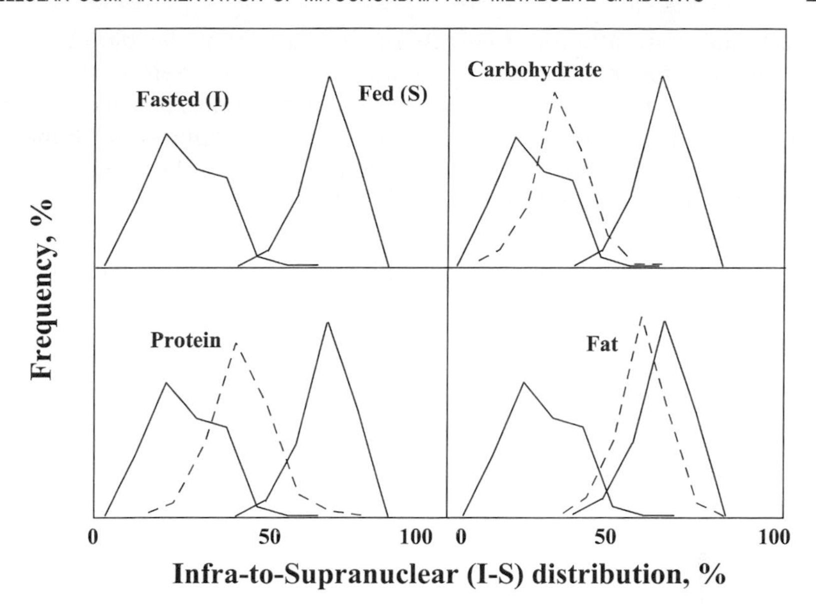

FIG. 2 Frequency of distribution of jejunal enterocytes with varied intracellular infra-to-supranuclear (I-to-S) distribution of mitochondria under different nutritional status. Rats are subjected to fed or 24 h fasted conditions, or conditions of fasting followed by refeeding with carbohydrate, protein, or fat diets. The number of mitochondria within the supranuclear (upper) or infranuclear (lower) regions of the cells was quantified and expressed as a percentage of total mitochondria in the cells. The number of enterocytes with different mitochondrial (I-to-S) distribution profiles was also quantified. The results are given as distribution frequency of cells with various defined I-to-S mitochondrial distribution.

Within cells, mitochondria are responsible for most of the O_2 consumption. Isolated mitochondria have been studied extensively with regard to their O_2 dependence as a function of O_2 concentration. The results show that the P_{50} value for mitochondrial function, defined as the O_2 concentration to give half-maximal activity, is dependent on mitochondrial respiratory activity, and is typically in the range of 0.1 to 0.3 μM (Oshino *et al.,* 1974). Because mammalian cells are normally exposed to O_2 in the range of 20–100 torr (i.e., 28–140 μM), one would not expect limitations in O_2 supply for maximal mitochondrial function in cells given an excess of O_2 availability in blood. In reality, this margin of O_2 excess is small. Studies in perfused kidney and brain show that, despite the relative abundance of O_2 in blood, a small decrease in arterial pO_2 can impair mitochondrial function (Balaban and Sylvia, 1981; Epstein *et al.,* 1982). A reasonable explanation for this apparent paradox was proposed by Tamura *et al.,* (1978) about 20 years ago. Based on analyses of the O_2 dependence of mitochondrial cytochrome oxidase and myoglobin in isolated perfused rat heart, these investigators suggest that concentration gradients of O_2 exist from the blood to the mitochondrial inner membrane (Tamura *et al.,* 1978). Their conclusion is that O_2 can readily equilibrate with the nonmitochondrial space, but the O_2 in the nonmitochondrial space does not readily equilibrate with the O_2 in the mitochondria. Studies from our laboratory and others support this conclusion (Jones and Kennedy, 1982a,b; Jones, 1984, 1986; Jones *et al.,* 1987).

A generalized model that can account for the existence of intracellular O_2 gradients is based on the heterogeneous distribution of mitochondria in cells. The fundamental concept is that mitochondrial aggregates create a localized region with a high O_2 consumption rate, and the O_2 diffusion into this region will differ dramatically from O_2 diffusion in other regions of the cell (Fig. 3). The magnitude of O_2 gradient is likely to be variable in different cell types, depending on mitochondrial cluster size, intracellular O_2 diffusion coefficient, and O_2 consumption rate. Furthermore, the steepness of O_2 gradients would be exaggerated under conditions of O_2 limitation. Thus, the distribution of mitochondria within cells would be an important contributor to the cellular O_2 dependence. Indeed, the extensive clustering of mitochondria contributes to an order of magnitude higher O_2 concentrations (15–25 μM) that is required to maintain maximal mitochondrial function in liver, heart, and kidney cells as compared to isolated mitochondria.

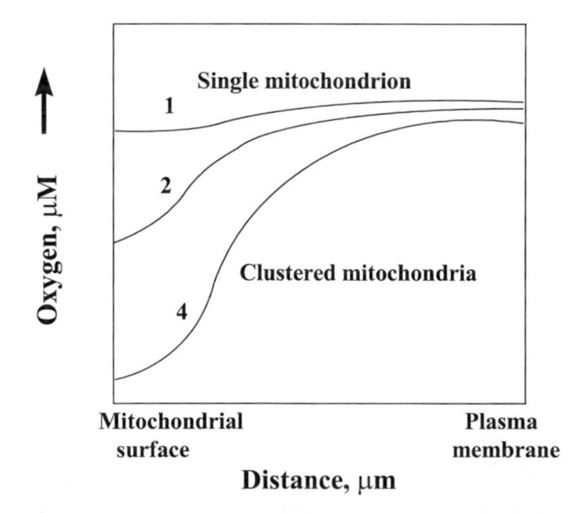

FIG. 3 O_2 concentration gradient in the region of a single mitochondrion or a cluster of mitochondria 2 or 4 across. O_2 consumption parameters were as described by Jones (1984), and the O_2 diffusion coefficient, D_i, was taken as 2×10^{-6} $cm^2 sec^{-1}$. The results show that significant O_2 gradients can occur in the vicinity of mitochondrial clusters.

A. Mitochondrial O_2 Dependence and Heterogeneity in Subcellular O_2 Delivery

1. Mitochondrial O_2 Dependence in Isolated Mitochondria and Cells

The liver is one of the best studied cell types with regard to oxygenation characteristics and mitochondrial O_2 requirement. Earlier studies of O_2 dependence of cell bioenergetics in isolated hepatocytes show that half-maximal changes in ATP/ADP and lactate/pyruvate occurred around 10 μM O_2 (Jones and Mason, 1978a,b; Jones *et al.,* 1983), a range that is 50-fold higher than the P_{50} for mitochondrial respiration and oxidation of mitochondrial cytochromes, ~0.2 μM (Oshino *et al.,* 1974). This finding indicates that a substantial O_2 concentration gradient occurred from the extracellular, nonmitochondrial space to the mitochondria themselves. Comparison of studies of O_2 dependence in isolated liver mitochondria and in liver cells reveals that the P_{50} for mitochondrial cytochrome *c* oxidation in cells was an order of magnitude higher than the P_{50} needed for cytochrome *c* oxidation in a well-mixed suspension of isolated mitochondria (Jones and Kennedy, 1982a), consistent with the existence of a dramatic O_2 gradient from the rest of the cytoplasm to the mitochondria. Another well-characterized cell with regard to O_2 dependence is the isolated proximal tubule cell. As expected from the clustering of mitochondria in the basal region of the cell, the half-maximal O_2 concentration for cytochrome, *c* oxidation is around 4 μM, 10-fold higher than the same measurement in isolated kidney mitochondria (Aw *et al.,* 1987b). This high O_2 requirement at the cellular level indicates existence of intracellular O_2 gradients, and may explain why the kidney is functionally O_2-deficient despite a tissue pO_2 of 10 to 15 torr (Balaban and Sylvia, 1981; Epstein *et al.,* 1982; Aw *et al.,* 1987b).

The same response occurs in cardiac myocytes, in that the O_2 requirement for mitochondrial cytochrome oxidation in quiescent cells is an order of magnitude greater than that needed for cytochrome oxidation in isolated heart mitochondria (Wittenberg and Robinson, 1981; Jones and Kennedy, 1982b; Kennedy and Jones, 1986). Furthermore, comparison of O_2 dependence of myoglobin oxygenation with that of mitochondrial cytochrome oxidation shows that, within a cardiac myocyte, significant O_2 gradients occur from extracellular space into the immediate vicinity of mitochondrial clusters and not into the cytoplasmic region of the cell (Tamura *et al.,* 1978; Kennedy and Jones, 1986). Given a more uniformly spaced rows of mitochondria between the myofibril, oxygenation of skeletal muscle may be different from that of heart muscle (Gayeski and Honig, 1986). Although the O_2 dependence has not been studied in skeletal myocytes, based on the predictions of the clustered mitochondria model, the extent of intracellular O_2 gradients in skeletal muscle cells may be small and the cellular O_2 dependence may be largely determined by transcellular gradients rather than intracellular O_2 diffusion into mitochondrial clusters. This means that, as compared to cardiac myocytes and other cell types with higher density and clusters of mitochondria, the isolated skeletal myocyte will require relatively lower O_2 concentrations for mitochondrial function.

It is apparent from the foregoing discussion that cellular O_2 concentration dependence is a function of mitochondrial clustering. The fundamental changes in mitochondrial distribution, structure, and function that occur concurrently with development, differentiation, and transition to culture predicts that neonatal cells (Aw and Jones, 1987, 1988a,b; Aw, 1991) and adult cells in culture (Aw and Jones, 1988a,b), both of which have low mitochondrial density and a more uniform mitochondrial distribution, should have small intracellular O_2 gradients. Consistent with this supposition, the experimentally determined P_{50} for mitochondrial function in neonatal hepatocytes and adult hepatocytes in culture are in a range that is similar to isolated mitochondria (~0.5 μM O_2) (Aw and Jones, 1987; 1988a,b) and an order of magnitude lower than adult cells (Fig. 4; also see Section III, A above). Functionally, perinatal cells and cells in culture

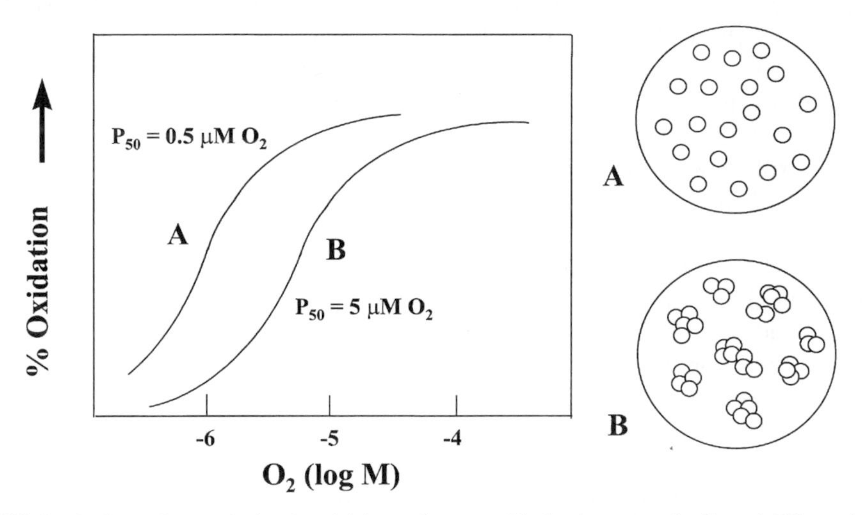

FIG. 4 O_2 dependence of mitochondrial cytochrome oxidation in neonatal cells and differentiated adult cell. (A) neonatal cells or adult cells in culture wherein mitochondria are fewer in number and are uniformly distributed. (B) differentiated adult cells wherein mitochondrial density is high and appear in clusters. The P_{50} for cytochrome oxidation in cells with mitochondrial arrangement (A) can be an order of magnitude less than that in cells with mitochondrial arrangements (B).

are expected to exhibit decreased oxidative capacity and achieve greater metabolic homogeneity.

To date, the clustering mitochondrial model appears to be adequate for explaining creation of intracellular O_2 gradients within cells. The diverse O_2 requirement of different tissues with different intracellular compartmentation of mitochondria is, in fact, consistent with this model. Such a model predicts that cells with mitochondria that are not present in clusters will remain functional at relatively low O_2 concentration, while cells with mitochondrial aggregates can be O_2 deficient even at physiologically relevant O_2 concentrations.

2. Functional Demand and Substrate Availability as Modulators of Cellular Mitochondrial O_2 Dependence

While mitochondrial distribution provides a primary mechanism to regulate steady-state cellular O_2 dependence, this O_2 dependence can be signifcantly altered by modulation of mitochondrial activity. In cardiac myocytes, electrical stimulation of quiescent cells increases the O_2 consumption rate and the P_{50} value for oxidation of mitochondrial cytochrome oxidase (Jones *et al.*, 1987). Similarly, stimulation of Na^+, K^+ ATPase activity in renal proximal tubule cells with nystatin increases O_2 consumption and the P_{50} for mitochondrial function, while decreasing ATP demand for pump activity with ouabain decreased both O_2 consumption and the P_{50} value (Aw *et al.*, 1987b, Fig. 5). A similar relationship of changes in P_{50} and respiration rate occurs for neonatal cells. In neonatal cells, wide variations in O_2 consumption can be obtained by varying substrate availability, such as succinate supply (Aw and Jones, 1987; 1988b). Comparison of O_2 dependences at different substrate concentrations show that the P_{50} varies as a function of the succinate-dependent O_2 consumption rate (Aw and Jones, 1987; 1988b). Taken together, these studies show that the cellular O_2 consumption rate is an important determinant of the O_2 concentration requirement for the mitochondria. Furthermore, the results suggest that mitochondrial O_2 dependence in excitable tissues, such as the brain, and in absorptive and transport tissues, such the small intestine and kidney, may be particularly sensitive to modulation by variation in workload and availability of specific substrates.

B. Mapping of Intracellular O_2 Gradients

1. Modeling Subcellular O_2 Diffusion

The model of Boag (1969) provides a reasonable working model to estimate the subcellular O_2 delivery. In the simplest case, a cell can be assumed to

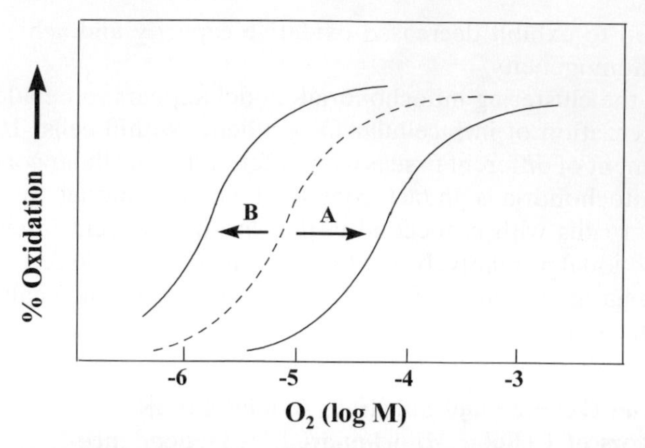

FIG. 5 O_2 dependence of mitochondrial cytochrome oxidation in a cell such as the kidney proximal tubule (dashed line). (A) represents stimulation of cellular O_2 consumption rate, e.g., increased Na^+, K^+-ATPase activity, causing a right shift in the O_2 dependency and a higher P_{50} for mitochondrial function. (B) represents a decrease in cellular O_2 demand by inhibition of pump activity with a resultant left shift in the O_2 dependence curve and a lower P_{50} for mitochondrial function.

be a sphere with uniform O_2 consumption throughout. In this idealized cell, O_2 gradients can occur in the unstirred layer surrounding the cell, at the plasma membrane, and in the cytosolic region within the cell. In cells of known radii and O_2 consumption rate, such as in Ehrlich ascites, and assuming the diffusion coefficient for O_2 in physiological salt solution, Boag (1970) found that the contribution of the unstirred layer is small. Similarly, the plasma membrane does not impose a significant barrier to O_2 diffusion; hence, large O_2 concentration gradient across the plasma membrane is unlikely to occur. If O_2 consumption is uniform in the cytoplasm, radial diffusion into the O_2-consuming sphere will result in a uniform decrease in O_2 concentration from the plasma membrane to the cell center (see Fig. 3). Based on the measured cellular O_2 consumption rate, cell volume and an estimated cytosol O_2 diffusion coefficient (D_i) to be that of O_2 in water (2×10^{-5} cm^{2}sec^{-1}), an intracellular O_2 gradient can be calculated for the cell (Huxley and Kutchai, 1981). In reality, this idealized cell model does not exist. Results from different studies show that the experimentally determined O_2 diffusion coefficient in cell cytoplasm is considerably less than that in water, in the range of 2×10^{-6} to 4×10^{-6} cm^{2}sec^{-1} (Boag, 1970; Clark *et al.,* 1987). A small fractional volume occupied by water, a high solute concentration, a high intracellular viscosity, a highly structured cytoskeleton and associated water, and the possible association of O_2 with cellular components all contribute to a slower O_2 diffusion within cell cytoplasm. Thus, with the consideration that the diffusion coefficient of O_2 is only 0.1 to 0.2 times that in water, the contribution of a gradient outside the cell is proportionately less than that inside the cell at a given flux.

The mapping of intracellular O_2 gradient is complicated by heterogeneous O_2 diffusion and asymmetric O_2 consumption that occurs within the cytoplasm of the cell due to distribution of mitochondrial aggregates. As discussed above, heterogeneously distributed mitochondria creates foci of high O_2 consumption rates that are surrounded by regions of relatively low O_2 consumption rate. Consequently, the presence of clusters of mitochondria in a cell can create a substantially different regional O_2 distribution than that predicted by the simple model of Boag, which considers uniform distribution of O_2 consumption. Nevertheless, the radial diffusion model of Boag (1969, 1970) can reasonably be used to analyze the O_2 diffusion into a cluster of mitochondria. In this scenario, the O_2 consumption per unit volume occupied by the mitochondrial cluster is several-fold higher than a single mitochondrion. As a result, the O_2 concentration at other sites of the cytoplasm distant from the mitochondrial clusters is much higher than that in the center of the cluster itself, thereby creating a substantial O_2 gradient in the vicinity of the cluster. Thus, the magnitude and variability of O_2 gradients that are created due to radial diffusion of O_2 will depend on the cell type and pattern of distribution of mitochondria and the O_2 consumption rate of the mitochondria. Since O_2 concentration gradients in the unstirred layer and extramitochondrial cytoplasmic space are relatively small, the vulnerability of cells to O_2 deficiency may be a direct function of the steepness of O_2 gradients associated with the existence or absence of mitochondrial clusters. Furthermore, under conditions of high O_2 consumption and low O_2 availability, such as during hypoxia, significant O_2 gradients will occur. Thus, the creation of variable O_2 gradients is an important consideration with regard to optimal oxygenation at different functional sites and under different physiological conditions.

IV. Intracellular Gradients of Metabolites and Substrates

Because of the preferential partitioning of mitochondria to discrete regions of the cytoplasm of cells, a unique pattern of microzonation of ATP within the aqueous cytoplasm is created in accordance with its association with the mitochondria. The resultant microheterogeneities in metabolite and substrate distribution may be important in regulation of cell structure and function.

A. Heterogeneity in ATP Consumption and Supply

1. ATP Concentration Gradients in Liver Cells

a. Intracellular Enzymes as Probes for Regional Differences in ATP Despite the absence of membranal barriers, several earlier studies suggested the existence of different pools of ATP in the cytoplasm of mammalian cells (Paul *et al.,* 1979, 1989; Barry *et al.,* 1980; Bricknell *et al.,* 1981; Lipton and Robacker, 1983; Lynch and Paul, 1983), consistent with occurrences of heterogeneities in ATP concentrations within cells. Moreover, the fact that mitochondrial clusters are often associated with high ATPase activities suggests that, under certain conditions, ATP supply to support ATPase function must be limiting. Studies from our laboratory show that during hypoxia, a substantial ATP concentration gradient occurs in the cytoplasm of isolated liver cells (Aw and Jones, 1985). We found that two ATP-utilizing systems with distinct subcellular localization in the cytoplasm and plasma membrane are not exposed to the same average ATP concentrations. This finding can be explained by the existence of consumption-induced ATP gradients due to ATP flux from the mitochondrial site of ATP production to the peripheral site of ATP utilization (Fig. 6).

The concept that enzymes can serve as intracellular probes of cellular metabolites and substrates was first pioneered by Chance and associates to determine cellular O_2 (Chance *et al.,* 1973) and H_2O_2 concentrations (Sies *et al.,* 1973). The fundamental hypothesis is that if an enzymatic reaction rate is largely determined by the concentration of its substrate,

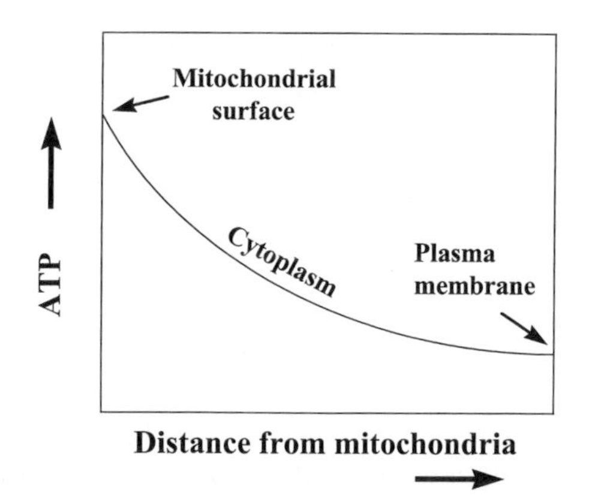

FIG. 6 Consumption-induced ATP concentration gradient occurs as a result of ATP flux from mitochondrial site of ATP production to the peripheral site of ATP utilization at the plasma membrane. Under limited ATP production, such as occurs during hypoxia, the supply radius of ATP will be markedly reduced.

then that defined enzymatic system can be used as a probe of the cellular substrate concentration (reviewed in Jones and Aw, 1990). Subsequent studies by others have also successfully employed specific enzyme–substrate coupling to estimate concentrations of cytoplasmic Ca^{2+}, H_2O_2, and mitochondrial matrix intermediates (Chance *et al.,* 1979; Williamson and Cooper, 1980; Bellomo *et al.,* 1984).

In our approach, we have employed two enzymes with different locations in liver cells, namely, the cytoplasmic ATP sulfurylase and the plasma membrane-bound Na^+, K^+-ATPase, whose activities are sensitive to ATP availability, to detect regional differences in ATP concentrations in the cytoplasm and plasma membrane, respectively (Aw and Jones, 1985). This approach to detect cellular ATP concentration is feasible because the reaction rate of the enzymes is largely determined by the concentration of ATP itself. Conceptually, this experimental approach of relating function with substrate availability is analogous to the use of mitochondrial cytochromes as sensors of O_2 concentration gradients as discussed above (Section III) for isolated cardiac myocytes, hepatocytes, and renal proximal tubule cells. ATP availability for ATP sulfurylase activity was determined by the rate of drug sulfation, while that for Na^+, K^+-ATPase activity was determined by cellular uptake of radiolabeled rubidium, Rb^+. The results show that sulfation rates decreased in parallel with decreases in average cell ATP over a wide range of ATP concentrations, while cellular Rb^+ uptake was more sensitive to decreases in cell ATP, indicating a reduced accessibility of ATP at the plasma membrane as compared to the cytoplasm (Aw and Jones, 1985; also see Fig. 6). Moreover, this differential in ATP availability was enhanced during hypoxia, wherein limited ATP production significantly curtailed ATP supply to peripheral sites of ATP consumption at the plasma membrane (Aw and Jones, 1985).

The general approach of exploiting the function of an enzyme to monitor substrate availability provides a specific and sensitive handle to define small differences in substrate concentrations in microenvironments surrounding organelles or membrane structures as evidenced by the novel studies of Aflalo and DeLuca (1987). In these studies, the investigators used firefly luciferase to continuously monitor ATP concentration in the microenvironment of immobilized ATP-producing (pyruvate kinase) or ATP-consuming enzyme (hexokinase) (Aflalo and DeLuca, 1987). They found that accessibility of ATP to luciferase either as ATP accumulation or ATP consumption depended on whether the two enzymes are associated with the particulate or soluble fractions. Notably, the ATP pool that is in the vicinity of immobilized enzyme does not readily equilibrate with the ATP pool in the bulk cytoplasmic phase, indicating the existence of two functionally distinct compartments. This heterogeneity of ATP concentration is relevant to the consideration of cellular ATP supply for function and regulation of membrane-bound enzymes. Importantly, the above noninvasive approach of enzyme-coupling for metabolite determination will have wide applicability for the determination of concentrations of other cellular substrates for a variety of different biochemical pathways in the cell.

b. Functional Significance of Heterogeneous ATP Availability A number of studies on cellular kinases and ATPase show that these ATP-dependent enzyme systems can be functionally compartmentalized, that is, independent of membranal partitioning. This compartmentation appears to occur in accordance with the localization of these enzymes proximal or distal to the mitochondria. In this regard, mitochondria with different subcellular distribution are functionally and metabolically distinct, particularly with regards to ATP supply for specific cellular activities. A decrease in mitochondrial ATP production without a proportionate decrease in cytosolic ATP consumption will pose a significant limitation on ATP availability to regions peripheral to the mitochondria. Enzymes at these sites are selectively susceptible to ATP depletion. Because the diffusion of ATP is low (Aw and Jones, 1985), juxtaposition of mitochondria to enzyme systems that utilize ATP minimizes diffusion limitations and allows the enzymatic machinery to respond directly to mitochondrial ATP production, thus ensuring maximal ATP delivery to zones with critical ATP requirements.

Several investigators have demonstrated a structure–function relationship between nonuniform distribution of glycolytic systems and heterogeneous ATP supply (Entman *et al.,* 1976; Barry *et al.,* 1980; Paul, 1983; Balaban and Bader, 1984). In these studies, it was found that glycolytically, rather than mitochondrially, derived ATP is the preferred energy source for membrane function. Examples include coupling to Ca^{2+} uptake in cardiac myocytes (Entman *et al.,* 1976), Na^+, K^+ transport in smooth muscle cells (Paul *et al.,* 1979; Lynch and Paul, 1983), and K^+ balance in neuronal cells (Lipton and Robacker, 1983). Thus, the heterogeneous and strategic localization of glycolytic and mitochondrial ATP sources can be an important determinant of ATP supply to different ATP-requiring reactions.

The association of ATP supply and ATP demand further represents an adaptive mechanism for metabolic regulation of cell function. The compartmentation of glucose metabolism is one such adapted system wherein hexokinase, which normally resides in the cell cytoplasm, specifically binds to the outer mitochondrial membrane and is activated under conditions of enhanced glycolytic activities (Wilson, 1980). This selective binding of hexokinase to the mitochondria eliminates limitation to ATP diffusion and allows preferential "trapping" of mitochondrially generated ATP for the kinase reaction. This finely controlled partitioning of hexokinase between the soluble and particulate fractions allows the cell to regulate function of the enzyme under varied conditions of glycolytic flux.

Another interesting example of subcellular heterogeneity of energy supply and energy transfer is the creatine phosphate system in skeletal muscle. Phosphorylation of creatine by the mitochondrially associated creatine kinase yields a readily diffusible high-energy compound, creatine phosphate (Bessman and Carpenter, 1985). Subsequent hydrolysis of creatine phosphate by myosin ATPase-associated creatine kinase provides an immediate ATP source at the site of ATP need. Thus, the function of mitochondrial and myofibrillar forms of creatine kinase overcomes limitations in ATP diffusion and provides an efficient means for energy transport over substantial intracellular distances within cells. The functional compartmentation of energy supply in intermediary metabolism will be considered in more detail in Section B2 below.

2. Modeling Subcellular ATP Diffusion

As with the mapping of intracellular O_2 gradients, the simple model of Boag (1969) can be applied to mathematically model the radial diffusion of ATP from the mitochondria into an increasing ATP-consuming volume. In the simplest consideration, which is based on two assumptions, namely, that mitochondria are spheres and that ATP consumption is zero order in ATP concentration (Aw and Jones, 1985), radial ATP flux and ATP utilization is defined by the following differential equation:

$$-4\pi Dr^2 \times dC/dr = Q_m - Q_c\,(r^3 - r_1^3/r_2 - r_2^3)$$

where D = diffusion coefficient for ATP in the cytosol; C = ATP concentration at radius r; r_1 = mitochondrial radius; r_2 = mean supply radius by a mitochondrion; Q_m = mitochondrial ATP production rate; and Q_c = ATP consumption rate in the region of the mitochondria. If the maximal ATP diffusion distance to the plasma membrane is represented by r_2, the ATP concentration at this distance is given by the simplified equation

$$C(r - r_2) = Q_m - Q_c/4\pi Dr_2$$

Thus, if ATP consumption in the cytoplasm of cells is independent of ATP concentration, a decrease in ATP production (Q_m) to equal that of ATP consumption (Q_c) would result in zero concentration at the plasma membrane (r_2). A further decrease in Q_m will result in a smaller ATP supply radius. From this simple analysis, one could predict that ATP concentration at the plasma membrane of cells will fall more drastically than the average cytoplasmic ATP concentration under ATP-limiting conditions, such as during hypoxia.

Given the above consideration, one could map the magnitude of ATP gradients in cells under different physiological conditions. Based on the high intracellular ATP concentration (mM) and an experimentally determined ATP diffusion coefficient of 10^{-6} cm/sec^{-1}, substantial ATP gradients are unlikely to occur in aerobic cells. However, during conditions of interrupted ATP production, such as in hypoxia, consumption-induced ATP gradients will be created due to the radial diffusion of a limited ATP supply into a large volume of continuous ATP consumption in the cytoplasm (Aw and Jones, 1985). Consequently, the function of ATPase at cell membranes will be selectively inhibited in accordance to the defined extent of decreased supply. In reality, the extent of ATP supply is difficult to quantify based on the simple model of Boag (1969, 1970) since mitochondria, whether singly or in clusters, are not perfect spheres and ATP consumption by ATPases does not always obey zero order kinetics.

B. Functional Compartmentation of DNA Replication and Biochemical Pathways of Carbohydrate Metabolism

The above consideration underscores the importance of metabolite gradients, such as ATP and O_2 in cellular regulation. Although the study of microcompartmentation of cellular metabolites has been limited, it can readily be appreciated that metabolite gradients can represent a common mechanism for regulation of biochemical reactions and metabolic pathways. Indeed, it would not be surprising that gradients of a variety of substrates and metabolites do exist in cells because of the rigidity that membrane partitioning can impose on cellular regulation. Therefore, modulation of substrate gradients may provide a more sensitive, versatile, and responsive means to fine tune metabolic performance in cells under varied physiological and pathophysiological conditions. In fact, creation of metabolic gradients and heterogeneous chemistry within cells may be a result of selective evolutionary driving force to optimize metabolic efficiency.

This next section is not intended to provide a comprehensive review of the literature on the subject of metabolite microcompartmentation. For this review, the reader is referred to an earlier excellent monograph that was devoted to covering many aspects of solute microcompartmentation and the metabolic implications within the structured environment of the cells (Jones, 1988). Rather, the following discussion will focus on two select topics of metabolic compartmentation that have general applicability and wide implications for cellular regulation in most adult mammalian cells. These are (a) the concept of functional compartmentation of DNA replication and (b) the concept that functional demand imposes a physical separation of metabolic pathways.

1. Compartmentation of DNA Precursor Pool

The microcompartmentation of DNA precursor has been documented and was first recognized in prokaryotic bacterial systems. In 1971, Werner found that, during DNA replication in *E coli,* thymidine labeling of DNA occurred at maximal rates before the dTTP pool has achieved maximal radioactivity (Werner, 1971), suggesting utilization of a preferential DNA precursor pool for DNA synthesis that is distinct from the average precursor pool. This suggestion was subsequently confirmed in studies by Manwaring and Fuchs (1979), who found that DNA replication in *E. coli* is fueled by a small and rapidly depleted pool which did not readily equilibrate with the bulk replication-inactive pool of dNTP. Because the average intracellular dNTP concentration in prokaryotes is low (~100 μM, ref), the V_{max} for DNA replication is high (~1000 nucleotides sec^{-1} per elongating strand), and the K_m for the replication apparatus for its substrate is low (Mathews, 1972; Mathews and Sinha, 1982), saturation of the replication machinery requires dNTP substrate concentration in the range of 0.2 to 1.0 mM (Mathews and Sinha, 1982; Mathews, 1985, 1993a). Therefore, it is reasonable to expect that dNTP concentration gradients must be generated in the vicinity of replication forks. In addition, dNTPs are needed in the cytoplasm for other processes, such as DNA repair and genetic recombination. Thus, it appears that there may exist two kinetically distinct dNTP pools for different metabolic purposes (Mathews, 1993b). This functional compartmentation may be represented as (a) a small, rapidly turning over, concentrated pool which supplies precursors specifically at replication sites for DNA replication and (b) a large, slow turning over, dilute cytoplasmic pool which supplies precursors for DNA repair and recombination. Again, as with ATP and O_2, the primary determinant of pool separation is functional rather than physical in nature.

The issue regarding DNA precursor heterogeneities and the functional implications of this inhomogeneity in eukaryotic cells is more complicated. Studies from several laboratories in the 1980s have demonstrated specific channeling of DNA precursor to replication sites in eukaryotes (Reddy, 1988; Kuebbing and Werner, 1975). Particularly, channeling of exogenous precursors for DNA replication appears to be preferred over endogenously synthesized substrates (Mathews and Slabaugh, 1986; Reddy, 1988). In general, the concept of metabolism channeling is consistent with the existence of multienzyme complexes in cells (Srere, 1987; Ovádi and Srere, this volume). However, the precise nature of this compartmentation has not been defined. An experimental approach to address DNA precursor compartmentation was by direct comparison of the turnover of the dTTP pool with its incorporation into DNA. Some investigators found that in cells like lymphocytes, dTTP turnover rate equals that for DNA incorporation (Taheri *et al.,* 1982), while others show that in cells like mouse fibroblast, dCTP turnover occurred at only a fraction of the rate of DNA synthesis (Nicander and Reichard, 1983, 1985). Moreover, in these latter studies, the rate of DNA synthesis in mouse fibroblasts equals that of dTTP turnover but was only one-third that of dCTP turnover (Nicarder and Reichard, 1985). Taken together, these results suggest that distinct metabolic DNA precursor pools can exist in mammalian cells, but the extent of inhomogeneities appears to be both cell- and nucleotide-specific.

Although less known and somewhat deviant from the consideration of intracellular compartmentation, the concept of temporal compartmentation warrants some mention. This notion was introduced in an earlier review by Moyer and Henderson (1985) based on the studies of Hordern and Henderson on cell-cycle specific changes in nucleotide-metabolizing enzymes in CHO cells (Hordern and Henderson, 1982). The term "temporal compartmentation" was later coined by Mathews and associates to make the distinction between intercellular, rather than intracellular, compartmentation. The basic concept is that selective exogenous ribonucleotide or deoxyribonucleotide pools are utilized for DNA synthesis, depending on whether cells in cultures, like CHO, are synchronized or not with respect to the replicative S or the quiescent G_1 phases (Leeds and Mathews, 1987). Whether DNA replication in different cell types is universally controlled by temporal compartmentation of DNA precursors is not known.

2. Functional Compartmentation of Carbohydrate Metabolism in Muscle

In contrast to the above treatise on specific metabolite gradients (Sections III and IV), the ensuing discussion deals with a more general subject of colocalization of enzyme or intermediary metabolic cascades for energy transformation with specific energy-dependent processes. One of the best-studied cell type with regard to functional compartmentation of carbohydrate metabolism is the vascular smooth muscle (Lynch and Paul, 1988; Paul *et al.,* 1989). Because of the large and rapid changes in energy demand in association with changes in the contractile state of the tissue, the smooth muscle is especially suited for studies of functional energy requirement with ATP synthesis by intermediary metabolism. Moreover, manipulation of smooth muscle cell energetics is relatively easy given the fact the oxidative and glycolytic pathways of intermediary metabolism can experimentally be altered independently and in opposing directions (Paul, 1983).

Early investigations show that function of the plasmalemma-associated Na^+, K^+-ATPase is correlated with ATP generated by aerobic glycolysis while actin–myosin interaction in the muscle fiber derived its ATP from mitochondrial oxidative phosphorylation (Paul, 1983). This finding suggests a physical separation of the Embden-Meyerhof pathway that is imposed by metabolic demand. This suggestion is substantiated by detection of two separate pools of glucose 6-phosphate, a common intermediate in glycolysis and glycogenolysis (Lynch and Paul, 1986). Using selective and independent manipulation of glycolytic and glycogenolytic activities as well as analyses of tracer kinetics using radiolabeled glucose, Lynch and Paul (1986) was able to show that exogenous glucose was the precursor of the membrane-associated glucose 6-phosphate pool for ATPase function, while the glucosyl moieties derived from glycogenolysis were the primary oxidative substrate for mitochondrial respiration for tension generation. Thus, the lack of mixing of intermediary metabolites and the physical separation of the two pathways strongly argue for cytoplasmic compartmentation independent of membrane barrier. Other investigators have also demonstrated a colocalization of glycolytic enzymes with plasma membrane in other cell types (Green *et al.,* 1965; Schrier, 1966; Paul *et al.,* 1989).

How this defined compartmentation is achieved and what advantages organizing enzyme clusters to biological interfaces and maintaining separate and distinct substrate pools would have for regulation of metabolism remain unclear. In this consideration, perhaps a more fundamental question to ask is whether functional compartmentation is selectively advantageous or a result of necessity. Either view has its merit. One could easily argue that the association of metabolic pathways or specific enzymes with structured cell components occurs by necessity in response to an imposed need. A good example is the partitioning of hexokinase during different glycolytic fluxes between the soluble and particulate fractions to allow for maximal regulation of hexokinase activity (Wilson, 1980). On the other hand, it is undeniable that functional compartmentation of metabolism provides a kinetic advantage wherein proximity of substrate pools for enzyme function minimizes diffusion limitations (Ovádi and Srere, this volume). The above discussion on juxtaposition of ATP production to ATP utilization systems as well as substrate channeling within multienzyme complexes provide clear illustrations of this fact. As the two adaptive responses are most likely to be inextricably linked, it is perhaps not as critical to resolve which mechanism predominates. Rather, in terms of understanding cellular regulation and homeostasis, it is more important to appreciate that metabolic compartmentation endows cells with a distinct edge for functional efficiency during normal metabolism and for optimal responses to enhanced metabolic stress.

V. Regional Compartmentation of Ions in Aqueous Cytoplasm in Mammalian Cells

A. Microzonation of pH in Renal Tubule Cells

1. Determinants of Regional pH Differences

The kidney plays a pivotal role in organismic acid–base balance and, because of the well-defined polarized distribution of mitochondria to the basal region of the proximal tubule cell, it offers an ideal model for studying regional difference in pH within mammalian cells. In this cell, transcellular movement of acid and base equivalents is effected by the function of the pH-controlling systems that are localized to the opposite poles of the cell (Boron, 1986). At the brush border membrane, the pH-sensitive Na^+/H^+ exchanger provides the driving force for luminal extrusion of H^+, while basal extrusion of HCO_3^- regulates base equivalents in the cell (Fig. 7). Thus, the coordinate function of H^+-removing and H^+-yielding systems controls the intracellular pH and contributes to regional differences in pH.

The polarized distribution of the mitochondria to the basal region plays an important role in the local generation of H^+ in that continuous pumping of H^+ into the narrow cytoplasmic space surrounding the mitochondria occurs in association with the chemiosmotic coupling of oxidative phosphorylation. Moreover, the high basal ATPase activity causes significant hydrolysis of ATP at physiological pH with concurrent H^+ release, which contributed to acidification of the basal region of the cell. Consequently, the immediate vicinity of the mitochondrial clusters in this basal region is expected to experience a lower pH than the rest of the bulk cytoplasmic

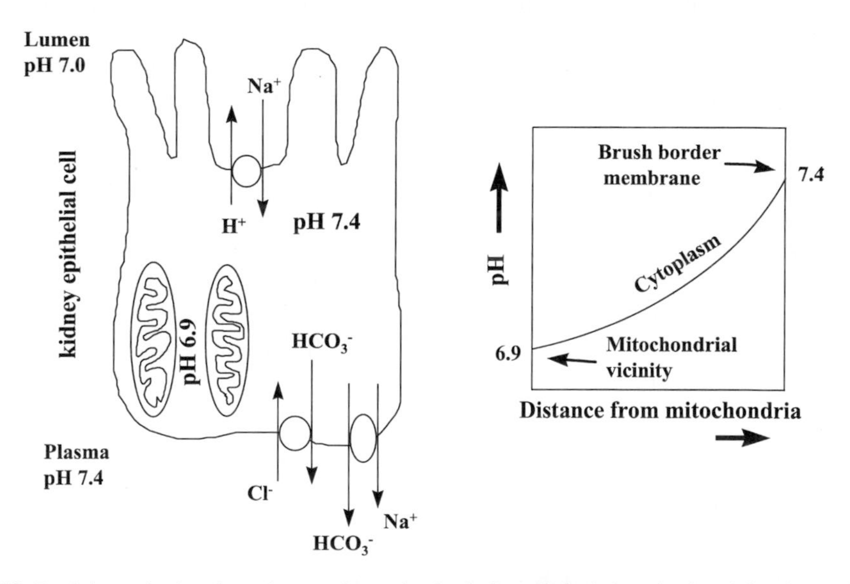

FIG. 7 Schematic drawing of a renal proximal tubule cell depicting the brush-border and basal membrane transport systems that function to regulate intracellular pH. The indicated pH are experimentally determined values for matrix and cytoplasmic compartments and for mitochondrial transmembranal ΔpH. The graph shows a pH gradient of 0.5 units from the microenvironment surrounding the mitochondria in the basal region of the cell (pH 6.9) to the bulk cytoplasmic mileu and the brush-border membrane (pH 7.4).

phase in the cell (Fig. 7). In earlier studies, detection of pH inhomogeneities in microenvironments was not possible using conventional techniques for measurements of average intracellular pH by microelectrode, pH sensitive dyes, ^{31}P-NMR, and distribution of weak acids. However, our development of a functional approach (see the following) has allowed the determination of heterogeneities within the kidney proximal tubule cells. Conceptually, this experimental approach to regional pH determination is analogous to the mapping of intracellular O_2 gradients using mitochondrial cytochromes as O_2 sensors and the mapping of intracellular ATP gradients using ATP-dependent enzymes as endogenous probes for ATP availability (Jones and Aw, 1990).

2. Detection and Experimental Determinations of pH Gradients and Proton Fluxes

The basic principle for development of the experimental approach for detection of pH gradients capitalizes on three known facts, namely, the average cellular pH of renal cells is around 7.2 to 7.5 (Radda *et al.,* 1980; Kleinman *et al.,* 1980; Chaillet *et al.,* 1985), kidney mitochondria are uniquely localized to the basal region of the cell (Sjöstrand and Rhodin, 1953), and the mitochondrial transmembranal ion accumulation requires a ΔpH with a lower pH on the outside of the mitochondria (LaNoue and Schoolwerth, 1984; Schoolwerth and LaNoue, 1985). Taken together, these observations predict that the pH in the narrow region surrounding the mitochondria must necessarily be lower than the rest of the cytoplasm (Fig. 7). The determination of mean cytosolic pH, the transmembranal ΔpH, and the mitochondrial matrix pH will allow for an estimate of the pH in the immediate vicinity of the mitochondria (Aw *et al.,* 1987a; Jones *et al.,* 1987). Furthermore, comparison of the value to the average pH in the bulk cytosolic phase provides a reasonable measure of the magnitude of the pH gradient from the cytoplasm to the mitochondrial cluster (see Fig. 7).

Experimentally, cytosolic and matrix pH can be determined by the distribution of weak acids, such as dimethadione (DMO), between the cytoplasmic and mitochondrial compartments (Hoek *et al.,* 1980; Anderson and Jones, 1985). A similar approach based on the matrix-to-cytosolic distribution of substrates, such as phosphate and pyruvate (Aw *et al.,* 1987a; Aw and Jones, 1989) whose mitochondrial matrix accumulation occurs by H^+-compensated electroneutral uptake transporters, can be used to determine transmembranal ΔpH (LaNoue and Schoolwerth, 1984). Studies from our and other laboratories show that in proximal tubule cells, the average cytosolic pH is 7.4, the matrix pH is 7.6 (Radda *et al.,* 1980; Kleinman *et al.,* 1980; Chaillet *et al.,* 1985; Aw and Jones, 1989), and the transmembrane ΔpH is at least in the order of 0.7 pH units (Aw and Jones, 1989).

Consequently, the pH in the cytoplasm in the narrow region surrounding the mitochondria is estimated to be around 6.9, or 0.5 pH units lower than the average cytosolic pH value. This means that a gradient of protons occurs between the bulk of the aqueous cytoplasmic compartment and the small zone of cytoplasmic space that is occupied by mitochondrial aggregates in the basal region of the cell (Fig. 7). Direct visualization of pH within single cells using pH sensitive dyes by video-enhanced methods substantiated the existence of pH heterogeneities in different regions of the cell in the same order of magnitude and distribution (Radda *et al.,* 1980; Balaban, 1982) as that obtained using the above functional approach.

At present, the quantitative diffusional fluxes of H^+ in cytoplasm of mammalian cells are unknown. In early studies, Robinson and Stokes (1976) estimated a diffusion coefficient for H^+ to be around 8–9 $\times 10^{-5}$ cm^2sec^{-1} in giant squid axoplasm assuming that axoplasmic H^+ diffusion is the same as that in water. Based on this estimate, a H^+ flux of 500 pmol $cm^{-2}sec^{-1}$ was calculated within the cell cytoplasm (Robinson and Stokes, 1976). In other studies in squid axon, other investigators show, not surprisingly, that pH gradients and H^+ fluxes can be dramatically altered by the buffering capacity of the cell cytoplasm (Junge and McLaughlin, 1987). Maximal buffering occurs with pK equaling the cell pH (Roos and Boron, 1981). While the squid axon is a relatively simple cell type, and the factors that control H^+ movement are likely to be different from one cell type to another, it seems reasonable that mammalian cells may be subjected to similar mathematical analysis for quantification of cytoplasmic H^+ fluxes under any given set of conditions.

3. Physiological Significance of Intracellular Regional pH Differences

How heterogeneities in pH within cell can be of physiological importance in regulation of cell function is not entirely clear. One interesting consideration is the possible role that a low pH microenvironment plays in the compartmentation of organelles or cellular macromolecules. For instance, if the average cytosolic pH in a cell is 7.4, and that ATP synthase activity and matrix accumulation of anions such as phosphate, pyruvate, and citrate require a transmembrane pH gradient in the order of at least 0.5 for oxidative phosphorylation, an extramitochondrial pH in the small region surrounding the mitochondria must be in the region of 7.0 or lower. Thus, mitochondria are likely to associate with one another in clusters such that the microzone in the vicinity of the mitochondria be kept at a pH lower than the remainder of the cytosol to support optimal mitochondrial function. Indeed, the mitochondrial arrangement in proximal tubule cells is consistent with this suggestion.

If pH does, in fact, favor clustering of mitochondria, it may explain the association of mitochondria to sites of high ATP demand. Because ATP hydrolysis generates H^+, regions of high ATPase activity should be expected to create microenvironments of low pH. Based on the above reasoning, it may be speculated that this low pH environment would promote mitochondrial aggregation and result in a region of high mitochondrial densities. While this hypothesis is highly speculative and remains to be experimentally tested, morphometric studies of many adult mammalian cell types appear to support such a contention.

B. Spatial Distribution of Calcium in Cell Cytoplasm

Given the fact that metabolite gradients play important regulatory roles in cells, it is reasonable to expect that ion gradients can similarly exist within cytoplasm of mammalian cells and contribute to cellular function and regulation. The challenge, however, lies not in the conceptual acceptance of a valid hypothesis, but, rather, in the experimental demonstration of microheterogeneities of different ions within cell cytoplasm. One of the better studied ions is Ca^{2+}, perhaps because of its central role in a variety of metabolic and biological processes, including cell signaling, plasma membrane electrical activity, cell regulation, and cell proliferation.

An earlier review by Abercrombie (1988) provides an excellent discussion of the diffusion of Ca^{2+} in free water and salt solutions *in vitro* (Wang, 1953). Within cell cytoplasm, the diffusion of Ca^{2+}ions is likely to differ from that in free solution. A structured cytoplasm, the interaction of Ca^{2+} with Ca^{2+}-binding proteins, and the sequestration of Ca^{2+} into organelles such as the mitochondria and endoplasmic reticulum can all contribute to decrease intracellular Ca^{2+} fluxes. Early studies by Hodgkin and Keynes (1957) in giant squid axon provided the first direct measurement of intracellular Ca^{2+} movement and Ca^{2+} gradients. Indeed, much of our current knowledge on Ca^{2+} inhomogeneities within cell cytoplasm is derived from studies with this cell type. Hodgkin and Keynes (1957) found that Ca^{2+} diffusion in the axoplasm was at least two-fold lower than that in water based on analyses of the Ca^{2+} distribution profile along the axon after axial injection of $^{45}Ca^{2+}$. The major impediment for diffusion of free Ca^{2+} was Ca^{2+} binding. It was estimated that the removal of $^{45}Ca^{2+}$ occurred at a rate of $\sim$0.1 sec^{-1} and that the diffusion coefficient was 0.3×10^{-6} cm^2sec^{-1} (Blaustein and Hodgkin, 1969). In other studies, Dipolo and co-workers (1976) used a Ca^{2+} sensitive probe, the protein aequorin, to demonstrate the presence of Ca^{2+} gradients from the surface membrane into the axoplasm. Moreover, Abercrombie and Hart (1986), using ruthenium red to inhibit mitochondrial Ca^{2+} uptake, further show that mitochondrial Ca^{2+} sequestration is an important determinant of the steepness of axoplasmic free Ca^{2+} gradients.

To date, demonstration of Ca^{2+} inhomogeneities within the cytoplasm of mammalian cells has come from studies using fluorescence digital imaging of intracellular Ca^{2+} concentrations in different regions of the cell cytoplasm. Ca^{2+}-sensitive fluorescent dyes, such as Quin-2 and Fura-2, have been used extensively by many investigators for experimental determination of intracellular Ca^{2+} concentrations in single cells (reviewed in Williford *et al.,* 1988). The first documentation of spatial heterogeneity of Ca^{2+} was reported by Williams and associates in isolated smooth muscle cells (Williams *et al.,* 1985; Williams and Fay, 1986). They found that nuclear and subsarcolemmal regions have higher Ca^{2+} concentration compared to other regions of the cytoplasm. Furthermore, the changes in Ca^{2+} levels in these compartments are not passively dependent on the cytoplasmic Ca^{2+}, but are regulated. These results are consistent with compartmentation of the free Ca^{2+} ion by membrane delineated compartments. Whether microinhomogeneities of Ca^{2+} occur within the cytoplasmic phase itself without membranal barrier remains unclear. However, it is expected that future development of new imaging tools, as well as novel functional approaches, should provide satisfactory resolution of this issue.

VI. Concluding Remarks

Intracellular asymmetry is a unique characteristic of adult mammalian cells. At a structural and organellar level, cytoplasmic inhomogeneity is best reflected in heterogeneous distribution of mitochondria, which often appear in clusters or distinct zones. At a functional level, mitochondrial clustering causes steep gradients of low molecular weight species like O_2, ATP, and pH, thereby resulting in metabolite inhomogeneities at different cytoplasmic sites without membranal delineation. Consequently, heterogeneity of substrate pools (e.g., ATP, DNA precursors, or glycolytic intermediates) or of ions (e.g., Ca^{2+}, H^+) will create spatially unique and defined functional microcompartments of cellular metabolism within the cytoplasm of cells. Although not widely appreciated, the establishment of microzones of metabolic activities in association with cellular organelles, membranal components, or a structured cytoplasm is likely to be integral to normal function and regulation of adult mammalian cells. There remains the burden of proof for future research endeavors.

Acknowledgments

Research in the author's laboratory was supported by grants from the National Institutes of Health GM-36538, GM-28176 (to D. P. Jones), and DK-44510, DK-43785 (to T. Y. Aw). T. Y. Aw is a recipient of an Established Investigatorship Award from the American Heart Association.

References

Abercrombie, R. F. (1988). Hydrogen and calcium ion diffusion in axoplasm. *In* "Microcompartmentation" (D. P. Jones, ed.), pp. 209–225. CRC Press, Boca Raton, FL.

Abercrombie, R. F., and Hart, C. E. (1986). Calcium and proton buffering and diffusion in isolated cytoplasm from *Myxicola* axons. *Am. J. Physiol.* **250,** C391–C405.

Aflalo, C., and DeLuca, M. (1987). Continuous monitoring of ATP in the microenvironment of immobilized enzymes by firefly luciferase. *Biochemistry* **26,** 3913–3920.

Andersson, B. S., and Jones, D. P. (1985). Use of digitonin fractionation to determine mitochondrial transmembrane ion distribution in cells during anoxia. *Anal. Biochem.* **146,** 164–172.

Aw, T. Y. (1991). Postnatal changes in pyridine nucleotides in rat hepatocytes: Composition and O_2 dependence. *Pediatr. Res.* **30,** 112–117.

Aw, T. Y., and Jones, D. P. (1985). ATP concentration gradients in cytosol of liver cells during hypoxia. *Am. J. Physiol.* **249,** C385–C392.

Aw, T. Y., and Jones, D. P. (1987). Respiratory characteristics of neonatal rat hepatocytes. *Pediatr. Res.* **21,** 492–496.

Aw, T. Y., and Jones, D. P. (1988a). Microzonation of ATP and pH in the aqueous cytoplasm of mammalian cells. *In* "Microcompartmentation" (D. P. Jones, ed.), pp. 367–377. CRC Press, Boca Raton, FL.

Aw, T. Y., and Jones, D. P. (1988b). Succinate and oxygen dependence of mitochondrial functions in neonatal hepatocytes. *In* "Integration of Mitochondrial Function" (J. J. Lemasters, C. R. Hackenbrock, R. G. Thurman, and H. V. Westerhoff, eds.), pp. 445–450. Plenum, New York.

Aw, T. Y., and Jones, D. P. (1989). Heterogeneity of pH in the aqueous cytoplasm of renal proximal tubule cells. *FASEB J.* **3,** 52–58.

Aw, T. Y., Andersson, B. S., and Jones, D. P. (1987a). Mitochondrial transmembrane ion distribution during anoxia. *Am. J. Physiol.* **252,** C356–C361.

Aw, T. Y., Wilson, E., Hagen, T.M., and Jones, D. P. (1987b). Determinants of mitochondrial O_2 dependence in kidney. *Am J. Physiol.* **253,** F440–F447.

Aw, T. Y., Tso, P., and Fuseler, J. W. (1992). Polarization of mitochondria in adult small intestine: Influence of nutritional status. *FASEB J.* **6,** A1840.

Balaban, R. S. (1982). Nuclear magnetic resonance studies of epithelial metabolism and function. *Fed. Proc., Fed. Am. Soc. Exp. Biol.* **41,** 42–47.

Balaban, R. S., and Bader, J. P. (1984). Studies on the relationship between glycolysis and (Na^++K^+)-ATPase in cultured cells. *Biochim. Biophys. Acta* **804,** 419–426.

Balaban, R. S., and Sylvia, A. L. (1981). Spectrophotometric monitoring of O_2 delivery to the exposed rat kidney. *Am. J. Physiol.* **241,** F257–F262.

Barry, W. H., Pober, J., Marsh, J. D., Frankel, S. R., and Smith, T. W. (1980). Effects of graded hypoxia on contraction of cultured chick embryo ventricular cells. *Am. J. Physiol.* **239,** H651–H657.

Bellomo, G., Nicotera, P., and Orrenius, S. (1984). Alterations in intracellular calcium compartmentation following inhibition of calcium efflux from isolated hepatocytes. *Eur. J. Biochem.* **144,** 19–23.

Bessman, S. P., and Carpenter, C. L. (1985). The creatine-creatine phosphate energy shuttle. *Annu. Rev. Biochem.* **54,** 831–862.

Blaustein, M. P., and Hodgkin, A. C. (1969). The effect of cyanide on the efflux of calcium from squid axons. *J. Physiol.* (*London*) **200,** 497–527.

Boag, J. W. (1969). Oxygen diffusion and oxygen depletion problems in radiobiology. *Curr. Top. Radiat. Res.* **5,** 141–195.

Boag, J. W. (1970). Cell respiration as a function of oxygen tension. *Int. J. Radiat. Biol.* **18,** 475–477.

Boron, W. F. (1986). Intracellular pH regulation in epithelial cell. *Annu. Rev. Physiol.* **48,** 377–388.

Bricknell, O. L., Davies, P. S., and Opie, L. H. (1981). A relationship between adenosine triphosphate, glycolysis and ischemic contracture in the isolated rat heart. *J. Mol. Cell. Cardiol.* **13,** 941–945.

Bulger, R. E., and Dobyan, D. C. (1982). Recent advances in renal morphology. *Annu. Rev. Physiol.* **44,** 147–149.

Chaillet, J. R., Lopes, A. G., and Boron, W. F. (1985). Basolateral Na-H exchange in the rabbit cortical collecting tubule. *J. Gen. Physiol.* **86,** 795–812.

Chance, B., Oshino, N., Sugano, T., and Mayevsky, A. (1973). Basic principles of tissue oxygen determination from mitochondrial signals. *Adv. Exp. Med. Biol.* **37A,** 277–292.

Chance, B., Sies, H., and Boveris, A. (1979). Hydroperoxide metabolism in mammalian organs. *Physiol. Rev.* **59,** 527–605.

Clark, A., Clark, P. A. A., Connett, R. J., Gayeski, T. E. J., and Honig, C. R. (1987). How large is the drop in PO_2 between cytosol and mitochondria? *Am. J. Physiol.* **252,** C583–C587.

Dawes, G. S., Mott, J. C., and Shelley, H. J. (1959). The importance of cardiac glycogen for the maintenance of life in foetal lambs and newborn animals during anoxia. *J. Physiol. (London)* **146,** 516.

DiPolo, R., Requena, J., Brinley, F. J., Jr., Mullins, L. J., Scarpa, A., and Tiffert, T. (1976). Ionized calcium concentrations in squid axons. *J. Gen. Physiol.* **67,** 433–467.

Entman, M. L., Kenichi, K., Goldstein, M., Nelson, T. E., Burnett, E. P., Futch, T. W., and Schwartz, A. (1976). Association of glycogenolysis with cardiac sarcoplasmic reticulum. *J. Biol. Chem.* **251,** 3140–3146.

Epstein, F. H., Balaban, R. S., and Ross, B. D. (1982). Redox state of cytochrome aa_3 in isolated perfused rat kidney. *Am. J. Physiol.* **243,** F356–F363.

Fawcett, D. W. (1981). "The Cell." Saunders, Philadelphia.

Garfinkel, D. (1963). A metabolic inhomogeneity of glycine in vivo. II. Computer simulation. *J. Biol. Chem.* **238,** 2435–2439.

Garfinkel, D., and Lajtha, A. (1963). A metabolic inhomogeneity of glycine in vivo. I. Experimental determination. *J. Biol. Chem.* **238,** 2429–2434.

Gayeski, T. E., and Honig, C. R. (1986). O_2 gradients from the sarcolemma to cell interior in a red muscle at maximal VO_2. *Am. J. Physiol.* **251,** H789–H799.

Gonzalez-Licea, A. (1970). Polarization of mitochondria in the absorptive cells of the small intestine of suckling rats. *Lab. Invest.* **23,** 163–167.

Gonzalez-Licea, A. (1971). A ultrastructural study of intestinal mitochondrial morphology during the absorption of various nutrients and water in suckling rats. *Lab. Invest.* **24,** 273–278.

Green, D. E., Murer, E., Hultin, H. O., Richardson, S. H., Salmon, B., Brierley, B. P., and Baum, H. (1965). Association of integrated metabolic pathways with membranes. I. Glycolytic enzymes of the red blood corpuscle and yeast. *Arch. Biochem. Biophys.* **112,** 635–647.

Heggeness, M. H., Simon, M., and Singer, S. J. (1978). Association of mitochondria with microtubules in cultured cells. *Proc. Natl. Acad. Sci. U.S.A.* **75,** 3863–3866.

Hirokawa, N. (1982). Cross-linker system between neurofilaments, microtubules, and membranous organelles in frog axons revealed by the quick-freeze, deep-etching method. *J. Cell Biol.* **94,** 129–142.

Hodgkin, A. L., and Keynes, R. D. (1957). Movements of labeled calcium in squid axons. *J. Physiol. (London)* **138,** 253–281.

Hoek, J. B., Nicholls, D. E., and Williamson, J. R. (1980). Determination of mitochondrial protomotive force in isolated hepatocytes. *J. Biol. Chem.* **255,** 1458–1464.

Hordern, J., and Henderson, J. F. (1982). Comparison of purine and pyrimidine metabolism in G1 and S phases of HeLa and Chinese hamster ovary cells. *Can. J. Biochem.* **60,** 422–433.

Huxley, V. H., and Kutchai, H. (1981). The effect of the red cell membrane and a diffusion boundary layer on the rate of oxygen uptake by human erythrocytes. *J. Physiol.* (*London*) **316,** 75–83.

Jones, D. P. (1984). Effect of mitochondrial clustering on O_2 supply in hepatocytes. *Am. J. Physiol.* **247,** C83–C89.

Jones, D. P. (1986). Intracellular diffusion gradients of O_2 and ATP. *Am. J. Physiol.* **250,** C663–C675.

Jones, D. P., ed. (1988) "Microcompartmentation". CRC Press, Boca Raton, FL.

Jones, D. P., and Aw, T. Y. (1988). Mitochondrial distribution and O_2 gradients in mammalian cells. *In* "Microcompartmentation" (D. P. Jones, ed.), pp. 37–53. CRC Press, Boca Raton, FL.

Jones, D. P., and Aw, T. Y. (1990). Use of enzymes and transport systems as in situ probes of metabolite and ion gradients. *In* "Structure and Organizational Aspects of Metabolic Regulation" (P. Srere, M. E. Jones, and C. Matthews, eds.), pp. 345–361. Alan R. Liss, New York.

Jones, D. P., and Kennedy, F. G. (1982a). Intracellular oxygen supply during hypoxia. *Am. J. Physiol.* **243,** C247–C253.

Jones, D. P., and Kennedy, F. G. (1982b). Intracellular O_2 gradients in cardiac myocytes. Lack of a role for myoglobin in facilitation of intracellular O_2 diffusion. *Biochem. Biophys. Res. Commun.* **105,** 419–424.

Jones, D. P., and Mason, H. S. (1978a). Gradients of O_2 concentration in hepatocytes. *J. Biol. Chem.* **253,** 4874–4880.

Jones, D. P., and Mason, H. S. (1978b). Metabolic hypoxia: Accumulation of tyrosine metabolites in hepatocytes at low pO_2. *Biochem. Biophys. Res. Commun.* **80,** 477–483.

Jones, D. P., Aw, T. Y., and Kennedy, F. G. (1983). Isolated hepatocytes as a model for the study of cellular hypoxia. In "Isolation, Characterization, and Use of Hepatocytes" (R. A. Harris, and N. W. Cornell, eds.), pp 323–332. Elsevier, New York.

Jones, D. P., Kennedy, F. G., and Aw, T. Y. (1987). Intracellular O_2 gradients and the distribution of mitochondria. *In* "Hypoxia: The Tolerable Limits" (C. Houston, and J. Sutton, eds.), pp. 59–69. Benchmark Press, Indianapolis, IN.

Jones, D. P., Aw, T. Y., and Sillau, A. H. (1990). Defining the resistance to oxygen transfer in tissue hypoxia. *Experientia* **46,** 1180–1185.

Jones, D. P., Aw, T. Y., Bai, C., and Sillau, A. H. (1991). Regulation of mitochondrial distribution: An adaptive response to changes in oxygen supply. *In* "Response and Adaptation to Hypoxia: Organ to Organelle" (S. Lahiri, and N. S. Cherniack, eds), pp. 25–35. Oxford University Press, London.

Junge, W., and McLaughlin, S. C. (1987). The role of fixed and mobile buffers in the kinetics of proton movement. *Biochim. Biophys. Acta* **890,** 1–5.

Jungermann, K. (1988). Metabolic zonation of carbohydrate metabolism in the liver. *In* "Integration of Mitochondrial Function" (J. J. Lemasters, C. R. Hackenbrock, R. G. Thurman, and H. V. Westerhoff, eds.), pp. 561–579. Plenum, New York.

Kennedy, F. G., and Jones, D. P. (1986). Oxygen dependence of mitochondrial function in isolated rat cardiac myocytes. *Am. J. Physiol.* **250,** C374–C383.

Kirkwood, S. P., Munn, E. A., and Brooks, G. A. (1986). Mitochondrial reticulum in limb skeletal muscle. *Am. J. Physiol.* **251,** C395–C402.

Kleinman, J. G., Brown, W. W., Ware, R. A., and Schwartz, J. H. (1980). Cell pH and acid transport in renal cortical tissue. *Am. J. Physiol.* **239,** F440–F444.

Krogh, A. (1919). The rate of diffusion of gases through animal tissues with some remarks on the coefficient of invasion. *J. Physiol.* (*London*) **52,** 391–415.

Kuebbing, D., and Werner, R. (1975). A model for compartmentation of *de novo* and salvage thymidine nucleotide pool in mammalian cells. *Proc. Natl. Acad. Sci. U.S.A.* **72,** 3333–3336.

LaNoue, K. F., and Schoolwerth, A. C. (1984). Metabolite transport in mammalian mitochondria. *In* "Bioenergetics" (L. Ernster, ed.), pp. 221–268. Elsevier, Amsterdam.

Leeds, J. M., and Mathews, C. K. (1987). Cell cycle-dependent effects on deoxyribonucleotide and DNA labeling by nucleoside precursors in mammalian cells. *Mol. Cell. Biol* **7,** 532–534.

Lipton, P., and Robacker, K. (1983). Glycolysis and brain function: $[K^+]_0$ stimulation of protein synthesis and K_+ uptake require glycolysis. *Fed. Proc. Fed. Am. Soc. Exp. Biol.* **42,** 2875–2880.

Lynch, R. M., and Paul, R. J. (1983). Compartmentation of glycolytic and glycogenolytic metabolism in vascular smooth muscle. *Science* **222,** 1344–1346.

Lynch, R. M., and Paul, R. J. (1986). Compartmentation of carbohydrate metabolism in vascular smooth muscle: Evidence for at least two functionally independent pools of glucose-6-phosphate. *Biochim. Biophys. Acta* **887,** 315–318.

Lynch, R. M., and Paul, R. J. (1988). Functional compartmentation of carbohydrate metabolism. *In* "Microcompartmentation" (D. P., Jones ed.), PP 17–35. CRC Press, Boca Raton, FL.

Manwaring, J. D., and Fuchs, J. A. (1979). Relationship between deoxyribonucleoside triphosphate pools and DNA synthesis in an *nrdA* mutant of *Escherichia coli. J. Bacteriol.* **138,** 245–248.

Mathews, C. K. (1972). Biochemistry of DNA-defective mutants of bacteriophage T4. III. Nucleotide pools. *J. Biol. Chem.* **247,** 7430–7438.

Mathews, C. K. (1985). Enzymatic channeling of DNA precursors. *Basic Life Sci.* **31,** 47–66.

Mathews, C. K. (1993a). The cell—bag of enzymes or network of channels? *J. Bacteriol.* **175,** 6377–6381.

Mathews, C. K. (1993b). Enzyme organization in DNA precursor biosynthesis. *Prog. Nucleic Acid Res. Mol. Biol.* **44,** 167–203.

Mathews, C. K., and Sinha, N. K. (1982). Are DNA precursors concentrated at replication sites? *Proc. Natl. Acad. Sci. U.S.A.* **79,** 302–306.

Mathews, C. K., and Slabaugh, M. B. (1986). Eukaryotic DNA metabolism. Are deoxyribonucleotides channeled to replication sites? *Exp. Cell Res.* **162,** 285–295.

Moyer, J. D., and Henderson, J. F. (1985). Compartmentation of intracellular nucleotides in mammalian cells. *CRC Crit. Rev. Biochem.* **19,** 45–61.

Nicander, B., and Reichard, P. (1983). Dynamics of pyrimidine deoxynucleoside triphosphate pools in relationship to DNA synthesis in 3T6 mouse fibroblasts. *Proc. Natl. Acad. Sci. U.S.A.* **80,** 1347–1351.

Nicander, B., and Reichard, P. (1985). Evidence for the involvement of substrate cycles in the regulation of deoxyribonucleoside triphosphate pools in 3T6 cells. *J. Biol. Chem.* **260,** 9216–9222.

Oshino, N., Sugano, T., Oshino, R., and Chance, B. (1974). Mitochondrial function under hypoxic conditions: The steady states of cytochrome $a+a_3$ and their relation to mitochondrial energy states. *Biochim. Biophys. Acta.* **368,** 298–310.

Papasozomenos, S. C., Yoon, M., Crane, R., Autilio-Gambetti, L., and Gambetti, P. (1982). Redistribution of proteins of fast axonal transport following administration of β,β'-iminodipropionitrile: A quantitative autoradiographic study. *J. Cell Biol.* **95,** 672–675.

Paul, R. J. (1983). Functional compartmentation of oxidative and glycolytic metabolism in vascular smooth muscle. *Am. J. Physiol.* **244,** C399–C409.

Paul, R. J., Bauer, M., and Pease, W. (1979). Vascular smooth muscle: Aerobic glycolysis linked to Na-K transport processes. *Science* **206,** 1414–1416.

Paul, R. J., Hardin, C. D., Raeymaekers, L., Wuytack, F., and Casteels, R. (1989). Preferential support of Ca^{2+} uptake in smooth muscle plasma membrane vesicles by an endogenous glycolytic cascade. *FASEB J.* **3,** 2298–2301.

Radda, G. K., Ackerman, J. J. H., Bore, P., Sehr, P., and Wong, G. G. (1980). ^{31}P-NMR studies on kidney intracellular pH in acute renal acidosis. *Int. J. Biochem.* **12,** 277–281.

Reddy, G. P. V. (1988). Functional compartmentation of deoxynucleotides for nuclear DNA replication in S phase mammalian cells. *FASEB J.* **2,** A1003.

Robinson, R. A., and Stokes, R. H. (1976). Electrolyte solution. *J. Gen Physiol.* **67,** 465.

Roos, A., and Boron, W. F. (1981). Intracellular pH. *Physiol. Rev.* **61,** 296–434.

Schoolwerth, A. C., and LaNoue, K. F. (1985). Transport of metabolic substrates in renal mitochondria. *Annu. Rev. Physiol.* **47,** 143–171.

Schrier, S. L. (1966). Organization of enzymes in human erythrocyte membranes. *Am. J. Physiol.* **210,** 139–145.

Sies, H., Bucher, T., Oshino, N., and Chance, B. (1973). Heme occupancy of catalase in hemoglobin-free perfused rat liver and of isolated rat liver catalase. *Arch. Biochem. Biophys.* **154,** 106–116.

Sillau, A. H., Aw, T. Y., and Jones, D. P. (1988). O_2 dependence of cytochrome *c* oxidation in hepatocytes isolated from hypoxic rats. *Physiologist* **31,**(4) A146.

Sjöstrand, F. S., and Rhodin, J. (1953). The ultrastructure of the proximal convoluted tubules of the mouse kidney as revealed by high resolution electron microscopy. *Exp. Cell Res.* **4,** 426–456.

Srere, P. A. (1987). Complexes of sequential metabolic enzymes. *Annu. Rev. Biochem.* **56,** 89–124.

Taheri, M. R., Wickremasinghe, R. G., and Hoffbrand, A. V. (1982). Functional compartmentation of DNA precursors in human leukaemoblastoid cell lines. *Br. J. Haematol.* **52,** 401–409.

Tamura, M., Oshino, N., Chance, B., and Silver, I. A. (1978). Optical measurements of intracellular oxygen concentration of rat heart *in vitro. Arch. Biochem. Biophys.* **191,** 8–22.

Wang, J. H. (1953). Tracer-diffusion in liquids. IV. Self-diffusion of calcium ion and chloride ion in aqueous calcium chloride solution. *J. Am. Chem. Soc.* **75,** 1769.

Werner, R. (1971). Nature of DNA precursors. *Nature (London), New Biol.* **233,** 99–103.

Williams, D. A., and Fay, F. S. (1986). Calcium transients and resting levels in isolated smooth muscle cells as monitored with quin-2. *Am. J. Physiol.* **250,** C779.

Williams, D. A., Forgarty, K. E., Tsien, R. Y., and Fay, F. S. (1985). Calcium gradients in single smooth cells revealed by the digital imaging microscope using fura-2. *Nature (London)* **318,** 558–561.

Williamson, J. R., and Cooper, R. H. (1980). Regulation of the citric acid cycle in mammalian systems. *FEBS Lett., Suppl.* **117,** K73–K85.

Williford, D. J., Walton, M. K., and Shen, S.-S. (1988). Fluorescence digital imaging microscopy-Spatial distribution of Ca^{2+} and H^{+} in single cells. *In* "Microcompartmentation" (D. P. Jones, ed.), pp. 227–249. CRC Press, Boca Raton, FL.

Wilson, J. E. (1980). Brain hexokinase, the prototype ambiquitous enzyme. *Curr. Top. Cell Regul.* **16,** 1–54.

Wittenberg, B. A., and Robinson, T. F. (1981). Oxygen requirements, morphology, cell coat and membrane permeability of calcium tolerant myocytes from hearts of adult rats. *Cell Tissue Res.* **216,** 231–251.

Macromolecular Compartmentation and Channeling

Judit Ovádi and Paul A. Srere†
Institute of Enzymology, Biological Research Center, Hungarian Academy of Sciences, Budapest, Hungary; †Department of Biochemistry, Veterans Affairs Medical Center, Dallas, Texas 75216

One of the accepted characterizations of the living state is that it is complex to an extraordinary degree. Since our current understanding of the living condition is minimal and fragmentary, it is not surprising that our first descriptions are simplistic. However, in certain areas of metabolism, especially those that have been amenable to experimentation for the longest period of time, the simplistic explanations have been the most difficult to revise. For example, current texts of general biochemistry still view metabolism as occurring by a series of independent enzymes dispersed in a uniform aqueous environment.

This notion has been shown to be deeply flawed by both experimental and theoretical considerations. Thus, there is ample evidence that, in many metabolic pathways, specific interactions between sequential enzymes occur as static and/or dynamic complexes. In addition, reversible interactions of enzymes with structural proteins and membranes is a common occurrence.

The interactions of enzymes give rise to a higher level of complexity that must be accounted for when one wishes to understand the regulation of metabolism. One of the phenomena that occurs because of sequential enzyme interactions is the process of channeling. This article discusses enzyme interactions and channeling and summarizes experimental and theoretical results from a few well-studied examples.

KEY WORDS: Metabolism, Channeling, Enzyme–enzyme interactions, Ambiquity, Enzyme structure protein interactions.

I. Introduction

Metabolism is the result of enzyme-catalyzed chemical reactions that, in sequence, break down the major nutrients, proteins, fats, and polysaccharides into small molecules which then produce chemical energy, mainly as ATP. These small molecules are also utilized, via various pathways, as precursors of macromolecules which are needed for the structure, function, and growth of the cell. To achieve an efficient regulated (homeostatic) system, a coordinated communication must exist between the metabolic processes. It has become apparent that, in addition to modulating the rates of individual enzyme steps to achieve regulation, physical separation of some pathways is required. One physical method employed by eukaryotic cells is that of membrane-bounded compartments in the form of organelles. These compartments contain specific sets of proteins specialized for particular functions. This kind of structural organization enables both synthetic and degradative processes to proceed simultaneously within the same cell but in different compartments, e.g., fatty acid synthesis takes place in the cytosol while fatty acid degradation occurs within the mitochondria.

There are many cases, however, in which both the synthetic and degradative processes occur in the same cellular compartment. For example, both the degradation and the synthesis of glucose proceed in the cytosol. The two processes are not completely the reverse of each other, in that several enzymes of gluconeogenesis bypass irreversible reaction steps in glycolysis. This is achieved by substituting phosphatases and carboxylases to catalyze the reverse reaction of the kinases. These enzymes are usually inversely regulated via "classical" regulatory mechanisms, like phosphorylation, allosteric activation/inhibition by certain metabolites, e.g., fructose 2,6-bisphosphate regulation of phosphofructokinase (PFK). Nevertheless, there are compelling observations which indicate a high degree of cellular organization within single compartments, which ensures the compartmentation of glycolytic intermediates in the cytosol of hepatocytes (Berry *et al.*, 1987, 1988) or in the cytosol of smooth muscle (Hardin and Kushmerick, 1994; Hardin and Robers, 1995). The nonhomogeneous distribution of the metabolic enzymes within a single cellular compartment is now accepted by most of the scientific community and is probably crucial for well-regulated cell functions. In this chapter, some new aspects of metabolic enzyme organization will be considered, with special emphasis on the relationship between organized structure and metabolic functions.

It is clear that each of the myriad of reactions that occurs in a cell affects many other reactions in the cell. Further, when one examines the metabolic systems composed of the individual enzymes, which also mutually sense and affect one another, one observes multiple interactions. Thus, the separations of these pathways are to a large extent artificial.

In an analogous manner, the sections of this paper are arbitrary. Topics overlap and interact, and many could have just as well been presented in another section.

II. Enzyme Interactions

A. Stable and Dynamic Complexes

There is a group of enzymes in which multiple enzyme activities occur on a single polypeptide chain; they may also have multiple activities on two different but interacting polypeptide chains. The chief interaction between active sites in such a group is clearly through their covalent linkage in the polypeptide chain. Such enzymes are multifunctional proteins and, when they catalyze a series of metabolic reactions, the metabolic intermediates are often covalently bound. An example is the fatty acid synthase of certain bacteria, yeast, and animals. Another group of multifunctional enzymes catalyzes nonsequential reactions, but we will not consider these here.

Heterologous enzyme associations are mainly stabilized by hydrophobic interactions, electrostatic forces, and hydrogen bonds. Depending on the strength of these "noncovalent" interactions, the macromolecular interactions can produce enzyme complexes of varying stability.

Static complexes of sequential enzymes are stable and time-invariant; they can be isolated in their complexed form since the enzymes remain bound to each other during extraction and purification procedures. These multienzyme complexes are a group of heteroenzyme associations since different functionally related enzymes are bound together by strong noncovalent forces into a highly organized structure. These static complexes require no energy to maintain structure and function. One often finds in these stable complexes that the intermediates are covalently bound, as in the case of the sequential multifunctional proteins. The α-keto acid dehydrogenase complexes are an example of this kind of static (stable) complex.

Dynamic heteroenzyme enzyme associations are stabilized by relatively weak forces. Therefore, these complexes are easily dissociated transient ones. These dynamic interactions are dependent on constant energy dissipation (Welch, 1977). Therefore, the terms "static" and "dynamic" are phenomenological descriptions of enzyme associations, and in nature there is a range of complexes with dissociation constants between the two extreme association states.

It is generally accepted that static enzyme complexes occur in highly processive pathways with no branches in the pathway, such as macromolecular biosyntheses and breakdown and nucleotide metabolism. Dynamic transitory complexes are seen in the amphibolic pathways, such as glycolysis and the citric acid cycle, which manifest numerous flow bifurcations along their metabolic route. At a metabolic crossroad where several enzymes compete for the substrate binding and conversion, the binding of one enzyme to another may direct the metabolic conversion of specific metabolites in one or other direction (Table I).

There are exceptions to this dogma, so that we must be careful with the generalization. As an example, in *Saccharomyces cerevisiae,* a bifunctional enzyme complex with carbamoyl phosphate synthetase and aspartate transcarbamoylase activities catalyzes the first two steps of the pyrimidine pathway. The time course of the coupled reactions catalyzed by the bifunctional enzyme showed virtually no lag phase in contrast to that measured with monofunctional enzymes (Hervé *et al.,* 1993). A more accurate analysis (Orosz and Ovádi, 1987) revealed the mechanism of the intermediate (carbamoyl phosphate) transfer and indicated that the channeling within this bifunctional enzyme complex was not perfect. A fraction of the intermediate diffused out of the complex (Penverne *et al.,* 1994). This structural arrangement of the active sites within the bifunctional enzyme complex can protect against rapid hydrolysis of carbamoyl phosphate. However, carbamoyl phosphate is also utilized by another metabolic pathway, the urea cycle. The efficiency of channeling may thus control flux of more than one pathway. Therefore, even in a multifunctional enzyme complex, the intermediate can escape from the "channel" and can have multiple conversions, as expected only in the cases of dynamic enzyme associations. This example illustrates that the stability of enzyme complexes does not determine unambiguously whether the intermediate has a single route or multiple conversions.

B. Sequential Enzyme Interaction

1. General Remarks

Specific interactions between sequential enzymes have been demonstrated for many different metabolic pathways. Most of the work in this area

TABLE I
Correlation between Organizational and Functional States of Metabolic Processes

Static (stable) systems
in anabolic (processive) pathways
one-way conversion
Dynamic (weakly interacting) systems
in amphibolic pathways
Multiple way for intermediate conversion
Direct metabolic flow in specific directions

has been previously reviewed (Ovádi, 1995; Srere, 1987). These pathways include nucleotide synthesis, Krebs urea cycle, Krebs TCA cycle, glycolysis, and the pentose phosphate shunt. Sequential metabolic complexes have been identified in bacteria, plants, and animals.

The techniques used to identify such complexes differ greatly. They include coprecipitation (Halper and Srere, 1977), fluorescent anisotropy (Tompa *et al.,* 1987), affinity chromatography (Persson and Srere, 1992), electrophoretic techniques (Beeckmans *et al.,* 1989), and countercurrent distribution procedures (Backman and Johansson, 1976). The specificity of interactions has been tested where possible by use of isozymes as controls. Other enzymes in the pathway one or two steps removed in the sequence are also used to assess specificity, and, finally, a series of unrelated proteins and enzymes can be used as controls.

Most studies are carried out with purified enzymes *in vitro*. For interactions with *large* dissociation constants, one often needs to use conditions which mimic the high protein concentrations in cellular compartments, including the addition of volume-excluding substances (see chapter by Johansson *et al.* in this volume) and low ionic strength solutions. The latter condition is considered to be "unphysiological," but one should remember that interactions studied *in vitro* are usually only between one or two enzymes, whereas *in vivo* multiple interactions which would enhance complex stability may well be the normal situation.

There is practically no experimental evidence to show that *in vivo* two sequential enzymes are spatially juxtaposed. This is an area in which sophisticated electron microscopic techniques ought to yield information. Channeling *in vivo* is the best, (though indirect) evidence currently available for this spatial contact between two sequential enzymes (see Section III).

2. Techniques for Detecting Interactions

It is not practical and would be redundant to cover all the techniques used to demonstrate sequential enzyme interactions for different pathways. However, for the purpose of illustration, the use of a variety of different techniques in such a study of the Krebs TCA cycle will be presented. Most of these techniques have also been applied with success to other metabolic pathways such as glycolysis, pentose phosphate shunt, amino acid synthesis, and nucleotide synthesis.

The interactions of sequential Krebs TCA cycle enzymes as measured with physical techniques will be discussed (Table II), starting with the malate dehydrogenase (MDH) and citrate synthase (CS) interaction and proceeding around the TCA cycle.

a. Physical Techniques Halper and Srere (1977) showed that pig CS and mitochondrial MDH (mMDH) precipitate in 14% poly(ethylene glycol)

TABLE II

Interaction between TCA Enzymes and TCA Related Enzymes Physical Evidence[a]

System	Technique	Interaction
CS+mMDH	precipitation/PEG	yes
CS+cMDH		no
CS+BSA		no
with pig and yeast enzymes		
CS+mMDH	anisotropy	yes
CS+mMDH	electrophoresis	yes
CS+mMDH	gel filtration	yes
CS+mMDH	pelleting	no
CS+mACO	preipitation/PEG	yes
mACO+NAD/ICDH		yes
NAD/ICD+KGDH	ultracentrifugation	yes
STK+KGDH	"	
FUM+mMDH	affinity column	
CS+FUM	affinity column	yes
CS+Thiolase	anisotropy	yes
CS+PDC	sedimentation	yes
CS+citrate transport	affinity chromatography	yes
mMDH+mAAT	countercurrent distribution (aqueous phase systems)	yes
mMDH+cAAT		no
cMDH+mAAT		no
cMDH+cAAT		small
mAAT+PC	Precipitation/PEG	yes
PC+KGDH	"	yes
GluDH+KGDH	"	yes/no
mAAT+mMDH	"	yes
PC+mMDH	"	yes
ternary and quateranary complexes	"	
TCA dehydrogenases +Complex I	pelleting	yes

[a] See text for references.

(PEG). Not cytosolic MDH (cMDH) nor bovine serum albumin nor many other proteins precipitate with CS in 14% PEG. This technique was also used to show that yeast enzymes, mitochondrial yeast CS (CS1), and mitochondrial yeast MDH (MDH1) behave in a similar manner.

The interaction between CS and mMDH has been demonstrated using a number of other physical techniques. Using fluorescently labeled proteins, it was shown that with CS and mMDH a change of anisotropy was observed. A K_{DISS} of 10^{-6} M was calculated using this technique (Tompa *et al.*, 1987). Beeckmans *et al.* (1989) used an affinity electrophoresis technique to show specific interaction between CS and mMDH. She also has immobilized CS or mMDH on Sepharose columns and has shown that the free enzyme of the pair bound to the immobilized enzyme (Beeckmans and Kanarek, 1981).

The technique of precipitation with PEG was employed to show specific interaction between CS and mitochondrial aconitase (mACO) and between mACO and NAD-specific isocitrate dehydrogenase (NAD-ICDH) (Tyiska *et al.,* 1986; Srere, 1987). Using one or more of these techniques outlined above, we know that NAD-ICDH interacts with α-KGDC (Porpaczy *et al.,* 1987), succinate thiokinase (STK) interacts with α-KGDC (Porpaczy *et al.,* 1983), and that FUM interacts with mMDH (Beeckmans *et al.,* 1989). Thus, six of the possible eight sequential interactions of the Krebs TCA cycle have been demonstrated. The remaining two sequential reactions are the putative interactions with succinate dehydrogenase (SDH). These two interactions, SDH with STK and SDH with FUM, have not been demonstrated with physical methods, but evidence from channeling experiments (see the following) indicates the presence of both interactions.

There are a number of other enzymes that are metabolically sequential to CS. Thiolase, the enzyme that produces acetyl-CoA from fatty acid oxidation, has been shown to interact with CS by anisotropic techniques (Sümegi *et al.,* 1985). Pyruvate dehydrogenase complex (PDC), which produces acetyl-CoA from glycolysis, has been shown to bind CS by centrifugation techniques (Sümegi and Alkonyi, 1983). It has also been shown that immobilized CS1 will bind mitochondrial citrate transporter, a membrane enzyme (Persson and Srere, 1992).

One of the first examples of a specific interaction of a Krebs TCA cycle enzyme was the demonstration by Backman and Johansson (1976) that mMDH and mitochondrial aspartate aminotransferase (mAAT) migrated together in a countercurrent distribution system, using immiscible polymer aqueous phases. This interaction was not observed using mMDH and cytosolic AAT (cAAT) or cMDH and mAAT. Some interaction was observed between cMDH and cAAT.

Fahien and Kmiotek (1979) have used the PEG precipitation technique to show specific interactions between auxiliary Krebs TCA cycle enzymes, such as AAT and pyruvate carboxylase (PC), and enzymes of the Krebs TCA cycle. He has demonstrated interactions between PC and α-ketoglutarate dehydrogenase complex (α-KGDC), between glutamate dehydrogenase (GDH) and α-KGDC, between mAAT and mMDH, and between PC and mMDH. Fahien and Chobanian (1997) have shown interactions with three or more auxiliary TCA cycle enzymes and TCA cycle enzymes.

There are several unexplained observations with regard to the specificity of CS interactions. First, it has been reported that fumarase (FUM) will bind to immobilized CS using affinity chromatography techniques. This interaction of nonsequential enzymes has not been tested with any other methodology as yet. The precipitation properties of FUM in PEG does not allow this method to be applied. Second, there is disagreement in the literature as to whether or not CS will bind to α-KGDC. Sümegi and Alkonyi (1983) reported that this nonsequential interaction does not occur while Fahien *et al.* (1989) reported that some interaction did occur. Third, a number of groups (unpublished results) have tested for the interaction of CS and mMDH using analytical ultracentrifugation. This method has not been successful with dynamically interacting systems. No interaction has been observed. These discrepancies and disagreements await further investigation.

There is evidence that dehydrogenases of the TCA cycle bind to Complex I of the inner membrane (Ovádi *et al.,* 1994; Sümegi and Srere, 1984b). All the enzymes of the Krebs TCA cycle have been shown to bind specifically to proteins of the matrix surface of the mitochondrial inner membrane (D'Souza and Srere, 1983). These interactions can be shown by incubating purified TCA cycle enzymes with inside-out inner membrane vesicles at low ionic strength and then collecting the vesicles by centrifugation. The enzymes are shown to sediment with the vesicles. In another approach, mitochondria were lightly sonicated to render them permeable to large molecules (Robinson and Srere, 1985). These preparations were shown to contain bound TCA cycle enzymes. This latter preparation was used to study coupled reactions of the Krebs TCA cycle.

A summary of the enzyme interactions and the techniques used to detect them appears in Table II.

b. Kinetic Techniques If a common metabolite of a sequential enzyme pair can be shown to be out of equilibrium with the bulk solution, that intermediate is said to be channeled (see Section III). One explanation for this phenomenon is that sequential enzymes are in close proximity, but other explanations are possible, depending on the system analyzed.

An early experiment co-immobilized CS and mMDH on Sepharose beads and entrapped them in acrylamide gels (Srere *et al.,* 1973). The immobilized enzyme couple had a shorter transient time than did the free enzymes. This indicates channeling, yet calculation showed that the average distance between enzyme molecules was quite large. Some other mechanism must have been operating, such as hindered diffusion of the intermediate due to unstirred layer effects.

In the permeabilized mitochondria containing bound CS and MDH, it was shown (Robinson *et al.,* 1987) that the coupled reaction was faster (3×) than that of equivalent free enzymes. Sümegi *et al.* (1992) have reported analogous results with other Krebs TCA cycle couples, such as fumarate oxidation which involves FUM, MDH, Complex I, and the electron transport chain. Again, enzyme interaction is a possible but not exclusive explanation. Ovádi *et al.* (1994) have shown channeling to occur between MDH and Complex I, and in this system, protein–protein interaction is the only possible explanation.

In another approach, Lindbladh *et al.* (1994a) have fused CS1 to MDH via a short amino acid linker by molecular biological techniques. The pure fusion protein showed channeling of oxalacetate (OAA) when a competing enzyme for OAA, AAT, was not as effective in trapping AAT from the fusion protein as it was when free CS1 and MDH1 was used (see Section III) (Lindbladh *et al.,* 1994b).

Using an isotope method, Sümegi *et al.* (1990) showed that when [2-^{13}C]propionate was given to respiring yeast cells, they observed a [2-^{13}C]alanine/[3-^{13}C]alanine ratio of greater than one. If the metabolism of propionate proceeds through succinate and fumarate of the Krebs TCA cycle, one would, classically, expect randomization of these two symmetrical intermediates, and the ratio of ^{13}C in the C2 and C3 of alanine should be one. One interpretation of a ratio greater than one is that the succinyl-CoA formed from propionate is converted to succinate by STK, but this intermediate is transferred directly to SDH with its orientation conserved. The fumarate formed from SDH is also passed to FUM with its orientation conserved. This would take special channeling of this intermediate. Similar results of orientation-conserved transfer can be seen with glutamate (Sümegi *et al.* 1993) in yeast. To observe this phenomenon, a short incubation period must be used, since multiple turns of the Krebs TCA cycle will result in randomization of the label, even with orientation-conserved transfer at each turn (Sherry *et al.,* 1994).

c. ***Genetic Techniques*** When enzymes of the Krebs TCA cycle are removed singly from yeast cells, the cells lose their ability to grow on acetate. Thus, if the gene for CS1 is disrupted, growth on acetate is lost, even though one might have expected that citrate made with the peroxisomal CS isozyme CS2 could provide adequate citrate for the mitochondrion and, thus, replace CS1 function. When cells devoid of CS1 are transformed with an inactive CS1, their ability to grow on acetate is partially restored. Thus, CS1 itself is necessary as well as its activity. This phenomenon was interpreted as support for the metabolon concept in that interactions between proteins to form supramolecular complexes yield a more efficient metabolic system with enhanced flux.

Hasselbeck and McAlister-Henn (1993) have observed that disruption of (NAD)ICDH1 results in an acetate$^-$ phenotype in spite of the presence in the mitochondrion of (NADP)ICDH. It should be recalled that (NADP)ICDH does not interact with ACO1 or α-KGDC and, thus, probably can not form a metabolon.

C. Enzyme-Structural Protein Interaction

The cytoplasm contains networks of the cytoskeleton composed of microtubules, actin filaments, and intermediate filaments. However, the ultrastructure of this network is more complex since many so-called "soluble enzymes" and nucleic acids transiently associate with these filaments and influence their structural and functional behavior (Fig. 1). The direct associations of enzymes plus indirect effects of multiple equilibria of enzyme subunits increase the complexity of cellular processes, which is probably a prerequisite for regulated optimal cell function (Fig. 2).

Association of glycolytic enzymes with the cytoskeletal network significantly contributes to the formation of the superstructure of cytomatrix. In muscle tissue, the enzymes are distributed in such a manner that high concentrations of glycolytic enzymes are found along the I band (Dolken

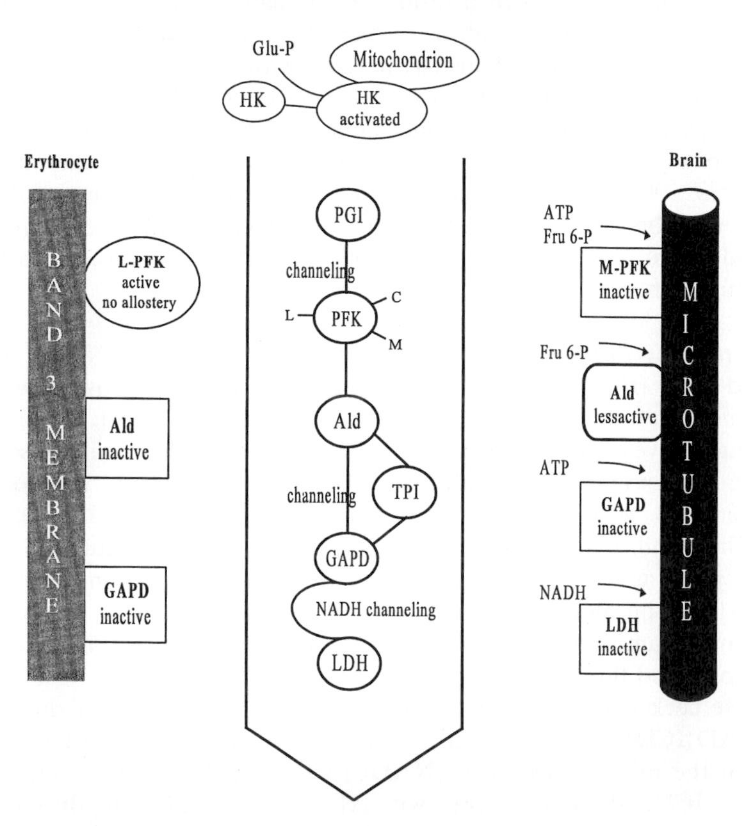

FIG. 1 Schematic for glycolytic enzyme interactions with structural proteins in erythrocytes and brain.

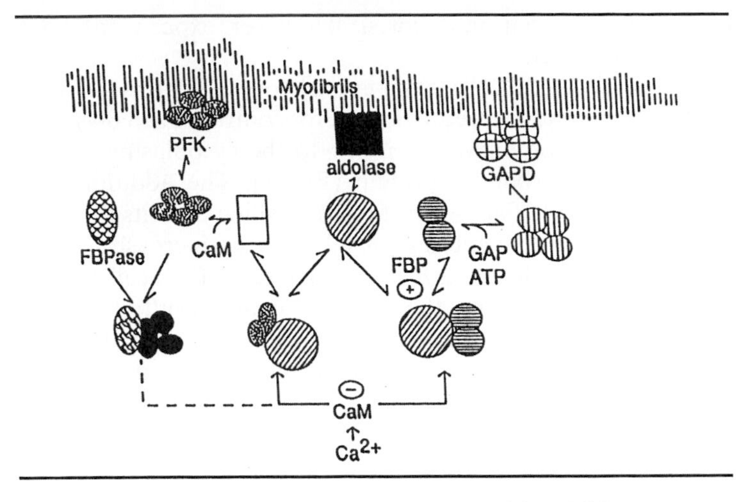

FIG. 2 Molecular model for the control of glycolysis. The activity of different enzyme species in reflected in the density of the symbols. From: Ovádi J. and Orosz F. in Current Topics in Cellular Regulation (E. R. Stadtman & P. Boon Chock, eds.), vol. 33, pp. 105–126, Academic Press, Inc., New York, 1992 with permission.

et al., 1975). Indeed, actin is one of the major cellular components in muscle that interacts with glycolytic enzymes (Masters, 1981; Clarke *et al.,* 1983). Distinct active sites and binding sites for PFK and glyceraldehyde-3-phosphate dehydrogenase on actin have been found (Humphreys *et al.,* 1986; Méjean *et al.,* 1989) (Fig. 2). Microcompartmentation of aldolase and glyceraldehyde-3-phosphate dehydrogenase was demonstrated also in cultured cell types (3T3 normal and transformed cells, fibroblast, embryo cardiomyocytes) (Minaschek *et al.,* 1992). The bound fraction of aldolase is released from I-actin *in vitro* when fructose 1,6-bisphosphate is added to a gelled aldolase F-actin mixture, demonstrating specificity and reversibility for the gelling activity. In brain, the microtubular system is of special importance. Transport of enzyme components along microtubule tracks in important in axons which comprise 99.9% of the neuronal volume and surface area. In *in vivo* experiments, some glycolytic enzymes, namely PFK and aldolase, have been identified in axonal transport as slow component b (Scb) with several other enzymes of intermediary metabolism which are organized as a discrete macromolecular cellular entity (Oblinger *et al.,* 1988). The glycolytic enzymes, glyceraldehyde-3-phosphate dehydrogenase and phosphofructokinase, produce cross links in tubules that alter their dynamics and ultrastructure. Glycolytic enzymes associate with microtubules or membranes depending on the cell type involved (Fig. 1). One would predict, on the basis of *in vitro* data, that a "surface enhanced" metabolic flux would occur in glycolysis. However, experimental data indicate the opposite effect.

Glycolysis in red blood cells is the energy source for this cell's function. Low *et al.,* (1993) provided direct evidence for control of glycolysis mediated by binding of glycolytic enzymes *in vivo* to the cytoplasmic extension of the anion transporter, band 3 protein (Fig. 1). The addition of specific antibodies (anti-peptide Fab) against band 3, which inhibits the association of the enzymes and the membrane, resulted in an increase in the rate of lactate production. On the other hand, enrichment of the cell with the band 3 peptide, that is, enrichment of the binding domain, resulted in a significant decrease in the rate of lactate accumulation. Therefore, the binding of the enzymes to a solid support decreases the glycolytic rate and does not produce a channeling effect.

D. Ambiquity

Ambiquitous enzymes have kinetically distinct subsets which partition between soluble and particle-bound forms in cells (Wilson, 1978, 1980). The association of the enzymes with particle is reversible and can be modulated by small ligands, other macromolecules, or changing conditions, e.g., by the metabolic state of the cell. This is true in most cases. However, the binding of enzymes to cellular surfaces (e.g., membrane, cytoskeleton) can induce structural changes, rendering the binding irreversible. The binding of PFK to microtubules shows typical ambiquitous behavior. The bound enzyme can be released specifically by ATP. If, however, bound enzymes cross-link tubules, the enzymes can no longer dissociate from the microtubules even in the presence of ATP. This appears to reflect a loss of ambiquity.

A number of scientists have hypothesized that the simultaneous associations of sequential enzymes on solid supports, e.g., cytoskeletal network and membranes, can enhance the direct transfer of a common intermediate, which could then favor increased metabolic flux just as channeling complex formation. According to theoretical predictions, the probability of intermediate transfer between active sites is higher in an organized structure. However, there is no experimental evidence so far to support this (Fig. 1).

It is apparent that ambiquity of "soluble" enzymes is an important aspect relating to their structure and function. In addition to an enzyme's ambiquitous behavior, other equilibria must be taken into account before assessing the physiological significance of the behavior.

E. Specific Characteristics of Heterologous Interactions

The dynamic associations or aggregations of enzymes are enhanced by macromolecular crowding (see chapter by Johansson *et al.,* this volume).

However, the macromolecular crowding effect itself can not promote *specific* enzyme interactions, since that requires biorecognition of the enzymes involved. Biorecognition arises as a result of surface complementarity, both spatial and electrostatic, of proteins. The specificity of enzyme associations is based upon this complementarity, which results in organized structures of physiological significance. The complementarity can be produced/diminished by the following mechanisms.

1. Conformational Changes of the Protein Induced by the Binding of Specific Ligands

This effect manifests itself by small conformational alterations. For example, the presence of fructose bisphosphate significantly increases the affinity of aldolase for glyceraldehyde-3-phosphate dehydrogenase, reflecting the essential role of protein conformation in specific associations (Vértessy *et al.,* 1991).

2. Direct or Indirect Competition of Ligands with Subcellular Structural Proteins for Enzyme Association

Aldolase binds to gelled F-actin of cell cytoplasm in a microdomain around stress fibers *in vivo* (Pagliaro *et al.,* 1989). An inhibitor of glycolysis, 2-deoxyglucose, specifically releases a significant fraction of the bound aldolase (Pagliaro and Taylor, 1992).

3. Dissociation or Association of Individual Oligomeric Enzymes

Different oligomeric forms of certain enzymes exhibit different affinities for the partner protein. If one oligomeric form specifically participates in macromolecular association and another oligomeric form exhibits different catalytic activities, the hetero association will result in altered overall activities due to a shift of equilibria between the oligomeric forms. Oligomer enzyme-form specific associations of muscle PFK to microtubules have been demonstrated. Since the dimeric form selectively binds to tubules, although it is inactive in both free and bound forms, the heterologous association results in a significant reduction in the overall activity of the kinase by increasing the amount of dimeric forms (Fig. 2).

4. Isoforms of an Enzyme with Different Associative Properties which Reside in the Same Cellular Compartment

If the association of enzymes alters the catalytic properties of an enzyme within the heterologous complex, then the specific complex formation can modulate enzyme activity. The association of PFK to microtubules is isoform-specific. The muscle enzyme (M type) associates with tubules, while the C-type isoform (which occurs in brain and tumor cells), and is a stable tetramer, shows little binding to tubules. Differences in intersubunit forces may, therefore, be primarily responsible for the selective associations of the isoforms of PFK and, thus, serve a function as a control mechanism in glycolysis.

5. Isoenzyme Specific Metabolite Channeling at a Metabolic Crossroad

This may be a means of sequestering intermediates from competing reactions by maintaining multiple pools of the same intermediate with isoenzyme–isoenzyme associations (Ureta, 1991). Different isoforms of hexokinase interact with sequential enzymes involved in glycogen metabolism, glycolysis, or the pentose phosphate pathway. The fate of a glucose 6-phosphate molecule is, thus, dependent on the specific interactions of the hexokinase isoforms.

III. Metabolic Channeling

A. General Concept

The rate of reactions in both chemical and biological systems depends on the encounter probabilities of the components. This probability is enhanced when the distance between the active sites of the sequential enzymes is reduced by juxtaposition of their respective active sites. This condition can be attained either by association of the sequential enzymes on subcellular particles or by association of the two enzymes catalyzing each consecutive reaction. In addition, linear and planar diffusion in two dimensions, compared to three-dimensional diffusion, results in significant increases in efficiency of transport of metabolites from one enzyme to another.

Channeling of metabolites in a reaction sequence is defined as the transfer of reaction product of one enzyme to the next enzyme in the metabolic sequence without the intermediate equilibrating with the bulk solution. The term "metabolic channeling" does not indicate a specific molecular mechanism but implies either stable or transient enzyme associations, resulting in dynamic microcompartmentation of metabolic intermediates.

A number of different conditions can alter the channel, which affects kinetic, functional, regulatory parameters. High local concentration of intermediates can occur due either to the physical barrier preventing the diffusion of intermediate into the bulk solution or to the close proximity of the active sites of sequential enzymes. A leaky channel is one in which a fraction of the intermediate leaks out, and a perfect channel one in which no escape of intermediate occurs.

Substrate channeling has been demonstrated in numerous systems (Agius and Sherratt, 1996) and in all parts of the cell and in microbes, plants, and animals. Compelling data exist on systems *in vivo, in situ,* and *in vitro*. Systems *in vivo* and *in situ* are important to demonstrate the existence, behavior, and biological relevance of substrate channeling. Studies *in vitro* are needed to clarify molecular mechanisms.

B. Molecular Mechanisms of Metabolite Channeling

1. The Intermediates Are Covalently Bound Sequentially to Active Sites

Pyruvate dehydrogenase complex is a multifunctional enzyme complex which catalyzes oxidative decarboxylation of pyruvate to acetyl-CoA, NADH, and CO_2. The overall reaction is carried out by three different enzymes (pyruvate dehydrogenase, dihydrolipoyl transacetylase, and dihydrolipoyl dehydrogenase), which form a tight multienzyme complex that is easily isolated as such by extraction of mitochondria. The covalently bound intermediates are transferred between individual active sites without leakage from the multienzyme assembly.

2. A Physical Barrier Tunnels the Intermediate between Active Sites

Tryptophan synthase catalyzes the biosynthesis of tryptophan from indol-3-glycerol phosphate and serine to tryptophan. In *Salmonella typhimurium,* the sequential reactions are catalyzed by a multienzyme complex ($\alpha_2\beta_2$) with indole as an enzyme bound intermediate. The three-dimensional crystal structure shows that each β monomer contains two structural domains of nearly equal size (Hyde *et al.,* 1988). The active site of each β monomer, which contains the bound coenzyme, pyridoxal phosphate, is sandwiched between these two domains. The structure of the enzyme complex shows that the two active centers in each α/β pair are 25 Å apart and are connected by a channel or "tunnel." This tunnel provides a pathway for the diffusion of indole from the active site of the α subunit to the active site of the β subunit. The action of tryptophan synthase complex shows that if the intermediate has just a single reaction fate in the cell, the perfect channeling of the intermediate can be fulfilled and that synthetic pathways are not necessarily the reverse of the degradative reaction. The physiological relevance of tunnel formation in this multienzyme complex is that the lipophilic indole molecule can not escape from the cell through its membranes. This could otherwise occur and necessitate the production of higher concentrations of indole to obtain the same flux to tryptophan.

3. Direct Site-to-Site Transfer of Noncovalently Bound Intermediates

Channeling of NADH between dehydrogenases of opposite chiral specificity appears to be one mechanism by which this common metabolite cofactor is transferred. This transfer occurs through formation of a ternary complex of the two enzymes and NADH (Srivastava and Bernhard, 1985). Direct transfer of NAD/NADH between alcohol dehydrogenase or lactate dehydrogenase and glyceraldehyde-3-phosphate dehydrogenase or glycerol phosphate dehydrogenase has been demonstrated (Srivastava and Bernhard, 1985; 1987). It is possible that, depending on enzyme chiral specificities, the rate of NADH transfer within enzyme complexes can be either substantially greater or less than the rate of coenzyme dissociation from the enzyme. Thus, enzyme–enzyme interactions, when occurring at relatively high physiological dehydrogenase concentrations, could modulate the rate of coenzyme transfer and control the direction of NADH conversion by the competition between dehydrogenases.

4. Transfer of Intermediates in an Unstirred Layer

Co-immobilization of enzymes catalyzing sequential metabolic reactions has been used as a technique to study proximity and microenvironmental effects on the coupled reaction rate. The analysis of the kinetics of surface-bound enzymes catalyzing coupled reactions showed that intermediates are converted to product with negligible loss of intermediate to the bulk phase. This was ascribed to an unstirred water layer surrounding the solid-state surface, which slowed the transport of the intermediate away from the surface (Goldman and Katchalski, 1971).

5. Electrostatic Channels—Covalent Coupling of Sequential Enzymes

To demonstrate this phenomenon, molecular biological techniques were used to produce fusion proteins of sequential enzymes (Bülow, 1987). This technique has been employed by Lindbladh *et al.* (1994b) to study Krebs TCA cycle enzymes. An in frame gene fusion between the C-terminus of CS1 and the N-terminal MDH1 was made and the fusion protein was expressed and transported into mitochondria of $CS1^-$ mutant yeast cells. The most remarkable observation was that when the fusion protein was assayed in the presence of a scavenger enzyme (AAT) for the OAA intermediate, it protected OAA when compared to the analogous system containing free CS1 and MDH1. A model of the crystal structure of pig heart CS and MDH as an end-to-end fusion protein showed a good spatial fit in the fusion protein, but the active sites of the two enzymes in the model were 60 Å apart. It was speculated that a positive "electrostatic channel" might exist between the two sites to facilitate the diffusion of the negatively charged OAA on the surface of the construct. This was confirmed by Brownian dynamic calculations of the electrostatic surface of the fusion protein (Elcock and McCammon, 1996). The calculations showed a continuous, positively charged surface between the two active sites and predicted a strong ionic strength effect on the channeling phenomenon. This prediction was recently found to hold (Morgunov and P. A. Srere, unpublished data). Thus, enzyme organization within a sequentially operating enzyme system can produce electrostatic pathways which sequester metabolites and enhance enzyme rates at relatively low bulk metabolite concentrations (Lindbladh *et al.,* 1992).

Another example of electrostatic effects in channeling is in the multienzyme complex which catalyzes nucleotide biosynthesis. It consists of a bifunctional enzyme containing both dihydrofolate reductase and thymidylate synthase activities. The crystal structure of this bifunctional enzyme from the protozoan *Leishmania major* showed that the enzymes are assembled around the thymidylate synthase core. The two intrasubunit active sites are located on the same side of the molecule and are separated by a distance (40 Å) which is too great for direct transfer of intermediate from one active site to the other. Direct transfer of folate intermediate is supported by kinetic data (Meek *et al.,* 1985), although the dihydrofolate reductase domain does not appear capable of flexing and, thus, bringing the two active sites closer together. These data, together with other experiments, suggest that the negatively charged folate travels on the strongly positive "electrostatic highway" across the surface that links the two active sites. Studies of the transient times for the reaction at different ionic strengths indicates the existence of an electrostatic channel (Trujillo *et al.,* 1996).

C. Consequences of Metabolic Channeling

The microscopic parameters (probability and lifetime) which characterize channeling can be converted into macroscopic parameters. These parameters (transient time, steady-state velocity) can be determined experimentally by analyzing the consecutive reactions catalyzed by sequential enzymes. The transient time (lag time) is a measure of the time required to establish a steady state in the concentrations of pathway intermediates. The decrease in transient time in interacting enzyme systems is characteristic of the extent of the channel. Analysis of this kind has been used extensively to investigate systems and identify channeling effects. This approach can be supplemented by an isotope dilution method with the appearance of labeled product after adding an external isotopic intermediate being a measure of the efficiency of the channel.

The catalytic advantages of channeling formation are summarized in Table III.

Most of the evidence for channeling cited above arises from *in vitro* studies, and many are from artificial biological constructs. The question remains as to whether or not channeling can be demonstrated *in vivo*. Is it possible to predict the occurrence of intermediate transfer on the basis of binding data of individual enzymes? To answer this question, systems need to be analyzed for which both binding and function data are available.

Another difference between *in vitro* and *in vivo* results is that *in vitro,* individual enzymes bind to distinct binding domains on the solid support while their binding is not simultaneous. Thus, the formation of metabolic channeling is not favored on the surface. Two glycolytic enzymes, PFK and aldolase, are involved in multiple equilibria. The individual enzymes bind to different domains of microtubules with PFK binding to the microtubule "body" and aldolase binding to the microtubule "tails." One would expect the simultaneous associations of these two enzymes to be possible. However, these enzymes form a hetero complex in solution which is unable to bind to microtubules. Therefore, no "surface-enhanced" metabolic transfer exists, although some authors (A. P. Minton, personal communication) predict a kinetic advantage for glycolysis by a surface-enhanced channeling effect.

It is difficult to envision the functional consequences of homo- and hetero-associations. It is known that the catalytic and allosteric properties of PFK

TABLE III

Catalytic Advantages of Metabolite Channeling

1. Prevention or impedance of a loss of intermediates by diffusion
2. Decreases the transit time required for an intermediate to reach the active site of the next enzyme
3. Reduces the transient time for the system to reach the new steady-state
4. Protects chemically labile intermediates
5. Circumvents unfavorable equilibria
6. Segregates the intermediates of competing chemical and enzymatic reactions

depend on its involvement in homologous and heterologous associations. The dimeric forms of PFK are involved in interactions with both aldolase and microtubules. However, the interaction of PFK with microtubules decreases its overall activity, while the formation of a complex with aldolase prevents the dissociation-induced inactivation of the kinase. Within an aldolase-PFK complex, aldolase is probably active because the aldolase stabilizes an "active" conformer of PFK which exists within the active tetrameric form (Vértessey *et al.,* 1997).

Another example of the difficulties encountered in applying *in vitro* studies to the *in vivo* situation comes from studies on the binding and kinetics of mitochondrial matrix enzymes to the mitochondrial membrane. The protein concentrations of the mitochondrial matrix are extremely high (nearly 600 mg protein/ml) (Srere, 1981). The binding of MDH to Complex I integrated into the membrane was characterized quantitatively and suggested that the enzyme had ambiquitous behavior *in vivo*. The specific binding of malate dehydrogenase to Complex I results in NADH channeling between the two active sites. The possible physiological advantage of this channeling formation is that direct transfer of NADH from mitochondrial matrix towards terminal oxidation can enhance both Krebs cycle flux and electron transport rates (Ovádi *et al.,* 1994). However, malate dehydrogenase can associate with other mitochondrial matrix enzymes, and these interactions may compete with MDH for binding to Complex I. Measurements with concentrated mitochondrial extract suggested that a significant amount of MDH was released from the membrane upon the addition of other matrix enzymes so that the direct transfer of NADH channeling was prevented (J. Ovádi and H. O. Spivey, unpublished results).

IV. Cross-Linking Processes

While the cytoskeleton presents major ultrastructural sites for interaction with glycolytic enzymes, ribosomes and polyribosomes have been implicated as cytoplasmic sites for the organization of enzymes involved in translation of nucleic acids (Spirin and Ajtkozhin, 1985; Ryazanov *et al.,* 1988). In the cytoplasm of eukaryotic cells, a fraction of the proteins (RNA-binding proteins) have an affinity for long-chain polynucleotides. This fraction contains the proteins of the translation machinery such as elongation and initiation factors and amino acyl-tRNA synthetases (Spirin and Ajtkozhin, 1985).

The protein biosynthetic machinery consists of a large number of protein and nucleic acid components. Extensive experimental data suggested a channeling mechanism for the transfer of tRNA in the mammalian translation system (Negrutskii and Deutscher, 1991). Exogenous aminoacyl-tRNAs are not able to enter the channel leading to protein synthesis. Only exogenous free amino acids were incorporated into protein. Many metabolic events were found to be limited when the tRNA is incompletely charged (e.g., in the case of amino acid deficiency). A rapid demise of cell function was suggested to be a consequence of the inhibition of PFK by free uncharged tRNA. This direct effect of tRNA on PFK activity was demonstrated in an *in vitro* system. However, a relation between this enzyme and its product, fructose bisphosphate, at the initiation stage of protein synthesis, has been shown in cell lysates. Stimulation of protein synthesis by fructose bisphosphate or the inhibition of protein synthesis by uncharged tRNA function as control mechanisms. A metabolic "crossover" point at the level of PFK has been determined in a variety of situations. Starvation, alloxan diabetes (equivalent to insulin withdrawal), or exposure to fatty acids leads to a decrease in glycolysis, which is associated with a rise in glucose-P and fructose-P concentrations and a fall in fructose bisphosphate concentration. This effect can be demonstrated in the perfused heart, skeletal muscle, and kidney cortex slice and clearly indicates that the decrease in glycolysis is related to a decrease in PFK activity (Rabinowitz, 1996).

The binding of glyceraldehyde-3-phosphate dehydrogenase to both tRNA and mRNA also has been demonstrated. The dehydrogenase selectively binds AU-rich RNA in the NAD(+) binding region (Rossman fold) (Nagy and Rigby, 1995). However, the cytoplasmic dehydrogenase was found in the polysomal fraction of lymphocytes. Although this association is relatively weak at physiological ionic strength, a considerable amount of the enzyme could be adsorbed on the ribosomes in the cell, where the concentrations of both glycolytic enzymes and ribosomes are very high. This and the specific binding of this enzyme to RNA support the view that glyceraldehyde-3-phosphate dehydrogenase has an important role in the regulation of mRNA turnover and translation. Compartmentation of the energy-supplying system on these structures can maintain a high local concentration of ATP and GTP near the sites of protein synthesis (Ryazanov *et al.,* 1988).

Ribosomes are present in an organized and repetitive fashion along the myofibrils and are associated with the myosin-containing A bands (Horne and Hesketh, 1990). mRNA species coding for actin, tubulin, and vimentin appear to be clustered around filamentous structures in the nonionic detergent-insoluble matrix (Singer *et al.,* 1989). The extent of the interaction of ribosomes/polysomes and cytoskeleton (microfilaments or myofibrils) varies according to different physiological conditions; more mRNAs and polysomes are associated with the cytoskeleton under conditions of increased protein synthesis. The association of mRNAs with the cytoskeleton might be important in the transport of mRNA from nucleus to cytoplasm

(Agutter, 1990). The relation between the translational apparatus and the filament system of cytoskeleton are of considerable significance with regard to how newly synthesized proteins are directed to their appropriate subcellular location (Hesketh and Pryme, 1991). If free polysomes, cytoskeletal-bound polysomes, and membrane-bound polysomes synthesize different proteins, then an association of a population of polysomes with the cytoskeleton could be a key feature in the way the cell sorts its newly synthesized proteins (Hesketh and Pryme, 1991). Alternatively, the association of mRNAs with the cytoskeleton might be important in the transport of mRNA from nucleus to cytoplasm (Agutter, 1990).

A considerable body of evidence indicates that many cytosolic enzymes are associated with different subcellular structures, such as membranes, the cytoskeleton, and ribosomes (Clarke and Masters, 1976; Masters, 1984; Ryazanov *et al.,* 1988; Keleti *et al.,* 1989; Knull and Walsh, 1992). This concept of enzyme organization includes the whole cytoskeletal network and other subcellular particles with many static and dynamic enzyme associations. One could envision metabolic networks whose connections are coupled and uncoupled like track switches in a railway network (Somero and Hand, 1990). New possibilities for metabolic coordination and channeling are emerging as a result of the channeling-based concept of metabolic compartmentation.

V. One Future Direction

There is no doubt that macromolecular interactions, especially those involving proteins, are widely accepted to be of primary importance in the balanced functioning of biological processes. However, the area of signal transduction has lagged in its investigation of physically integrated sequences. While there has been interest in individual interactions, there is little evidence on the relationships of each of the components in a transducing pathway to its sequential partners. When one considers the plethora of different cell signals, membrane receptors, and specific responses compared to the relatively small number of second messengers, it seems apparent that specificity can be achieved only by a mechanism analogous to the channeling process in metabolism. This relatively unexplored field holds great promise for the understanding of this crucial area of biology.

Acknowledgments

The authors wish to thank Penny Kerby for manuscript preparation. This work was supported by Hungarian grants OTKA T-17830 and MKM-FKFP 0158/97, 1023/97 (JO) and grants from the National Science Foundation and Department of Veterans Affairs (PAS)

References

Agius, L., and Sherratt, H. S. A. (1996). "Channelling in Intermediary Metabolism." Portland Press, London.

Agutter, P. S. (1990). "Between Nucleus and Cytoplasm." Chapman & Hall, London.

Backman, L., and Johansson, G. (1976). Enzyme-enzyme complexes between aspartate aminotransferase and malate dehydrogenase from pig heart muscle. *FEBS Lett.* **65,** 39–43.

Beeckmans S., and Kanarek, L. (1981). Demonstration of physical interactions between consecutive enzymes of the citric acid cycle and of the aspartate-malate shuttle: A study involving fumarase, malate dehydrogenase, citrate synthase and aspartate aminotransferase. *Eur. J. Biochem.* **117,** 527–535.

Beeckmans, S., Van Driessche, E., and Kanarek, L. (1989). The visualization by affinity electrophoresis of a specific association between the consecutive citric acid cycle enzymes fumarase and malate dehydrogenase. *Eur. J. Biochem.* **183,** 449–454.

Berry, M. N., Gregory, R. B., Grivell, A. R., Henly, D. C., Philips, J. W., Wallace, P. G., and Welch, G. R. (1987). Linear relationships between mitochondrial forces and cytoplasmic flows argue for the organized energy-coupled nature of cellular metabolism. *FEBS Lett.* **224,** 201–207.

Berry, M. N., Gregory, R. B., Grivell, A. R., Henly, D. C., Nobes, C. D., Philips, J. W., and Wallace, P. G. (1988). Intracellular mitochondrial membrane potential as an indicator of hepatocytes energy metabolism: Further evidence for thermodynamic control of mechanism. *Biochim. Biophys. Acta* **936,** 294–306.

Bülow, L. (1987). Characterization of an artificial bifunctional enzyme β-galactosidase/galactokinase prepared by gene fusion. *Eur. J. Biochem.* **163,** 443–448.

Clarke, F., Stephan, P., Morton, D., and Weidemann, J. (1983). The role of actin and associated structural proteins in the organization of glycolytic enzymes. *In* "Actin: Structure and Function in Muscle and Non-muscle Cells" (J. Barden, and C. Dos Remedios, eds.), pp. 249–257. Academic Press, Sidney.

Clarke, F. M., and Masters, C. J. (1976). Interactions between muscle proteins and glycolytic enzymes. *Int. J. Biochem.* **7,** 359–365.

Dolken, G., Leisner, E., and Pette, D. (1975). Immunofluorescent localization of glycogenolytic and glycolytic enzyme proteins and of malate dehydrogenase isozymes in cross-triated muscle and heart of the rabbit. *Histochemistry* **43,** 113–121.

D'Souza, S. F., and Srere, P. A. (1983). Cross-linking of mitochondrial matrix proteins *in situ. Biochim. Biophys. Acta* **724,** 40–51.

Elcock, A. H., and McCammon, J. A. (1996). Evidence for electrostatic channeling in a fusion protein of malate dehydrogenase and citrate synthase. *Biochemistry* **35,** 12652–12658.

Fahien, L. A., and Chobanian, M. C. (1997). Kinetic advantages of multienzyme complexes involving aminotransferases. *In* "Channeling in Intermediary Metabolism" (L. Agius and H. S. A. Sherratt, eds.), pp. 219–236. Portland Press, London.

Fahien, L. A., and Kmiotek, E. (1979). Precipitation of complex between glutamate dehydrogenase and mitochondrial enzymes. *J. Biol. Chem.* **254,** 5983.

Fahien, L. A., MacDonald, M. J., Teller, J. K., Fibich, B., and Fahien, C. M. (1989). Kinetic advantages of hetero-enzyme complexes with glutamate dehydrogenase and the α-ketoglutarate dehydrogenase complex. *J. Biol. Chem.* **264,** 12303–12312.

Goldman, R., and Katchalski, E. (1971). Kinetic behavior of a two-enzyme membrane carrying out a consecutive set of reactions. *J. Theor. Biol.* **32,** 243–257.

Halper, L. A., and Srere, P. A. (1977). Interaction between citrate synthase and mitochondrial malate dehydrogenase in the presence of polyethylene glycol. *Arch. Biochem. Biophys.* **184,** 529–534.

Hardin, C. D., and Kushmerick, C. M. (1994). Simultaneous and separable flux of pathways for glucose and glycogen utilization studied by ^{13}C NMR. *J. Mol. Cell. Cardiol.* **26,** 1197–1210.

Hardin, C. D., and Robers, T. M. (1995). Compartmentation of glucose and fructose 1,6-bisphosphate metabolism in vascular smooth muscle. *Biochemistry* **34,** 1323–1331.

Hasselbeck, R. J., and McAlister-Henn, L. (1993). Function and expression of yeast mitochondrial NAD- and NADP-specific isocitrate dehydrogenases. *J. Biol. Chem.* **268,** 12116–12122.

Hervé, G., Nagy, M., LeGouar, M., Penverne, B., and Ladjimi, M. (1993). The carbamoyl phosphate synthase-aspartate transcarbamoylase complex of Saccharomyces cerevisiae: Molecular and cellular aspects. *Biochem. Soc. Trans.* **21,** 195–198.

Hesketh, J. E., and Pryme, I. F. (1991). Interaction between mRNA, ribosomes and the cytoskeleton. *Biochem. J.* **277,** 1–10.

Horne, Z., and Hesketh, J. E. (1990). Immunological localization of ribosomes in striated rat muscle. *Biochem. J.* **268,** 231–236.

Humphreys, L., Reid, S., and Masters, C. J. (1986). Studies on the topographical localization of the binding sites for substrate and for actin on the enzymes, glyceraldehyde phosphate dehydrogenase and phosphofructokinase. *Int. J. Biochem.* **18,** 445–451.

Hyde, C. C., Ahmed, S. A., Padlan, E. A., Miles, E. W., and Davies, D. R. (1988). Three-dimensional structure of the tryptophan synthase a2b2 multienzyme complex from *Salmonella typhimurium. J. Biol. Chem.* **263,** 7857–7871.

Keleti, T., Ovádi, J., and Batke, J. (1989). Kinetic and physico-chemical analysis of enzyme complexes and their possible role in the control of metabolism. *Prog. Biophys. Molec. Biol.* **53,** 105–152.

Knull, H. R., and Walsh, J. L. (1992). Association of glycolytic enzymes with the cytoskeleton. *Curr. Top. Cell. Regul.* **33,** 15–30.

Lindbladh, C., Persson, M., Bülow, L., and Mosbach, K. (1992). Characterization of a recombinant bifunctional enzyme, galactose dehydrogenase/bacterial luciferase, displaying an improved bioluminescence in a three-enzyme system. *Eur. J. Biochem.* **204,** 241–247.

Lindbladh, C., Brodeur, R. D., Small, W. C., Lilius, G., Bülow, L., Mosbach, K., and Srere, P. A. (1994a). Metabolic studies on S. cerevisiae containing fused citrate synthase/malate dehydrogenase. *Biochemistry* **33,** 11684–11691.

Lindbladh, C., Rault, M., Hagglund, C., Small, W. C., Mosbach, K., Bülow, L., Evans, C., and Srere, P. A. (1994b). Preparation and kinetic characterization of a fusion protein of yeast mitochondrial citrate synthase and malate dehydrogenase. *Biochemistry* **33,** 11692–11698.

Low, P. S., Rathinavelu, P., and Harrison, M. L. (1993). Regulation of glycolysis via reversible enzyme binding to the membrane protein, Band 3. *J. Biol. Chem.* **268,** 14627–14631.

Masters, C. J. (1981). Interactions between soluble enzymes and subcellular structure. *CRC Crit. Rev. Biochem.* **11,** 105–143.

Masters, C. J. (1984). Interactions between glycolytic enzymes and components of the cytomatrix. *J. Cell Biol.* **99,** 222S–225S.

Meek, T. D., Garvey, E. R., and Santi, D. V. (1985). Purification and characterization of the bifunctional thymidylate synthase-dihydrofolate reductase from methotrexate-resistant *Leishmania tropica. Biochemistry* **24,** 678–686.

Méjean, C., Pons, F., Benyamin, Y., and Roustan, C. (1989). Antigenic probes locate binding sites for the glycolytic enzymes glyceraldehyde-3-phosphate dehydrogenase, aldolase and phosphofructokinase on the actin monomer in microfilaments. *Biochem. J.* **264,** 671–677.

Minaschek, G., Stewart, U. G., Blum, S., and Bereiter-Hahn, J. (1992). Microcompartmentation of glycolytic enzymes in cultured cells. *Eur. J. Cell Biol.* **58,** 418–428.

Nagy, E., and Rigby, W. F. (1995). Glyceraldehyde-3-phosphate dehydrogenase selectively binds AU-rich RNA in the NAD(+)-binding region (Rossmann fold). *J. Biol. Chem.* **270,** 2755–2763.

Negrutskii, B. S., and Deutscher, M. P. (1991). Channeling of aminoacyl-tRNA for protein synthesis *in vivo. Proc. Natl. Acad. Sci. U.S.A.* **88,** 4991–4995.

Oblinger, M. M., Foe, L. G., Kwiatkowska, D., and Kemp, R. G. (1988). Phosphofructokinase in the rat nervous system: Regional differences in activity and characteristics of axonal transport. *J. Neurosci. Res.* **21,** 25–34.

Orosz, F., and Ovádi, J. (1987). A simple approach to identify the mechanism of intermediate transfer: Enzyme system related to triose phosphate metabolism. *Biochim. Biophys. Acta* **915,** 53–59.

Orosz, F., Christova, T. Y., and Ovádi, J. (1987). Aldolase decreases the dissociation-induced inactivation of muscle phosphofructokinase. *Biochem. Biophys. Res. Commun.* **147,** 1121–1128.

Ovádi, J. (1995). "Cell Architecture and Metabolic Channeling," Molecular Biology Intelligence Unit. R.G. Landes and Springer-Verlag, New York and Berlin.

Ovádi, J., Huang, Y., and Spivey, H. O. (1994). Binding of malate dehydrogenase and NADH channeling to Complex I. *J. Mol. Recognition,* **7,** 265–273.

Pagliaro, L., and Taylor, D. L. (1992). 2-Deoxyglucose and cytochalasin D modulate aldolase mobility in living 3T3 cells. *J. Cell Biol.* **118,** 859–863.

Pagliaro, L., Kerr, K., and Taylor, D. L. (1989). Enolase exists in the fluid phase of 3T3 cells. *J. Cell Sci.* **94,** 333–342.

Penverne, B., Belkaid, M., and Hervé, G. (1994). *In situ* behavior of the pyrimidine pathway enzymes in *Saccharomyces cerevisiae*. 4. The channelling of carbamylphosphate to aspartate transcarbamylase and its partition in the pyrimidine and arginine pathways. *Arch. Biochem. Biophys.* **309,** 85–93.

Persson, L.-O., and Srere, P. A. (1992). Purification of the mitochondrial citrate transporter in yeast. *Biochem. Biophys. Res. Commun.* **183,** 70–76.

Porpaczy, Z., Sümegi, B., and Alkonyi, I. (1983). Association between the α-ketoglutarate dehydrogenase complex and succinate thiokinase. *Biochim. Biophys. Acta* **749,** 172–179.

Porpaczy, Z., Sümegi, B., and Alkonyi, I. (1987). Interaction between NAD-dependent isocitrate dehydrogenase, α-ketoglutarate dehydrogenase complex, and NADH:ubiquinone oxidoreductase. *J. Biol. Chem.* **262,** 9509–9514.

Rabinowitz, M. (1996). Uncharged tRNA-phosphofructokinase interaction in amino acid deficiency. *Amino Acids* **10,** 99–108.

Robinson, J. B., Jr., and Srere, P. A. (1985). Organization of Krebs tricarboxylic acid cycle enzymes in mitochondria. *J. Biol. Chem.* **260,** 10800–10805.

Robinson, J. B., Jr., Inman, L., Sümegi, B., and Srere, P. A. (1987). Further characterization of the Krebs Tricarboxylic acid cycle metabolon. *J. Biol. Chem.* **262,** 1786–1790.

Ryazanov, A. G. (1988). Organization of soluble enzymes in the cell. Relay at the surface. *FEBS Lett.* **373,** 1–4.

Ryazanov, A. G., Ashmarina, L. I., and Murinetz, V. I. (1988). Association of glyceraldehyde-3-phosphate dehydrogenase with mono-and polyribosomes of rabbit reticulocytes. *Eur. J. Biochem.* **171,** 301–305.

Sherry, A. D., Sümegi, B., Miller, B., Cottam, G. L., Gavva, S., Jones, J. G., and Malloy, C. R. (1994). Orientation-conserved transfer of symmetric Krebs cycle intermediates in mammalian tissue. *Biochemistry* **33,** 6268–6275.

Singer, R. H., Langevin, G. L., and Lawrence, J. B. (1989). Ultrastructural visualization of cytoskeletal mRNAs and their associated proteins using double-label *in situ* hybridization. *J. Cell Biol.* **108,** 2343–2353.

Somero, G. N., and Hand, S. C. (1990). Protein assembly and metabolic regulation: Physiological and evolutionary perspectives. *Physiol. Zool.* **63,** 443–471.

Spirin, A. S., and Ajtkozhin, M. A. (1985). Informosomes and polyribosome-associated proteins in eucaryotes. *Trends Biochem. Sci.* **10,** 162–165.

Srere, P. A. (1981). Protein crystals as a model for mitochondrial matrix proteins. *Trends Biochem. Sci.* **6,** 4–7.

Srere, P. A., and Mosbach, K. (1974). Metabolic compartmentation: Symbiotic, organellar, multienzymic, and microenvironmental. *Annu. Rev. Microbiol.* **28,** 61–83.

Srere, P. A., Mattiasson, B., and Mosbach, K. (1973). An immobilized three-enzyme system: A model for microenvironmental compartmentation in mitochondria. *Proc. Natl. Acad. Sci. U.S.A.* **70,** 2534–2538.

Srivastava, D. K., And Bernhard, S. A. (1985). The mechanism of transfer of NADH dehydrogenases. *Biochemistry* **24,** 623–628.

Srivastava, D. K., and Berhard, S. A. (1987). Biophysical chemistry of metabolic reaction sequences in concentrated enzyme solution and in the cell. *Annu. Rev. Biophys. Biophys. Chem.* **16,** 175–204.

Sümegi, B., and Alkonyi, I. (1983). A study on the physical interaction between the pyruvate dehydrogenase complex and citrate synthase. *Biochim. Biophys. Acta* **749,** 163–171.

Sümegi, B., and Srere, P. A. (1984b). Complex I binds several mitochondrial NAD-coupled dehydrogenases. *J. Biol. Chem.* **259,** 15040–15045.

Sümegi, B., Gilbert, H. F., and Srere, P. A. (1985). Interaction between citrate synthase and thiolase. *J. Biol. Chem.* **260,** 188–190.

Sümegi, B., Sherry, A. D., and Malloy, C. R. (1990). Channeling of TCA cycle intermediates in cultured *Saccharomyces cerevisiae. Biochemistry* **29,** 9106–9110.

Sümegi, B., Porpaczy, Z., McCammon, M. T., Sherry, A. D., Malloy, C. R., and Srere, P. A. (1992). Regulatory consequences of organization of citric acid cycle enzymes. *Curr. Top. Cell. Regul.* **33,** 249–260.

Sümegi, B., Sherry, A. D., Malloy, C. R., and Srere, P. A. (1993). Evidence of orientation-conserved transfer in the TCA cycle in *Saccharomyces cerevisiae:* ^{13}C NMR studies. *Biochemistry* **32,** 12725–12729.

Tompa, P., Batke, J., Ovádi, J., Welch, G. R., and Srere, P. A. (1987). Quantitation of the interaction between citrate synthase and malate dehydrogenase. *J. Biol. Chem.* **262,** 6089–6092.

Trujillo, M., Donald, R. G. K., Roos, D. S., Greene, P. J., and Santi, D. V. (1996). Heterologous expression and characterization of the bifunctional dihydrofolate reductase-thymidylate synthase enzyme of *Toxoplasma gondii. Biochemistry* **35,** 6366–6374.

Tyiska, R. L., Williams, J. S., Brent, L. G., Hudson, A. P., Clark, B. J., Robinson, J. B., Jr., and Srere, P. A. (1986). Interactions of matrix enzyme activities with mitochondrial inner membranes. *NATO* Series A: Life Sciences, vol. 127, pp. 177–189, Plenum Press, New York.

Ureta, T. (1991). The role of isoenzymes in metabolite channeling. *J. Theor. Biol.* **152,** 81–84.

Vértessy, B. G., Orosz, F., and Ovádi, J. (1991). Modulation of interaction between aldolase and glycerol-phosphate dehydrogenase by fructose phosphates. *Biochim. Biophys. Acta.* **1078,** 236–242.

Vértessy, B. G., Orosz, F., Kovács, J., and Ovádi, J. (1997). Alternative binding of two sequential glycolytic enzymes to microtubules: Molecular studies in phosphofructokinase/aldolase/microtubule system. *J. Biol. Chem.* **272,** 25542–25546.

Welch, G. R. (1977). On the role of organized multienzyme systems in cellular metabolism: A general synthesis. *Prog. Biophys. Mol. Biol.* **32,** 103–191.
Wilson, J. E. (1978). Ambiquitous enzymes: Variation in intracellular distribution as a regulatory mechanism. *Trends Biochem. Sci.* **3,** 124–125.
Wilson, J. E. (1980). Brain hexokinase, the prototype ambiquitous enzyme. *Curr. Top. Cell. Regul.* **16,** 1–54.

The State of Water in Biological Systems

Keith D. Garlid
Department of Biochemistry and Molecular Biology, Oregon Graduate Institute of Science and Technology, Portland, Oregon 97291

This paper addresses the issue of how the aqueous cytoplasm is organized on a macroscopic scale. Mitochondria were used as the experimental model, and a unique experimental approach was used to probe the properties of water in the mitochondrial matrix. The results demonstrate aqueous phase separation into two distinct phases with different osmotic activity and different solute partition coefficients. The larger phase, designated "normal water," is osmotically active and behaves in every respect like a bulk, dilute salt solution. The smaller phase, designated "abnormal water," is osmotically inactive and comprises the water of hydration of matrix proteins. It is, nevertheless, solvent water, with highly selective partition coefficients, and behaves like a Lewis base.

KEY WORDS: Water, Mitochondria, Partition coefficient, Osmotic, Membranes, Protein hydration.

I. Introduction*

This review deals with the time-averaged, equilibrium properties of water in cells. In this domain, one is dealing with familiar macroscopic properties such as osmotic activity and solute activity coefficients.

In the early 1970s, I put forward a novel hypothesis for the macroscopic state of water in biological systems: that biological water spontaneously separates into two (or more) phases with distinct solvent properties (Garlid, 1976, 1978, 1979). This hypothesis was both unusual and, it still seems to me, inescapable. Thus, a straightforward application of the scientific method conclusively excluded the alternative hypothesis that biological water com-

* Abbreviations: BSA, bovine serum albumin; CCCP, carbonyl cyanide m-chlorophenylhydrazone; PEG, polyethylene glycol.

International Review of Cytology, Vol. 192
0074-7696/00 $30.00

prises a single, homogeneous phase. This experimental evidence will be reviewed.

Isolated rat liver mitochondria were used as the experimental model for this investigation. The mitochondrial matrix is a single, membrane-bounded compartment that is very rich in proteins. A thermodynamically complete description of matrix water requires no more than two different phases,[1] which I call "normal" and "abnormal." The *normal* aqueous phase behaves in every respect like a bulk aqueous solution of similar composition. The *abnormal* aqueous phase is osmotically inactive and comprises the water of hydration of mitochondrial membranes and proteins. Nevertheless, the abnormal phase dissolves small solutes, with solute activity coefficients that are very different from those in bulk aqueous solutions.

The decisive results came from a novel, but perfectly straightforward, experimental approach, in which the effects of matrix volume on nonelectrolyte distribution coefficients were compared with distribution coefficients measured at constant volume. It remains surprising to me that this powerful, but conceptually simple, approach has not been applied to other biological systems.

The two-phase hypothesis contains important implications for understanding the aqueous cytoplasm. The proteins of the cell are neither distributed randomly nor do they impart the aqueous phase separation in a manner analogous to that observed with macromolecules in a test tube, as suggested by Walter and Brooks (1995). Rather, they are directed by the biological machinery of the cell to form close associations with each other and the cytoskeleton and membranes of the cell. The association serves the purpose of minimizing the water of hydration associated with all proteins. Based on hydration estimates of serum albumin, it has been suggested that, if it did not occur, all of the water of the cell could be osmotically inactive (Cameron *et al.,* 1997). Proteins are also specifically localized within the cell. This localization not only serves the purpose of mechanical function, as in muscle, but also the purpose of metabolic channeling (Saks *et al.,* 1994). The picture that emerges is that cell proteins exist in a semi-solid, gel-like state, and their water of hydration possesses unique solvent properties as a consequence of its organization. The remainder, comprising the major portion of cell water, is a dilute salt solution, probably containing few proteins. This bulk solution, through which metabolites and signal ligands flow, is dispersed in streams and pockets among the organized matrices of proteins and membranes.

[1] A small fraction of cell water is irrotationally bound to proteins and, therefore, cannot participate as a solvent. Attempts to quantitate this fraction lead to values on the order of 0.1% of total cell water (Cooke and Kuntz, 1974). All of the comments in this article are meant to exclude this irrotationally bound water.

II. Osmotic Behavior of Polar Solutes—Introduction to the Osmotic Intercept

The thermodynamic activity of water in cells, a_w, normally ranges between 0.990 and 0.997. The preferred measure of water activity is the osmolality, ϕ, which avoids the awkwardness of these numbers, giving a range between 166 and 558 mosM. Osmolality of a solution containing a single solute is defined as follows (Kirkwood and Oppenheim, 1961, discussed in Garlid, 1998):

$$\phi = gm = -\frac{1000}{M_w}\ln a_w \tag{1}$$

where m is molality of the solute and M_w is the molecular weight of water; g is the osmotic coefficient and is equal to 1 for an ideal solution. Note that osmolality, which can be measured by freezing point depression, osmotic pressure, or vapor pressure lowering, is a thermodynamically rigorous measure of water activity.

Polar solutes exhibit nonideal behavior in aqueous solutions, as illustrated by the osmotic behavior of sucrose and PEG400, plotted in Fig. 1. Both curves deviate strongly from the line of ideal behavior ($g = 1$), but extrapolate to zero at infinite dilution. BSA exhibits similar behavior (not shown).

As shown nearly 80 years ago by Scatchard (1921), nonideal behavior of polar solutes can be rationalized by considering their hydration. In solution, sucrose is unlikely to undergo thermal motion as the dry molecule. Rather, it will move as the hydrate, and the water available for freezing or vaporization will be correspondingly reduced by an amount equal to the molal hydration, W_h. A new concentration, m', may thus be defined:

$$m' = \frac{m}{(1 - W_h m)} \tag{2}$$

where W_h is the water of hydration, in kg H_2O/mol solute. We may write

$$\phi = g'm' \tag{3}$$

where g' is the osmotic coefficient of the newly defined solution. To linearize the expression, we take the inverse of Eq. (2) and obtain

$$1/m = W_h + g'/\phi \tag{4}$$

Two aspects of Eq. (4) are noteworthy. First, an entirely analogous relationship between inverse quantities is used for osmotic studies in cells and organelles. Second, the intercepts of such plots represent the approach

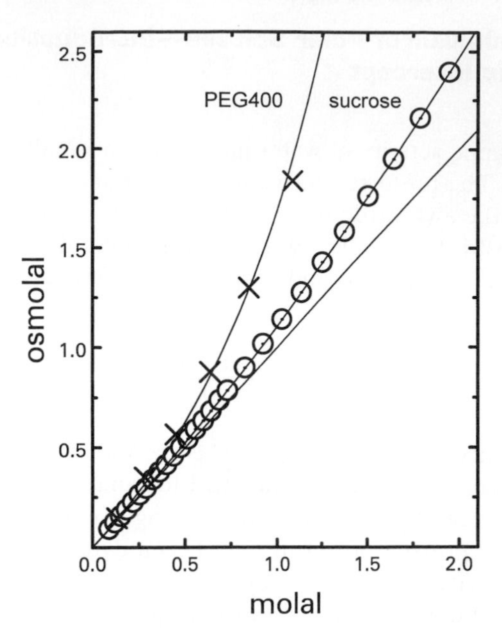

FIG. 1 Osmotic behavior of polar molecules in solution. Osmolality is plotted as a function of molal concentration of PEG400 and sucrose. The straight line, osmolal = molal, is included to demonstrate the behavior of an ideal solute. The values for PEG400 were converted from the data of Parsegian *et al.* (1995). The values for sucrose were converted from the data of Wolf and Brown (1965). The curves drawn through the data points used $\phi = m/(1 - W_h m)$ (see Eq. (2) in text), where W_h is the constant water of hydration. Use of this equation implies that sucrose and PEG400 behave as ideal solutes when hydration is taken into account ($g' = 1.0$ in Eq. (1)).

to infinite concentration, rather than infinite dilution. Thus, *the osmotic intercepts cannot equal zero except for ideal solutions.*

Figure 2 contains data for sucrose, PEG400, and BSA, plotted according to Eq. (4). Note that all three sets of data are linear with positive intercepts, W_h. More impressively, the slopes of the lines equal 1.0 within experimental error. This means that each of these solutes behaves as an *ideal* solute ($g' = 1$) when a constant molar hydration is taken into account. Thus, hydrated sucrose behaves ideally to 2 molal.

The hydration values are instructive. In mol H_2O/mol solute, they are 5 for sucrose, similar to the value obtained by Scatchard (1921), 22.7 for PEG400, and 1.05×10^4 for BSA. The range is still large in g H_2O/g solute: 0.28 for sucrose, 1 for PEG400, and 2.86 for BSA. These numbers indicate that sucrose is hydrated with one water per –OH group, whereas the large surface of BSA permits building up of water multilayers.

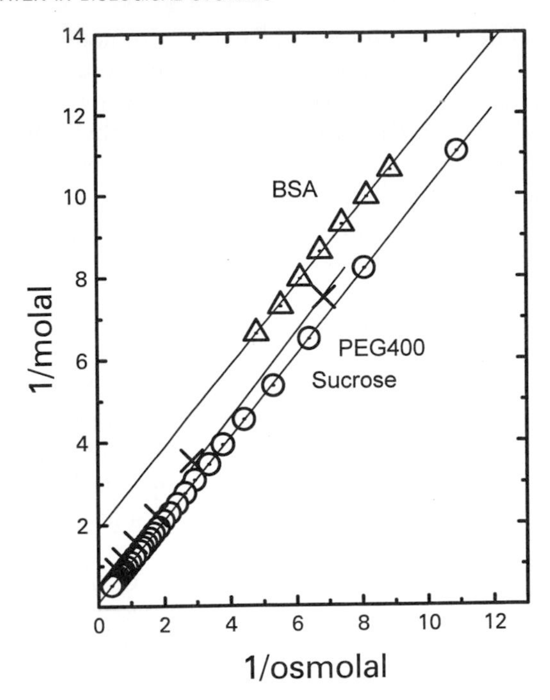

FIG. 2 Ideal osmotic behavior of hydrated polar solutes and macromolecules. Inverse molality is plotted versus inverse osmolality, as in Eq. (4). Molal and osmolal values for BSA were multiplied by 100 to fit in the same figure. The intercepts (W_h) and slopes (g') were obtained by linear regression, with correlation coefficients > 0.999 for each curve. Sucrose: 0.0956, 1.003; PEG400: 0.4094, 1.043; BSA: 189.48, .989. The values for BSA were converted from those of Zimmerman *et al.* (1995), after correcting observed osmolalities for the Donnan effect (Garlid, 1998).

III. Mitochondria—The Experimental Model

The mitochondrion is an excellent experimental model for studies of biological water. It is structurally simple and contains a single osmotically active compartment. Under controlled experimental conditions, this compartment retains its ionic contents during wide volume changes. Mitochondria behave as osmometers (Tedeschi and Harris, 1955; Bentzel and Solomon, 1967; Beavis *et al.,* 1985; Garlid and Beavis, 1985), and they can undergo very large amplitude swelling without rupture, a consequence of the extensive folding of the inner membrane (Stoner and Sirak, 1969).

Distributions of water and nonelectrolytes in isolated rat liver mitochondria at 4°C were measured according to exacting protocols, which are described elsewhere (Garlid, 1998). Briefly, mitochondria were added to assay media at a final concentration of 5 mg/ml. For nonelectrolyte distributions and osmotic studies, parallel assays contained the isotope pairs ^3H_2O/^{14}C-sucrose and ^3H-sucrose/^{14}C-nonelectrolyte. The nonelectrolytes studied were ethanol, glycerol, urea, antipyrine, and dimethylsulfoxide (DMSO). Incubations were carried out at 0–2°C for 5–10 minutes, followed by centrifugation. The pellet was extracted by heating at the average isoelectric point (pH 4.6). Pellet extract and supernatant were then analyzed by scintillation counting and atomic absorption spectroscopy. All dilutions were gravimetric.

Solution osmolalities were measured by freezing point depression. Solution densities, used to convert molar to molal concentrations, were measured with a Mettler–Paar model 02C Density Meter. All radioactive probes except THO, ethanol, and DMSO were purified by paper chromatography prior to use, a procedure which proved to be essential for accurate measurements. We also measured the distribution of antipyrine by chemical analysis (Mendelsohn and Levin, 1960), and the results confirmed the distributions measured isotopically.

These protocols led to quantitative estimates of matrix water content, W_i (mg H_2O/mg dry wt); matrix nonelectrolyte content, X_i (nmol solute/mg dry wt); and the corrected pellet dry weight, W^o_{dry}, which was used to normalize the data. It is necessary to emphasize the fact that these protocols measure matrix *water* content, and not matrix *volume*.

Over a period of years, rat liver mitochondria proved to be highly reproducible. For example, total K^+ and Mg^{2+} in the stock suspension were almost invariant over 6 years of measurement, at 153 ± 1 nmol/mg and 40 ± 0.3 nmol/mg (n = 20), respectively. Some K^+ was predictably lost upon dilution, and matrix values for K^+ and Mg^{2+} were 125 ± 8 nmol/mg and 38 ± 0.3 nmol/mg, respectively. Matrix Mg^{2+} was unaffected by osmotic swelling at 0°C.

IV. Osmotic Equilibria in Mitochondria

A. Theory

Osmotic equilibrium is the state in which, in the absence of pressure gradients, water activity, a_w, is equal in the two phases. Thus, osmolality, ϕ, is equal in the two phases, and

$$g_i\, s_i = \phi_o \tag{5}$$

where s_i designates the ideal osmolality of the internal phase; $s_i = \Sigma v_j\, m_j$ where v_j = the number of particles into which solute j dissociates. The variable ϕ_0 is the (measured) osmolality and g is the osmotic coefficient. (It must be emphasized that g is a property of the *solution* and not of the individual solutes, as sometimes claimed.) Endogenous matrix solutes are not susceptible to direct measurement, but they may be written

$$S_i \equiv s_i W_i \tag{6}$$

where S_i is the amount (nosmol/mg protein) of osmotically active solute, and W_i is matrix water content. Combining Eqs. (5) and (6), we achieve the general expression of osmotic equilibrium

$$W_i = g_i S_i/\phi_0. \tag{7}$$

As discussed in Section I, this proportionality cannot hold at infinite osmolality, where the influence of macromolecular hydration will be most extreme. Thus, the correct equation for a perfect osmometer is

$$W_i = W_2 + g' S_1/\phi_0 \tag{8}$$

where W_2 is the osmotic intercept (the water of hydration) and g' and S_1 are the osmotic coefficient and amount of solute, respectively, in the osmotically active phase. According to Eq. (8), water content should vary linearly with inverse osmolality, provided that there is no solute movement during the measurement (S_1 is constant), and there should be a positive intercept.

B. Osmotic Swelling in Mitochondria

Figure 3 contains plots of matrix water content, W_i, vs inverse osmolality. The data, which cover a 9-fold range of volume, are in complete agreement with the predictions of Eq. (8). Each curve is linear, and the slopes increase with increasing solute content. The curves have a common, nonzero intercept (W_2), equal to 0.28 mg H_2O/mg mitochondrial protein. Thus, when hydration is taken into account, the mitochondrion behaves as a perfect osmometer.

These results indicate that matrix water partitions into two aqueous phases, an osmotically active phase, W_1, and an osmotically inactive phase, W_2. The question whether W_2 behaves as a solvent or represents "bound," nonsolvent water will be addressed in the section on nonelectrolyte distributions.

C. Osmotically Active Matrix Solutes and the Osmotic Slope

The major solutes of freshly isolated mitochondria are potassium salts of phosphate and organic anions (Gamble and Hess, 1966). Experimentally,

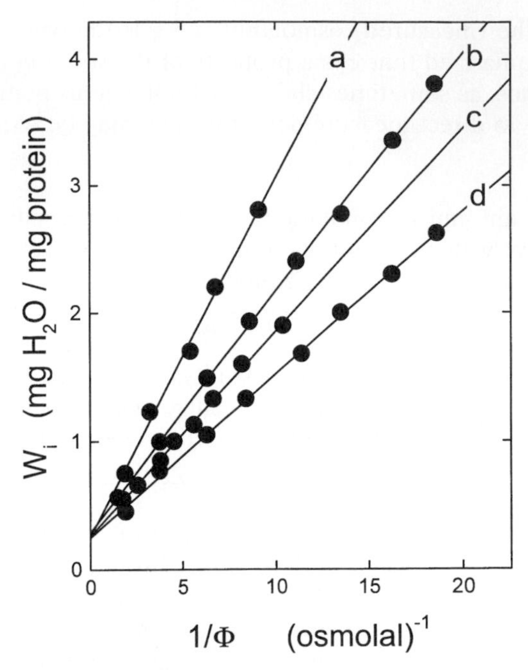

FIG. 3 Osmotic swelling in mitochondria. Matrix water content, W_i (mg H_2O/mg dry weight), is plotted versus inverse osmolality, ϕ^{-1}. The data shown were obtained on four separate preparations with different matrix K^+ contents. Each preparation began with mitochondria suspended in 0.25 M (0.272 osmolal) sucrose. **Curve a,** mitochondria were preincubated at 25°C in the presence of substrate, O_2 and K^+. Prior to the assay, the preparation was treated with rotenone and subjected to a 5-min 0°C wash in K^+ malate to block the K^+/H^+ antiporter. They were resuspended in sucrose following a large volume wash in 0.25 M sucrose. K^+ content was 165 nmol/mg dry weight. **Curve b,** no pretreatment. K^+ content was 125 nmol/mg dry weight. In order to maintain constant K^+, aliquots were incubated for 10 s prior to centrifugation. Longer incubations cause curvature due to swelling-induced K^+ loss (Garlid and Beavis, 1985). **Curve c,** mitochondria were incubated in 80 mosmolal sucrose for 15 min, during which they lost K^+ via the K^+/H^+ antiporter. They were then reisolated and suspended in 0.25 M sucrose. K^+ content was 82 nmol/mg dry wt. **Curve d,** mitochondria were added to 54 mosmolal sucrose for 15 min, reisolated and suspended in 0.25 M sucrose. K^+ content was 75 nmol/mg dry weight. These results are representative of 10 osmotic swelling curves. In each case, the slopes depended on K^+ content and the curves were linear if matrix K^+ was constant.

it is a simple matter to alter the amount of K^+ and the amount and type of anion in respiring mitochondria. It was essential to maintain constant amounts of endogenous solutes through subsequent volume changes, so that the quantity S_1 remained constant. By manipulating the Mg^{2+}-regulated K^+/H^+ antiporter (Garlid, 1980, 1988), we succeeded in preparing mitochondria which contained widely varying amounts of K^+ salts and which did not lose these solutes during swelling.

Figure 4 contains a plot of the osmotic slopes, S_1, versus matrix K^+, using the data in Fig. 3. The slope of the line is 1.52 nosmol/nmol K^+, and the intercept is 15 nosmol/mg. Uptake and loss of K^+ salts constituted the only significant change in matrix solutes in these experiments, and any net change in matrix K^+ was accompanied by anions of average valence, z_a. The relationship between osmotic slope and K^+ may be written

$$S_1 = g'\{s_1{}^0 + (1 + 1/z_a)\, K_i^+\} \tag{9}$$

where $s_1{}^0$ represents internal solutes not included in the second term in the equation. We have estimated g', the osmotic coefficient of the "normal" phase, to be 0.94, so the slope of 1.52 yields an estimated anion valency, z_a, of 1.62. This is in good agreement with expectations, since the labile anions are primarily phosphate and divalent anions under these conditions. This is an approximate analysis, because the distribution of K^+ between phases 1 and 2 is unknown.

D. Osmotically Inactive Matrix Water and the Osmotic Intercept

Four facts can be noted about the osmotic intercept: (1) W_2 represents osmotically inactive *water,* as should be clear from the way the measurements were carried out. The suggestion frequently made that the osmotic

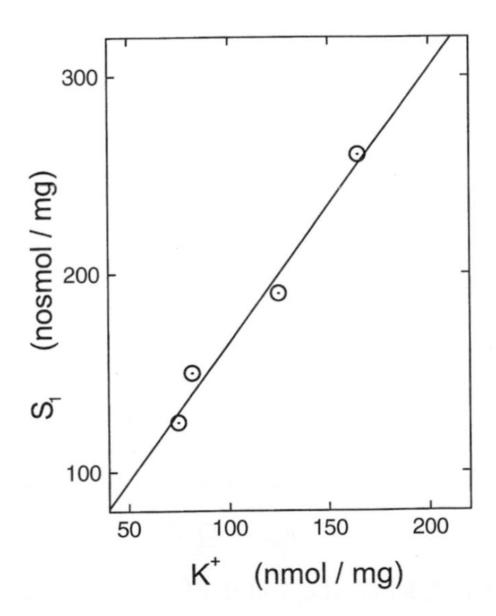

FIG. 4 The effect of matrix K^+ on the osmotic slope. The osmotic slope, S_1, is plotted versus matrix K^+ content. The values plotted correspond to the four curves in Fig. 1.

intercept represents proteins and lipids is incorrect; (2) W_2 is unaffected by changes in matrix solute concentrations; (3) A positive osmotic intercept is characteristic of all biological systems, as well as of most inanimate systems, ranging from small nonelectrolytes to gels and coacervates; (4) The size of W_2 is about an order of magnitude less than the water of hydration of proteins in solution (compare W_2 with W_h for BSA).

V. Nonelectrolyte Distributions in Mitochondria

A. Theory

When solute x is at equilibrium between two homogeneous phases, the following relationship holds:

$$m_{x1} = f\, m_{x0} \tag{10}$$

where f is the thermodynamic *partition coefficient,* given by

$$f \equiv Y_{x0}/Y_{x1} \tag{11}$$

where the Ys are molal activity coefficients. The internal concentration, m_{x1}, cannot be measured directly. Instead, we know internal solute content, X_i, and total matrix water content, W_i, which leads to an average concentration, $[X_i]$:

$$[X]_i \equiv X_i/W_i \tag{12}$$

Note that $[X]_i$ is a *derived,* rather than measured, quantity. Solute distribution is then given by

$$[X]_i = Qm_{x0} \tag{13}$$

where Q is the experimental *distribution coefficient.* Note that $Q = f$, if, and only if, the matrix contains a single aqueous phase.

The key to the experimental proof, developed below, is that $[X_i]$ contains two independent variables. Thus, Q can be estimated at constant volume from the slopes of data plotted according to Eq. (13). In fact, this was the only way that solute distributions had previously been studied. The variable Q can also be estimated from slopes of data plotted according to Eq. (14)

$$X_i/m_{x0} = QW_i \tag{14}$$

which is equally valid, since the water of distribution, X_i/m_{x0}, is a measured quantity.

B. Nonideal Distribution Coefficients for Simple Nonelectrolytes

In the 1930s, there was considerable controversy over the extent of normal and bound water in cells. In a very nice review entitled, "Water, Free and Bound," Blanchard (1940) suggested that permeant nonelectrolytes should be useful as probes of "free" water, since one would not expect them to penetrate "bound" water. In particular, polar nonelectrolytes should all have equal volumes of distribution and, if all water is normal, their distribution coefficients should be close to 1.0.

These expectations are strikingly unfulfilled in mitochondria. This is demonstrated by the constant volume data in Fig. 5, plotted according to Eq. (13). Note that Q for each solute is independent of concentration up to 0.5 M, the highest level measured. The results in Fig. 5 are qualitatively

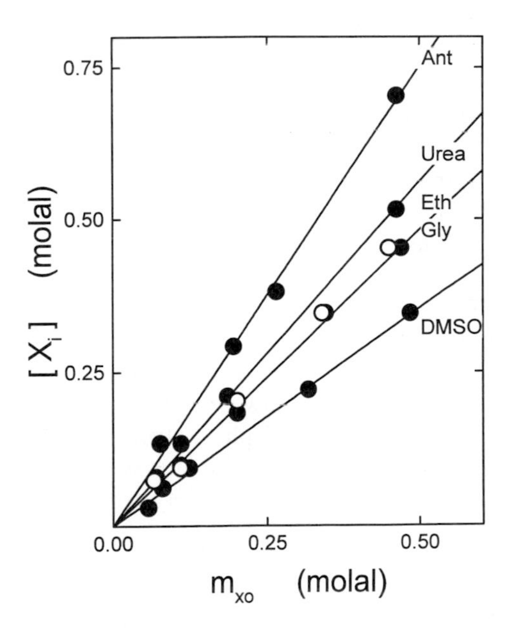

FIG. 5 Nonelectrolyte distribution in mitochondria at constant volume. Apparent molal concentration in the matrix $[X_i]$ is plotted versus medium concentration, m_{xo}, for various nonelectrolytes: Ant. = antipyrine; Urea = urea; Eth. = ethanol; Gly. = glycerol; and DMSO = dimethylsulfoxide. Matrix water was held constant at 1.0 ± 0.05 mg/mg dry weight. Results are representative of 3 (ethanol, glycerol) or 4 (antipyrine, urea, DMSO) experiments. Ethanol and glycerol always exhibited a volume of distribution equal to that of water. These are true phase distributions; there was no saturation in the concentration range between 10^{-4} and 10^{-1} M.

similar to the results of Ling (1970) and Troshin (1966) in various gels, tissues, and cells.

The results with the waterlike solutes, ethanol and glycerol, which have distribution coefficients of 1.0, strongly indicate that all of matrix water is solvent, including the hydration water, W_2.

The other probes confound expectations in that they differ from 1.0 and differ from each other. From such results, Ling (1970) and Troshin (1966) were led to conclude that all of biological water is abnormal. The logical fallacy of this conclusion lies in the implicit assumption that all of matrix water constitutes a single phase. To get around this problem, it is necessary to perform experiments in a different way, as shown in the next section.

VI. Solute Distribution as a Function of Matrix Volume

A. Experimental Evidence for Phase Heterogeneity in Mitochondria

Equation (14) provides a rationale for measuring distributions as a function of matrix volume, and it contains no assumptions about aqueous phase structure in the matrix. Equation (14) also provides a direct test of the single-phase theory. If one homogeneous phase exists, then the slopes, Q, of data plotted according to Eq. (14) must equal the slopes plotted according to Eq. (13) (Fig. 5).

The results plotted in Fig. 6 clearly demonstrate that this is not the case. The slopes of Fig. 6 differ from those of Fig. 5, and, moreover, all solute distributions fall on parallel curves with slopes approximately equal to 1.0.

The inescapable conclusion to draw from these results is that the mitochondrial matrix has a heterogeneous phase structure. This conclusion does not depend on the absolute values of the slopes or intercepts, but rather on the large and unmistakable differences in the slopes of Figs. 5 and 6.

B. Phase Heterogeneity—Theory

The general expression for multiphase solute distribution is

$$X_i/m_{x0} = \Sigma f_j W_j \quad (15)$$

where f_j are the thermodynamic partition coefficients in each phase. It is sufficient to define two hypothetical classes of water: Phases of variable

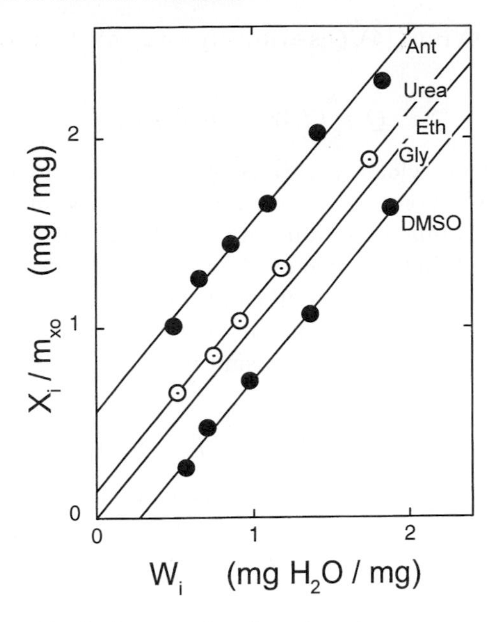

FIG. 6 Nonelectrolyte distributions as a function of matrix water content. The symbols and conditions are as described in the legend to Fig. 5. Multiple points for urea represent data obtained at different urea concentrations ranging from 10^{-4} to 0.5 M. All other probe concentrations were 50 mM. Matrix water (W_i) was varied osmotically by varying medium sucrose plus buffer concentration from 0.08 M to 0.33 M. All lines have slopes of 1.0 ± 0.04 by linear regression. Ethanol and glycerol distributions were always found to be within 4% of W_i, and their data points have been omitted for clarity. Antipyrine distributions were confirmed by chemical analysis.

(osmotically active) water content, designated by W_1, and phases of constant (osmotically inactive) water content, designated by W_2. Thus,

$$W_i = \mathrm{W}_1 + W_2 \tag{16}$$

If two solvent phases coexist in the matrix, then solutes must partition into both phases. This is in contrast to the commonly held assumption that osmotically inactive water must simultaneously be nonsolvent (Ling, 1972; Dick, 1971). To express total matrix solute, X_i, in terms of measured quantities, we begin with the following:

$$X_i = W_1 m_{x1} + W_2 m_{x2} \tag{17}$$

Introducing thermodynamic partition coefficients $f_1 = \mathrm{m}_{x1}/m_{x0}$ and $\mathrm{f}_2 = \mathrm{m}_{x2}/m_{x0}$, we obtain

$$X_i = (f_1 W_1 + f_2 W_2) m_{x0} \tag{18}$$

By comparison with Eq. (14), Q is evidently a weighted sum of partition coefficients:

$$Q = (f_1 W_1 + f_2 W_2)/W_i \tag{19}$$

Equation (19) makes clear why Q is insensitive to phase heterogeneity when determined at constant volume. At constant volume, compartment sizes (W_1, W_2, W_i) remain constant; moreover, the partition coefficients, f, remain constant within the range of Henry's Law. Under these conditions, even if Q is not a true partition coefficient, a constant slope will be obtained from experiments such as those in Fig. 5, as was observed.

We can now rearrange Eq. (18), using Eq. (16), in terms of experimental variables:

$$X_i/m_{x0} = f_1 W_i + (f_2 - f_1)W_2 \tag{20}$$

It should now be clear why the comparison of slopes between Eqs. (13) and (14) provides a decisive test of the homogeneous phase model. Thus, Eq. (20) reduces to Eq. (14) (and $Q = f$), only when the matrix is a single aqueous phase; that is, only when $W_2 = 0$.

The separation into two phases, W_1 and W_2, represents what we are *capable* of knowing about the compartmentation of water and solute. This separation is thermodynamically complete, because it accounts for all phase types that can be distinguished experimentally.

C. Phase Heterogeneity—Quantitative Results

Table I contains a summary of the results of the nonelectrolyte experiments, interpreted using Eq. (20). Some additional inferences may be drawn from these data:

1. The theoretical development does not preclude the possibility of water shifts between phases; however, the fact that the data conform to the linear relationship of Eq. (20) suggests that such shifts do not occur to any significant extent.
2. The results are independent of the means used to vary W_i. The variable W_i may be changed by experimental alterations in two independent parameters: internal solute content and medium concentration of impermeant solute. Solute distributions are dependent on W_i and independent of the parameter or parameters used to achieve a given matrix volume (data not shown).
3. $f_1 = 1.00 \pm 0.02$ for all nonelectrolytes studied and remained constant over a 5-fold range of matrix water content. The variable f_1 is independent

TABLE I

Properties of Mitochondrial Water

	Abnormal phase	Normal phase
Extent (mg H_2O/mg dry wt)	0.28 (30%)	0.68 (70%)[a]
Osmotic coefficient	inactive	0.94[b]
Partition coefficients		
antipyrine	3.0	1.0
urea	1.5	1.0
glycerol	1.0	1.0
ethanol	1.0	1.0
dimethyl sulfoxide	0.04	1.0

[a] Value given is for isolated mitochondria in equilibrium with 0.272 osmolal sucrose. The abnormal phase is estimated to be 15% of total water *in vivo,* because total water content is about 1.93 mg/mg *in vivo.*

[b] By inference.

of probe concentration, showing that Henry's Law is obeyed. Phase I may therefore be considered to be a normal phase indistinguishable in its solvent properties from the medium.

4. DMSO, antipyrine, and urea have nonzero intercepts. From Eq. (20), it may be concluded that W_2 is different from zero. That is, mitochondria contain a fraction of abnormal water that is also osmotically inactive. It is noteworthy that this finding is deduced without consideration of osmotic swelling curves.

5. We may also conclude that partition coefficients in W_2 differ for urea, DMSO, and antipyrine. The f_2s are independent of concentration, so solutes in phase 2 also obey Henry's Law.

6. Whereas the f_1s are the same for all solutes, the f_2s are solute specific. This behavior, together with osmotic inactivity, characterizes phase 2 as an abnormal aqueous phase.

7. The quantitative distribution of water between normal and abnormal phases cannot be resolved by nonelectrolyte probes; however, limits can be deduced from Fig. 6. DMSO is evidently excluded, to a considerable extent, from W_2. If f_2 were equal to zero for DMSO, W_2 would equal the negative intercept of the DMSO curve, since $f_1 = 1$. This value is 0.27 g H_2O/g dry wt, representing the *minimum* value of W_2. This is very close to the independent estimate of W_2 (0.28 g H_2O/g dry wt) from the osmotic studies of Fig. 1.

8. A further implication of these results is that matrix solutes, as well as matrix water, must be divided into osmotically inactive and active portions. That is, solutes in W_2 do not contribute osmotically to phase 1.

VII. Discussion

A. The Evidence for Aqueous Phase Separation in the Mitochondrial Matrix

The hypothesis that biological water separates into two distinct phases was introduced more than 20 years ago (Garlid, 1976, 1978, 1979) and has recently been revived (Walter and Brooks, 1995). Despite an extensive literature of studies on cell water, the experimental approach used to test the hypothesis remains unique. This approach is also powerful, because it provides decisive evidence. It was possible to exclude the alternative hypothesis that matrix water comprises a single phase, and thus prove, in a Popperian sense, that phase separation occurs. The evidence is simple and straightforward:

First, mitochondria behave as perfect osmometers, including a modest osmotic intercept (Fig. 3).

Second, the aqueous distributions of simple nonelectrolytes exhibit strikingly nonideal behavior (Fig. 5).

Third, the *combination* of osmotic swelling and solute distribution excludes a single-phase system (Fig. 6). In a single-phase model of matrix water, the slopes of Figs. 5 and 6 must be identical. They are not, and this simple fact is proof of aqueous phase heterogeneity in the mitochondrial matrix.

As a consequence of this evidence, the osmotic intercept, which constitutes a separate, osmotically inactive fraction of matrix water, may be identified as being responsible for the abnormal solute distributions.

B. The Nature of the Normal Aqueous Phase in Mitochondria

The normal phase exhibits all the characteristics of a dilute solution of potassium salts. It is a perfect osmometer, and the osmotic coefficient appears to be constant over the rather wide range studied, which is typical of a dilute salt solution. It expands and contracts with transport of salts across the inner membrane (Fig. 4). It dissolves nonelectrolytes with partition coefficients of 1.0. This behavior leads me to speculate that the normal phase contains no proteins, and that most matrix proteins are sequestered in the abnormal phase.

C. The Nature of the Abnormal Aqueous Phase in Mitochondria

1. Interpretation of Osmotic Curves and the Osmotic Intercept

Osmotic swelling curves, such as those in Figs. 2 and 3, must necessarily break down toward the origin as external osmolality approaches infinity (dehydration). Apropos the discussion of polar solutes in solution, we infer that hydration begins at extremely high osmolalities and achieves a constant saturation value at water activities much lower than those of dilute solutions. There is considerable evidence that this is an accurate description.

Ingenious studies on hemoglobin crystals by Perutz (1946) strongly suggest the presence of "bound" water at low water activities. The quantitative relationship of this fraction to the osmotic intercept in red cells was demonstrated by Drabkin (1950). Thus, the value for bound water in hemoglobin suggests that the linear part of the osmotic curve extends very close to the y axis ($1/\phi_0 \rightarrow 0$). This is also seen in studies of red cell membrane preparations in which hydration saturates at about 0.70 g H_2O/g dry membrane (Schneider and Schneider, 1972). This works out to a layer of water with an average thickness of 50 Å, very close to the minimum value for abnormal water associated with the mitochondrial membrane, as will be discussed.

One of the most thoroughly studied systems is that of the brine shrimp, *Artemia salina,* and this brief summary cannot do justice to the valuable work of Clegg and coworkers (Clegg, 1978, 1979, 1992), who have characterized the hydration state of *Artemia* using metabolic studies, nmr, differential scanning calorimetry, and dielectric measurements. Encysted brine shrimp gastrula commonly undergo complete desiccation during their development, and the concomitant cessation of metabolism is completely reversed by rehydration. Up to 0.3 g H_2O/g dry weight, all of the water entering the desiccated cyst is used in primary hydration of proteins and membranes. Below this value, no bulk water exists. Remarkably, several metabolic pathways become active at 0.3 g/g, indicating that metabolites can diffuse within this limited water phase. Sufficient water to support full-blown metabolism is only achieved at 0.6 g H_2O/g dry weight after which further hydration has little effect.

2. Solvent Properties of the Abnormal Phase

The abnormal aqueous phase of the mitochondrial matrix exhibits widely varying solute activity coefficients (Table I). The pattern that emerges from the limited number of solutes studied is that the abnormal phase preferentially attracts solutes with excess hydrogen bond donor groups (urea, antipyrine) and excludes aprotic solutes (DMSO). Thus, abnormal water behaves as if it contains excess electron acceptor groups relative to bulk water, that is, it has properties of a Lewis base. It follows that this phase should exclude other Lewis bases, such as anions of weak acids, and should concentrate Lewis acids, including protonated acids and amines, perhaps including tetraphenylphosphonium and other probes of membrane potential.

How does a solute such as urea, which is already extensively hydrogen-bonded in the normal phase, preferentially partition into abnormal water? Transfer to a region of more structured water would be energetically unfavorable, so it is unlikely that abnormal water is icelike and extensively hydrogen-bonded. If water structure begins with hydration of hydrogen-bond acceptor groups on membranes and proteins, subsequent layers of water would have the same polarization, inhibiting formation of three-dimensional clusters. Such a structure may well have a higher entropy than that of bulk water and would tend to exhibit properties of a Lewis base, as observed. Some of these predictions could be tested by studying the effects of temperature on the nonelectrolyte distributions.

3. Location of the Abnormal Phase and Distribution of Proteins

"Abnormal water" is a phenomenological term, introduced to describe physicochemical solution properties (Garlid, 1979). "Vicinal water" is a morphological term, introduced to describe water near surfaces (Drost-Hansen, 1969). I have suggested that abnormal matrix water and its associated proteins are located next to the inner membrane (Garlid, 1976). Srere (1982, 1985) has also reviewed a considerable body of evidence indicating that matrix proteins are adjacent to the inner membrane.

Studies on a variety of systems indicate that water hydrates the surfaces of proteins and membranes, leading to rapid saturation at low vapor pressures (high osmolalities). Further increases in water activity have little further effect on this surface phase. Thus, it seems reasonable to postulate that abnormal water is vicinal water—hydration water whose structure is induced by macromolecular surfaces.

The observation that proteins are extensively hydrated in solution, and that W_h is osmotically inactive, requires that the abnormal phase contain proteins. The question is, how much? What fraction of matrix proteins associates with the abnormal phase of mitochondria?

The answer to these questions may be found in the 10-fold disparity between hydration of macromolecules in solution and hydration of cells and organelles, as revealed by osmotic behavior. If cellular proteins were randomly located in the cytosol, then hydration water could occupy all of the cell, as pointed out by Cameron *et al.* (1997). The simplest explanation for this disparity is that cellular proteins are not in solution, but rather in a quasi-solid gel phase. Thus, soluble enzymes and other proteins preferentially associate with each other, presumably because protein–protein interactions are more favorable than water–protein interactions. I favor the view that virtually all proteins in the mitochondrial matrix are associated with this phase. Similarly, virtually all proteins in the cytosol, and their associated water, are associated with internal membrane surfaces and cytoskeleton. The effect of these interactions will be to minimize the water of hydration in cells and, thus, to minimize the osmotic intercept.

Because the hydration water behaves as a Lewis base, its time-averaged structure must differ from that of bulk water. Accordingly, the phase boundary is likely to be sharp rather than gradual.

4. The Extent of Abnormal Water in the Matrix

Taking the value of abnormal water at 0.28 g H_2O/g dry weight, I have calculated the thickness of this phase, assuming it forms a layer along the inner surface of the inner mitochondrial membrane (Garlid, 1998). I obtained two values, depending on the amount of matrix protein in this phase. The average thickness of the abnormal phase layer would be about 59 Å if it were protein-free. Its thickness would be about 189 Å if it were to contain all of matrix proteins. Because I believe that most of the proteins are associated with the abnormal phase, I favor estimates near the latter value.

D. Biological Consequences of Aqueous Phase Heterogeneity

A theme of this analysis is that proteins are segregated into a quasi-solid gel. This results in a minimization of the water of hydration, which would otherwise occupy all of the cell. There are now instances in biology which suggest that minimization of the hydration water may be variable and under regulation. Kelly *et al.* (1995) have shown that the immature frog oocyte exhibits osmotic behavior typical of most cells. The mature egg, however, is osmotically unresponsive, despite having a normal permeability to water and ions. Since the mature egg must survive in distilled (fresh) water, this is a marvelous adaptive mechanism. Having excluded other alternatives, Kelly *et al.* (1995) postulate that, during maturation, increased cross-linking of cytoskeletal polymers acts to increase the degree of cytoplasmic gelation. It is interesting to speculate that the opposite occurs; that the cytoskeleton *releases* proteins from the gel domain. This exposure of hydration surfaces would have the effect of converting all the bulk water to osmotically inactive hydration water, making the cell osmotically unresponsive and able to survive in pure water.

Similar regulatory changes appear to occur with single proteins that exist in different conformations, such as hemoglobin and the delayed rectifier K^+ channel of squid axon (Zimmerberg *et al.,* 1990; Colombo *et al.,* 1992). It is to be expected that the two conformations may have different hydration values, and these authors have devised methods for quantitating the difference. Parsegian and Rau (1984) propose that protein–water interactions (hydration) play a major role in both biological mechanisms and intracellular protein associations.

There are preliminary indications that the extent of abnormal water in mitochondria depends on the redox state of the electron transport chain. Although W_2 is normally invariant, I have observed that the uncoupler, CCCP causes a significant and reproducible reduction from 0.28 to 0.19 mg H_2O/mg protein, without changing the osmotic slope (Garlid, 1998). It would be interesting to know the effects of other membrane-localizing drugs on the osmotic intercept.

VIII. Concluding Remarks

Walter and Brooks (1995) have suggested that cellular proteins segregate into two or more phases, much like macromolecules in the test tube, as described by Albertsson (1971, 1986). Similar forces may be involved in cells, but I do not believe that proteins segregate randomly, by a purely physicochemical process. Rather, I believe that proteins undergo "channeling" of their own, on their passage from the ribosomes to their target destinations. In this view, the overwhelming majority of proteins are never "dissolved" in the bulk water of cytoplasm, but rather form a segregated domain more akin to a hydrated solid, or gel. In cytosol, this domain is associated with endoplasmic reticulum and cytoskeletal filaments. In the mitochondrial matrix, this domain is concentrated along the inner surface of the inner membrane.

Water in the hydration phase is osmotically inactive and separate from bulk cellular water. Both phases dissolve small solutes, but the partition coefficients may differ appreciably from one.

Localization of cellular proteins must be organized if bulk water is to be present in the cell. Protein localization is also a prerequisite for metabolic channeling, which appears to be ubiquitous in nature. The metabolic consequences of such microcompartmentation have been discussed extensively by Srere (1985) and Clegg (1978, 1979, 1992), and the present analysis supports Clegg's model of enzyme segregation (1978). Clegg (1992) has raised the additional questions of whether proteins are specifically localized and, if so, what determines this localization. In my view, the existence of organized enzyme complexes within the cell (Srere, 1987; Saks *et al.*, 1994) is now well established. Such assemblies would indeed require biological machinery, beyond the physicochemical forces causing aqueous phase separation, to provide the spatial organization and localization of proteins necessary for metabolic channeling.

Acknowledgments

The author wishes to acknowledge the expert technical assistance of Mr. Craig Semrad and to express special gratitude to James S. Clegg for his encouragement of these efforts in the early years. This work was supported in part by NIH grants GM31086 and GM55324 from the National Institute of General Medical Sciences.

References

Albertsson, P.-Å. (1971). "Partition of Cell Particles and Macromolecules." Wiley-Interscience, New York.

Albertsson, P.-Å. (1986). "Partition of Cell Particles and Macromolecules." Wiley-Interscience, New York.

Beavis, A. D., Brannan, R. D., and Garlid, K. D. (1985). Swelling and contraction of the mitochondrial matrix. *J. Biol. Chem.* **260,** 13424–13433.

Bentzel, C. J., and Solomon, A. K. (1967). Osmotic properties of mitochondria. *J. Gen. Physiol.* **50,** 1547–1563.

Blanchard, K. (1940). Water, free and bound. *Quantv. Biol.* **8,** 1–8.

Cameron, I. L., Kanal, K. M., Keener, C. R., and Fullerton, G. D. (1997). A mechanistic view of the non-ideal osmotic and motional behavior of intracellular water. *Cell Biol. Int.* **21,** 99–113.

Clegg, J. S. (1978). Hydration-dependent metabolic transitions and the state of cellular water in *Artemia* cysts. *In* "Dry Biological Systems" (J. H. Crowe and J. S. Clegg, eds.), pp. 117–153. Academic Press, New York.

Clegg, J. S. (1979). Metabolism and the intracellular environment. *In* "Cell-Associated Water" (W. Drost-Hansen and J. S. Clegg, eds.), pp. 363–413. Academic Press, New York.

Clegg, J. S. (1992). Cellular infrastructure and metabolic organization. *Curr. Top. Cell. Regul.* **33,** 3–14.

Colombo, M. F., Rau, D. C., and Parsegian, V. A. (1992). Protein solvation in allosteric regulation: A water effect on hemoglobin. *Science* **256,** 655–659.

Cooke, R., and Kuntz, I. D. (1974). The properties of water in biological systems. *Annu. Rev. Biophys. Bioeng.* **3,** 95–126.

Dick, D. A. T. (1971). Water movements in cells. *In* "Membranes and Ion Transport" (E. E. Bittar, ed.), Vol. 3, p. 211. Wiley-Interscience, London.

Drabkin, D. L. (1950). Hydration of macrosized crystals of human hemoglobin, and osmotic concentrations in red cells. *J. Biol. Chem.* **185,** 231–245.

Drost-Hansen, W. (1969). Structure of water near solid interfaces. *Ind. Eng. Chem.* **61,** 10–47.

Gamble, J. L., Jr., and Hess, R. C., Jr. (1966). Mitochondrial electrolytes. *Am. J. Physiol.* **210,** 756–770.

Garlid, K. D. (1976). Free and bound water in mitochondria: Two distinct aqueous phases with different solution properties. *In* "L'Eau et les Systèmes Biologiques" (A. J. Berteaud and A. Alfsen, eds.), pp. 317–321. CNRS, Paris.

Garlid, K. D. (1978). Overview of our understanding of intracellular water in hydrated cells. *In* "Dry Biological Systems" (J. H. Crowe and J. S. Clegg, eds.), pp. 3–19. Academic Press, New York.

Garlid, K. D. (1979). Aqueous phase structure in cells and organelles. *In* "Cell-Associated Water" (W. Drost-Hansen and J. S. Clegg, eds.), pp. 293–361. Academic Press, New York.

Garlid, K. D. (1980). On the mechanism of regulation of the mitochondrial K^+/H^+ exchanger. *J. Biol. Chem.* **255,** 11273–11279.

Garlid, K. D. (1988). Mitochondrial volume control. *In* "Integration of Mitochondrial Function" (J. J. Lemasters, C. R. Hackenbrock, R. G. Thurman, and H. V. Westerhoff, eds.), pp. 257–276. Plenum, New York.

Garlid, K. D. (1999). Unpublished results.

Garlid, K. D., and Beavis, A. D. (1985). Swelling and contraction of the mitochondrial matrix. *J. Biol. Chem.* **260,** 13434–13441.

Kelly, S. M., Butler, J. P., and Macklem, P. T. (1995). Control of cell volume in oocytes and eggs from *Xenopus laevis. Comp. Biochem. Physiol. A* **111A,** 681–691.

Kirkwood, J. G., and Oppenheim, I. (1961). "Chemical Thermodynamics." McGraw-Hill, New York.

Ling, G. N. (1970). The physical state of water in living cells and its physiological significance. *Int. J. Neurosci.* **1,** 129–152.

Ling, G. N. (1972). Hydration of macromolecules. *In* "Water and Aqueous Solutions" (R. A. Horne, ed.), p. 663. Wiley-Interscience, New York.

Mendelsohn, D., and Levin, N. W. (1960). A colorimetric micromethod for the estimation of antipyrine in plasma or serum. *S. Afr. J. Med. Sci.* **25,** 13–18.

Parsegian, V. A., and Rau, D. C. (1984). Water near intracellular surfaces. *J. Cell Biol.* **99,** 196s–200s.

Parsegian, V. A., Rand, R. P., and Rau, D. C. (1995). Macromolecules and water: Probing with osmotic stress. *In* "Methods in Enzymology" (M. L. Johnson and G. K. Ackers, eds.), **259,** pp. 43–93. Academic Press, San Diego, CA.

Perutz, M. D. (1946). The composition and swelling properties of haemoglobin crystals. *Trans. Faraday Soc. London* **42B,** 187–197.

Saks, V. A., Khuchua, Z. A., Vasilyeva, E. V., Belikova, O. Yu., and Kuznetsov, A. V. (1994). Metabolic compartmentation and substrate channeling in muscle cells. *Mol. Cell. Biochem.* **133/134,** 155–192.

Scatchard, G. (1921). Hydration of sucrose in water solution as calculated from vapor-pressure measurements. *J. Am. Chem. Soc.* **43,** 2406–2418.

Schneider, M. J. T., and Schneider, A. S. (1972). Water in biological membranes: Adsorption isotherms and circular dichroism as a function of hydration. *J. Membr. Biol.* **9,** 127–140.

Srere, P. A. (1982). The structure of the mitochondrial inner membrane-matrix compartment. *Trends Biochem. Sci.* **7,** 375–378.

Srere, P. A. (1985). Organization of proteins within the mitochondrion. *In* "Catalytic Facilitation in Organized Multienzyme Systems" (G. R. Welch, ed.), pp. 1–61. Academic Press, New York.

Srere, P. A. (1987). Complexes of sequential metabolic enzymes. *Annu. Rev. Biochem.* **56,** 89–124.

Stoner, C. D., and Sirak, D. J. (1969). Osmotically induced alterations in volume and ultrastructure of mitochondria isolated from rat liver and bovine heart. *J. Cell Biol.* **43,** 521–538.

Tedeschi, H., and Harris, D. L. (1955). The osmotic behavior and permeability to nonelectrolytes of mitochondria. *Arch. Biochem. Biophys.* **58,** 52–67.

Troshin, A. S. (1966). "Problems of Cell Permeability." Pergamon, Oxford.

Walter, H., and Brooks, D. E. (1995). Phase separation in cytoplasm, due to macromolecular crowding, is the basis for microcompartmentation. *FEBS Lett.* **361,** 135–139.

Wolf, A. V., and Brown, M. G. (1965). Concentrative properties of aqueous solutions. *In* "Handbook of Chemistry and Physics," 46th ed., pp. D163–D164. Chemical Rubber Co., Cleveland, OH.

Zimmerberg, J., Bezanilla, F., and Parsegian, V. A. (1990). Solute inaccessible aqueous volume changes during opening of the potassium channel of the squid giant axon. *Biophys. J.* **57,** 1049–1064.

Zimmerman, R. J., Kanal, K. M., Sanders, J., Cameron, I. L., and Fullerton, G. D. (1995). Osmotic pressure method to measure salt induced folding/unfolding of bovine serum albumin. *J. Biochem. Biophys. Methods* **30,** 113–131.

Mechanisms for Cytoplasmic Organization: An Overview

Len Pagliaro
Cerep, Inc., Redmond, Washington 98052

One of the basic characteristics of life is the intrinsic organization of cytoplasm, yet we know surprisingly little about the manner in which cytoplasmic macromolecules are arranged. It is clear that cytoplasm is not the homogeneous "soup" it was once envisioned to be, but a comprehensive model for cytoplasmic organization is not available in modern cell biology. The premise of this volume is that phase separation in cytoplasm may play a role in organization at the subcellular level. Other mechanisms for non-membrane-bounded intracellular organization have previously been proposed. Some of these will be reviewed in this chapter. Multiple mechanisms, involving phase separation, specific intracellular targeting, formation of macromolecular complexes, and channeling, all could well contribute to cytoplasmic organization. Temporal and spatial organization, as well as composition, are likely to be important in defining the characteristics of cytoplasm.

KEY WORDS: Cytoplasm, Compartmentation, Macromolecular interactions, Cytoplasmic organization, Targeting, Trafficking.

I. Introduction

"It is believed that the directing effect of the internal surfaces is displayed predominantly by an organized network of protein molecules, forming a three-dimensional mosaic extending throughout the cell. The enzymes would form part of this structure, their activity being largely controlled by the mosaic. This conception of living matter reconciles some difficulties; for instance, independent chemical reactions could proceed simultaneously in various parts of the cell but, at the same time, the mosaic could react as a whole to stimuli transmitted from the cell surface" (Peters, 1930, pp. 797–798).

International Review of Cytology, Vol. 192
0074-7696/00 $30.00

Almost 70 years ago, Sir Rudolph Peters considered cells from a three-dimensional, systems perspective in his elegant paper entitled "Surface Structure in the Integration of Cell Activity." As important as this perspective was, there are still gaps in our understanding of biology at the whole cell level, in part due to the reductionist tradition of biochemistry and molecular biology, and in part due to the difficulty of studying whole, individual, living cells. Nevertheless, defining the mechanisms by which the many molecular components of metabolism are organized and integrated into the highly regulated three-dimensional system of the cell is an essential challenge for biologists. In this chapter, I will argue that the scheme Peters proposed is basically correct and that, in the intervening years, several mechanisms have been identified which together could direct intracellular organization.

II. Cytoplasm as a Biochemical Environment: Compartmentation is a Fundamental Characteristic

All life is characterized by the ability to establish and maintain order (Schrödinger, 1944). Within a living cell, order is established by compartmentation. The most widely applied definition of intracellular compartmentation is that of a membrane-bound compartment that is topologically discontinuous with the cytoplasmic space from which is it separated. Examples of this include the nucleus, mitochondrion (both inner and outer membranes), Golgi apparatus, endoplasmic reticulum, and a variety of endocytic and secretory vesicles. Clearly, this form of structural compartmentation is important, but there is substantial evidence that traditional biochemical and ultrastructural approaches tend to emphasize and/or preserve this kind of compartmentation preferentially, perhaps at the expense of other less obvious—but perhaps no less physiologically significant—mechanisms for intracellular compartmentation. Even simple macromolecular complexes can constitute effective mechanisms for intracellular compartmentation on a limited scale, but these are difficult to demonstrate experimentally.

Other compartmentation mechanisms may be transient, existing in a fully functional form only for microseconds (Kaprelyants, 1988). Again, while it is difficult to obtain a biochemical or ultrastructural "snapshot" of this kind of organization, it may still play an essential role in cell physiology. Characterizing and understanding weak and/or transient regimes of intracellular organization continue to be an important area of research.

Cytoplasm is generally defined rather loosely as the substance, in the space outside of the nucleus and inside the plasma membrane, that is not part of a distinct organelle such as the mitochondria, Golgi, or endoplasmic reticulum. A similar lack of detail is conveyed by operational definitions, such as the term "cytosol," first used by Lardy (1965) or more recently the "aqueous cytoplasm" (Clegg, 1984a). While these descriptions have been useful from a microanatomical point of view, they do little to convey the functional characteristics of the substance contained in the cytoplasmic space. As our understanding of cytoplasm improves, it is becoming clear that an important characteristic of cytoplasm is not only its molecular composition or physical properties but also the complex information network which defines cytoplasm and distinguishes it from other cellular regions. This type of "megamacromolecular" or "submicroscopic" level of organization was termed a "no man's land" by DeDuve (1984) because this level of biological organization is below the resolution limit of conventional microscopy, yet of larger scale and more complex than can be easily elucidated by conventional chemical approaches. In spite of these challenges, a number of organizational schemes have been identified in cytoplasm and there is abundant evidence that cytoplasm contains organization that is ultrastructurally and physiologically significant (Bhargava, 1985).

To study the physical chemical basis for the formation of cytoplasmic structure, it is necessary to consider the properties of the cytoplasmic proteins themselves. Among the factors which may be important are the high concentration of intracellular proteins (Srere, 1967) and the large size of enzymes which provide adequate surface for protein interaction as well as the carefully structured active site (Srere, 1984). Much of the mass of many enzymes is not directly involved in their catalytic roles but contributes other characteristics essential for cytoplasmic organization, including protein folding, protein–protein interactions, and intracellular targeting. These secondary protein characteristics provide a unique cytoplasmic environment quite distinct from the mixture of activities that one might measure if the respective proteins were simply mixed to homogeneity *in vitro*. Further, the high concentration of many cytoplasmic enzymes and structural proteins means that relatively weak molecular interactions can be significant under physiological conditions. The potential importance of weak binding can be seen in the interactions between many of the glycolytic enzymes and either the actin cytoskeleton, the tubulin cytoskeleton, or both (Arnold and Pette, 1968; Bronstein and Knull, 1981; Karkhoff-Schweizer and Knull, 1987; Walsh *et al.*, 1989). It is often assumed that a K_D in the micromolar range, as is the case for actin and aldolase (Walsh *et al.*, 1977), would not be physiologically important. However, the intracellular concentrations of both aldolase and actin are in the tens of micromolar (Srivastava and Bernhard, 1986), i.e., much higher than generally used for *in vitro* biochemical studies. Thus, conditions *in vivo* make this weak interaction physiologically significant.

In addition to high concentrations of individual molecular species, the high total macromolecular concentration in cytoplasm leads to crowding that influences protein–protein interactions (Zimmerman and Minton, 1993). One must also consider the effect of volume exclusion in a confined system. Thus, there exist diffusional constraints due to molecular "meshing" or "sieving" in cytoplasmic structure (Gershon *et al.,* 1985; Luby-Phelps *et al.,* 1987, 1988). Confinement of molecules in the pores of such a network would be an additional driving force for association of macromolecular species (Minton, 1992).

While water is abundant in cytoplasm, the high concentration of solvated molecules may lead to a significant amount of this water on/around some macromolecular species to be in an ordered state (Clegg, 1984b, 1991). The water in this zone may be relatively discontinuous with bulk-phase water in the cell, leading to what physiologists refer to as an "unstirred layer." A domain containing ordered water could result in diffusive characteristics of reduced dimensionality, including two-dimensional and one-dimensional diffusion (Berg and von Hippel, 1985), with corresponding changes in diffusion times. Such constrained or reduced dimensionality could itself represent a relatively nonspecific mechanism for cytoplasmic compartmentation. Osmotically active and inactive water have also been described (see chapter by Garlid in this volume).

Our understanding of advantages of cytoplasmic organization has evolved over the past few decades. One explanation, based on simple membrane-bound compartments, is that the kinetics of a given series of enzymes packaged together into a smaller volume with the necessary substrates and cofactors will lead to faster catalytic throughput for the system. While true in simple systems, and clearly important in more complicated systems, the complexity of cytoplasm is several orders of magnitude greater than systems that can be replicated *in vitro* or described with computer models. Concepts for metabolic regulation, such as the existence of a single rate-limiting step in a metabolic pathway, do not adequately describe the complex metabolic systems in a cell. More recent approaches such as metabolic control analysis (Fell, 1992), in which control is shown to be distributed at various steps of the pathway, provide a description of metabolic regulation that is much better adapted to the systems context of complex pathways. They are also in agreement with recent metabolic experiments. In turn, compartmentation of metabolism could provide part of the explanation for fine metabolic regulation in a complex milieu.

III. Mechanisms, Models, and Dynamics for Establishing and Maintaining Cytoplasmic Organization

Some molecules and molecular complexes can self-assemble, and the terms "self-assembly" and "self-organization" have been used to describe these poorly understood complex molecular interactions (Hess and Mikhailov, 1994). There is theoretical evidence that self-organization plays important roles in the generation of molecular networks in cytoplasm (Hess and Mikhailov, 1995) or flow-based intracellular traffic (Mikhailov and Hess, 1995). At present, the rules which govern the ability of molecules to organize and contribute to cytoarchitecture are largely unknown.

Some important mechanisms proposed as organizational elements of cytoplasm are given in Table I. They share important features. First, these schemes do not involve a membrane-delineated compartment (as in the case of a membrane-bounded organelle, e.g., mitochondrion or vesicle). Second, they describe transient events that occur on time scales that are short relative to cellular life span, but long enough to have significant physiological consequences. It is, thus, a biology of metastable organization. Third, these interactions are generally dependent on secondary, weak binding activities of molecules that are known for a primary (typically catalytic) activity. Weak binding activities can become important when the molecular species involved are present in high numbers and/or high local concentrations—a characteristic of cytoplasm. While there is clearly some overlap between the categories below, I have tried to assign a concept to a category based on its predominant feature.

A. Membranes and Surfaces

Apart from the basic form of membrane-bounded compartmentation discussed above, membranes and other intracellular surfaces (e.g., those provided by macromolecules) play an organizational role by "reduction of dimensionality" of reactions that are primarily diffusion-driven (Adam and Delbrück, 1968). Such surfaces reduce the number of dimensions in which molecules can diffuse from three to two in the case of membranes, or to one in the case of DNA-binding proteins diffusing on a DNA molecule

TABLE I
Suggested Bases for macromolecular and Cellular Organization

Mechanism	Event	Reference
Reduction of dimensionality	Diffusion is constrained at/near surfaces.	Adam and Delbrück (1968)
Reversible adsorption	Reversible adsorption of enzymes as a possible allosteric control mechanism.	Masters *et al.*, (1969)
Ambiquity	Structural allostery; variation in intracellular distribution as a regulatory mechanims.	Wilson (1978)
Quinary structure	Transient macromolecular interactions *in vivo*.	McConkey (1982)
Metabolon	Supramolecular complex of sequential metabolic enzymes and cellular structural elements.	Srere (1985)
Metabolite transfer	Direct enzyme–enzyme transfer of metabolites without involvement of the aqueous environment.	Srivastava and Bernhard (1986)
Surface relay	Transient surface adsorption resulting in local accumulation and functional compartmentation of enzymes.	Ryazanov (1988)
Pairwise dynamic associations	Metabolite-modulated dynamic enzyme assembly.	Ovádi (1988)
Diazymes	Mobile clusters of enzymes interacting via common metabolic products.	Dillon and Clark (1990)

(Berg and von Hippel, 1985). Such surface interactions define and limit the microenvironment in which these molecules interact, creating a functional compartment and reducing diffusive distances and times. The surface may, thus, play a role that is no less important for maintaining intracellular organization than an enclosing membrane is for a vesicle, but the less well circumscribed nature of the compartment formed makes it more difficult to define rigorously.

Surface interactions and reduced dimensionality are also key components in other schemes that have been proposed. The concept of "reversible adsorption"—now almost 30 years old—was based on a relatively weak interaction between enzyme molecules and a surface (Masters *et al.,* 1969). According to this model, the interaction between an enzyme and a surface could do two things. First, it could localize the enzyme near a surface, and second, interaction between the enzyme and the surface could modify the activity of the enzyme, with the surface taking on some of the characteristics of an allosteric subunit for the enzyme. This phenomenon was termed ambiquity by Wilson (1978). Experimental evidence for the second of these consequences was easier to generate. A number of cytoplasmic enzymes, including aldolase (Walsh *et al.,* 1977), hexokinase (Wilson, 1980), and phosphofructokinase (Andrès *et al.,* 1996), has been shown to have modified activities when they interact with cytoskeletal surfaces under physiological conditions. In addition, a fraction of aldolase is concentrated on or near the surface of the actin cytoskeleton (Pagliaro and Taylor, 1988) and the localization of aldolase in cytoplasm is metabolically sensitive (Pagliaro and Taylor, 1992), consistent with physiological relevance of this enzyme localization. Together, reduction of dimensionality, reversible adsorption, and ambiquity describe a regime of cytoplasmic organization that is consistent with that proposed by Peters in 1930.

An excellent model for a dynamic, surface-based mechanism organizing a sequential metabolic pathway, termed "surface relay," was developed by Ryazanov (1988). A series of enzymes interact with a surface. Binding of substrate to one enzyme tends to release that enzyme from the surface. The resulting ES complex has a greater affinity than the free enzyme for the next enzyme in the series which is still surface-bound. When the ES complex binds to the next enzyme in the series, the first enzyme is released. The reduced dimensionality of the zone near the surface increases the probability that the free enzyme will reassociate with the surface, thus maintaining a domain enriched in enzymes and substrates near the surface. An important aspect of this model underscores the importance of studying these phenomena at the system level. Although the substrate for any one enzyme may inhibit surface binding of that enzyme, metabolic flux through the system (which mandates the presence of substrate for all reactions) will tend to maintain the integrity of a non-membrane-bound compartment near the surface. Even a simple system can exhibit behavior not predicted by the action of any of its components.

For a physical chemical analysis, it is easiest to consider a given molecule as either bound or free, relative to the surface with which it can interact. For a complex ensemble of molecules dimensionally constrained near a surface, any individual molecule may be either bound or free on a time scale of microseconds (Sperelakis, 1995). Establishing the state of molecules is difficult even when time scales of the order of seconds are involved. The finding that, *on average,* a certain fraction of a given population of molecules is bound leaves us with only a low resolution description of the system. One of the best approaches to this challenge has come from work in the chemical engineering field on enzymes interacting with substrate that is immobilized on surfaces (Gaspers *et al.,* 1994). This study used total internal reflection fluorescence photobleaching to monitor the diffusive characteristics of fluorescently labeled enzymes near surfaces bearing differing patterns and densities of tethered substrate molecules. It was shown that enzymes diffused in two dimensions on a surface containing substrate, and that both diffusion and catalytic rates of the enzyme could be regulated by the pattern and density of the substrate presentation on the surface, consistent with two-dimensional diffusion. While this approach is useful for characterizing macromolecular organization and behavior in a complex system, it cannot be adapted readily to studies of living cells or tissues.

B. Macromolecular Complexes

A variety of macromolecular complexes have been described, including the tricarboxylic acid cycle complex (Robinson and Srere, 1985), the fatty acid synthase complex (Wakil *et al.,* 1983), and the pyruvate dehydrogenase complex (Reed, 1969; Patel and Roche, 1990; Mattevi *et al.,* 1992). Complexes such as these are considered to be relatively stable, i.e., can be isolated intact from cells, and are generally the exception, rather than the rule, in metabolic biochemistry. Srere (1985) coined the term "metabolon" to describe functional units (such as the TCA cycle complex) which integrate a segment of a metabolic pathway in a functional macromolecular complex which can be isolated. One of the characteristics of a metabolon is specific binding interactions between/among the molecules in the complex (Small *et al.,* 1995). These interactions may appear to be primarily structural at the molecular level in that they do not affect the kinetics of the individual enzymes in solution. However, the effects of such interactions on metabolic flux is often difficult to predict based on the behavior of the individual constituent molecules.

A related concept was proposed in an elegant paper by McConkey (1982), in which he defined the level of organization above quaternary structure as "quinary structure." This concept is very similar to the term "megamacromolecular" used by DeDuve (1984) to describe high order molecular organization. However, McConkey arrived at his conclusion using an approach quite different from that used by DeDuve. McConkey argued that the small differences in isoelectric point and size in the polypeptides of distantly related species cannot be explained except by the evolutionary conservation of surface charge and function. He proposed that the surface characteristics of proteins play an important role in determining cytoplasmic, tissue, and ultimately organismal organization based on protein–protein interactions. Thus, unless there is retention of surface properties, relatively small sequence differences that do not affect catalytic activity at the molecular level might influence higher order cytoplasmic organization and have major effects on the tissue and organism.

C. Intracellular Targeting and Trafficking

In many ways, mechanisms for cytoplasmic organization via intracellular targeting of macromolecules form a continuum with macromolecular complex formation. The class of macromolecular complexes discussed above are relatively stable (half-lives on the order of minutes to many hours) and, generally, involve only a few neighboring molecules. Intracellular sorting and targeting mechanisms, on the other hand, are generally short lived (half-lives on the order of milliseconds to minutes) and may involve fairly extensive intracellular domains. Targeting and trafficking pathways are generally divided into two major categories: secretory and endocytic (Alberts *et al.,* 1994; Stryer, 1995). The secretory pathway encompasses protein production and modification including synthesis, processing, targeting, transport, and secretion; the endocytic pathway includes receptor-mediated endocytosis, vesicle formation, internalization, and degradation/processing in the endosome/lysosome and ubiquitin/proteasome mediated systems. Which of the components of these two systems becomes involved in the life cycle of a given protein is governed by the nature of the protein and cellular localization. Secreted (e.g., insulin), cytoplasmic (e.g., glycolytic enzymes), nuclear (e.g., transcription factors) proteins, and so on, will thus involve different pathway components or combinations of components.

The secretory and endocytic pathways have been extensively studied and many of the molecular components and interactions involved in intracellular targeting have been characterized. It is clear that a host of specific macromolecular interactions is involved in targeting, trafficking, and maintaining the polarity in cells. One of the earliest signal sequences identified was that present on nascent proteins to permit their transport across the endoplasmic reticulum membrane. The signal sequence is subsequently removed by cleavage (Blobel, 1980; Walter and Lingappa, 1986). The signal sequence is necessary for targeting protein to the rough ER so that the newly formed protein can be processed (posttranslational modifications such as glycosylation) and packaged for transport to a desired intracellular location or to the cell surface for an exocytic secretory event (Rapoport, 1992). While specific molecular interactions are involved in the sorting and targeting of proteins, the extensive ER–Golgi membranous network is also an essential part of the system, enclosing proteins for secretion in a space that is topologically distinct from that of the bulk of the cytoplasm. Thus, this kind of intracellular organization is, ultimately, membrane-dependent but in a way that uses specific molecular interactions and active cytoskeletal transport to achieve organization (Nelson, 1991; Rothman, 1994).

Enzymes of other metabolic pathways in the cytoplasm depend on another-kind of intracellular targeting. When no membrane transport is involved, proteins lack signal sequences but have binding domains for cytoplasmic structural components. A number of glycolytic enzymes have binding activities targeted to the actin or tubulin cytoskeleton (or to both). In addition, binding domains for enzymes may exist on cytoskeletal proteins (Carr and Knull, 1993; Volker and Knull, 1993). There is experimental evidence that the intracellular localization of aldolase described above is driven, at least in part, by specific molecular binding to F-actin. A number of site-directed mutants of aldolase were used and specific mutants found which affect either catalytic or actin-binding activity (Wang *et al.,* 1996). Fluorescent analogs of such mutants were then microinjected into cells and localized differentially using ratio imaging (Wang *et al.,* 1997). A single amino acid substitution which affected actin-binding activity on aldolase also significantly affected its intracellular localization, thus providing evidence that a specific molecular interaction is responsible for targeting this enzyme to a domain on/around the actin cytoskeleton. It is not yet known what the effect of binding is on the enzyme's activity.

D. Functional Compartmentation; Channeling

One of the obvious consequences of cytoplasmic organization is the process known as channeling [see Ovádi (1995) for review and chapter by Ovádi and Srere, this volume), a process independent of the actual nature of cytoplasmic organization. Organization of metabolic enzymes (e.g., in metabolons) may result in high local concentrations of metabolic intermediates which ensure high metabolic rates, even when the bulk concentration of the metabolites is relatively low. These functional compartments increase the probability of certain preferred reactions taking place without the need for the membrane-bound structures associated with physical compartments. Such compartments clearly do not have the structural integrity of a membrane-bound organelle but there is good evidence that they can segregate biochemical pathways with common intermediates effectively. Hardin and Kushmerick (1994) used ^{13}C-NMR studies to show that cytoplasmic glucose and glycogen consumption remain cleanly separated in smooth muscle, even though common enzymes of glycolysis and glycogenolysis are present in the cytoplasmic compartment. More recent work has demonstrated that glycolysis and glycogenolysis can be regulated independently by exogenously applied substrates (Hardin and Roberts, 1997), establishing a high degree of separation of the two pathways in the same topological compartment. Clearly, the environmental volumes accessible to even identical intracellular substrates is quite different from that deduced from conventional ultrastructural studies.

A number of related models for functional compartments or "channels" have been developed, and they share some common features. Most involve relatively few kinds of molecules that interact specifically. These interactions provide an environment in which a given intermediate is extremely likely to be captured by the next molecule in the series, and the probability that it will diffuse away into the bulk cellular pool of intermediates is small. Specific examples include "metabolite transfer," originally based on substrate "handoffs" in dehydrogenase complexes (Srivastava and Bernhard, 1986), "pairwise dynamic associations," a similar mechanism in which metabolites mediate the formation of transient E-S-E complexes (Ovádi, 1988), and "diazymes," proposed for the coupling of pyruvate kinase and creatine kinase (Dillon and Clark, 1990). These mechanisms are representative of dynamic channels in which the channel is transient with a very short half-life (Mendes *et al.,* 1992). The transient nature of dynamic channels does not necessarily make them any less important physiologically. Transient events that occur in the time domain of molecular dynamics can be critically important for a metabolic pathway whose reactions exist in this time domain but can not be isolated biochemically.

E. Multiple Mechanisms for Cytoplasmic Organization

Multiple weak interactions can have major effects (Pagliaro, 1993). It is likely that the role of large-scale, weak interactions is to provide the molecular mechanism(s) which drive cytoplasmic organization. Widespread weak interactions, coupled with molecular crowding (Minton, 1983, 1992; Zimmerman and Minton, 1993; Zimmerman, 1993), clearly have the potential to establish non-membrane-bounded cytoplasmic organization. The weakness of the interactions causes them sometimes to be disregarded, but at high concentrations of reactants in crowded conditions, weak interactions can not be neglected (see chapter by Johansson *et al.,* this volume).

The dynamic organization of the cytoskeleton is well established and there is evidence that both the actin and tubulin cytoskeleton occur as liquid crystals *in vivo.* Studies on microtubule solutions, based on both packing dynamics (Brown and Berlin, 1985) and optical birefringence (Hitt *et al.,* 1990), have shown that polymerized microtubules can form nematic liquid crystal phases *in vitro,* and that microtubule-associated proteins (MAPs) can regulate formation of these phases. Similarly, actin filaments can form nematic liquid crystals *in vitro* (Kerst *et al.,* 1990), and the actin-binding protein gelsolin can play a role in modulating liquid crystal formation (Furukawa *et al.,* 1993). These studies suggest that a role of cytoskeletally associated proteins (both MAPs and actin-binding proteins) is to regulate the liquid crystalline phase of the cytoskeleton. While not yet established *in vivo,* liquid crystalline cytoskeletal dynamics is an intriguing possibility that bears a resemblance to the "three-dimensional mosaic" proposed by Peters in 1930.

There are important biological cell phenomena that are difficult to explain by conventional mechanisms. For example, it is well established that cytoplasm has an inherent mesh size that excludes diffusive translocation of molecules larger than that mesh size (Gershon *et al.,* 1985; Luby-Phelps *et al.,* 1987, 1988). What is less clear is how this mesh is regulated in three-dimensional space over time. Swanson and McNeil (1987) injected mammalian cells with differentially labeled large and small fluorescent dextrans and observed the behavior of the dextrans as the cells underwent mitosis. The large dextrans were excluded from the nucleus in interphase, but as the nuclear envelope broke down and the cells entered mitosis, the two dextrans intermingled and surrounded the condensed chromosomes. The surprising result, however, was that the large dextran was again excluded from the nascent nucleus at the end of mitosis, before the nuclear envelope reformed. The fact that these were neutral, inert dextrans argues that specific molecular interactions were not involved, and that molecular sieving excluded the larger particles.

IV. Concluding Remarks

The genomes of yeast and bacteria have been sequenced, and the sequence of the human genome will also be available soon. Analogously, many proteins have been characterized by their sequence and three-dimensional structure. The fields of genomics and proteomics have grown and become established on the basis of technologies which permit sequencing the genome, establishing amino sequences and relating protein structure and function. We know much about the stepwise logic and linear sequences of metabolic pathways. In spite of this information, however, our understanding of metabolic regulation in cytoplasm remains incomplete. Organizational mechanisms for metabolism could have a significant bearing on regulation, and there is good evidence that regulation *in vivo* is far more complex than simple models based on individual rate-limiting steps. Metabolic control is probably widely distributed among constituent members of a given pathway (Fell, 1992). This kind of regulation is consistent with metabolic organization in cytoplasm that resembles Peters' description. Specific interactions among the macromolecules of intermediary metabolism could be the mechanism that establishes and maintains this level of cellular organization.

References

Adam, G., and Delbrück, M. (1968). *In* "Structural Chemistry and Molecular Biology" (A. Rich and N. Davidson, eds.). Freeman, San Francisco.

Alberts, B., Bray, D., Lewis, J., Raff, M., Roberts, K., and Watson, J.D. (1994). "Molecular Biology of the Cell," 3rd ed., Chapters 12 and 13. Garland Publishing, New York and London.

Andres, V., Carreras, J., and Cusso, R. (1996). Myofibril-bound muscle phosphofructokinase is less sensitive to inhibition by ATP than the free enzyme, but retains its sensitivity to stimulation by bisphosphorylated hexoses. *Int. J. Biochem. Cell Biol.* **28,** 1179–1184.

Arnold, H., and Pette, D. (1968). Binding of glycolytic enzymes to structure proteins of the muscle. *Eur. J. Biochem.* **6,** 163–171.

Berg, O. G., and von Hippel, P. H. (1985). Diffusion-controlled macromolecular interactions. *Annu. Rev. Biophys. Biophys. Chem.* **14,** 131–160.

Bhargava, P. M. (1985). Is the "soluble" phase of cells structured? *BioSystems* **18,** 135–139.

Blobel, G. (1980). Intracellular protein topogenesis. *Proc. Natl. Acad. Sci. U.S.A* **77,** 1496–1500.

Bronstein, W. W., and Knull, H. R. (1981). Interaction of muscle glycolytic enzymes with thin filament proteins. *Can. J. Biochem.* **59,** 494–499.

Brown, P. A., and Berlin, R. D. (1985). Packing volume of sedimented microtubules: Regulation and potential relationship to an intracellular matrix. *J. Cell. Biol.* **101,** 1492–1500.

Carr, D., and Knull, H. (1993). Aldolase-tubulin interactions: Removal of tubulin C-terminals impairs interactions. *Biochem. Biophys. Res. Commun.* **195,** 289–293.

Clegg, J. S. (1984a). Properties and metabolism of the aqueous cytoplasm and its boundaries. *Am. J. Physiol.* **246,** R133–R151.

Clegg, J. S. (1984b). Intracellular water and the cytomatrix: Some methods of study and current views *J. Cell Biol.* **99,** 167s–171s.

Clegg, J. S. (1991). Metabolic organization and the ultrastructure of animal cells. *Biochem. Soc. Trans.* **19,** 985–991.

DeDuve, C. (1984). "A Guided Tour of the Living Cell," Vol. 1, p. 17. Scientific American Book, New York.

Dillon, P. F., and clark, J. F. (1990). The theory of diazymes and functional coupling of pyruvate kinase and creatine kinase. *J. Theor. Biol.* **143,** 275–284.

Fell, D. A. (1992). Metabolic control analysis: A survey of its theoretical and experimental development. *Biochem. J.* **286,** 313–330.

Furukawa, R., Kundra, R., and Fechheimer, M. (1993). Formation of liquid crystals from actin filaments. *Biochemistry* **32,** 12346–12352.

Gaspers, P. B., Robertson, C. R., and Gast, A. P. (1994). Enzymes on immobilized substrate surfaces: Diffusion. *Langmuir* **10,** 2699–2704.

Gershon, N. D., Porter, K. R., and Trus, B. L. (1985). The cytoplasmic matrix: Its volume and surface area and the diffusion of molecules through it. *Proc. Natl. Acad. Sci. U.S.A.* **82,** 5030–5034.

Hardin, C. D., and Kushmerick, M. J. (1994). Simultaneous and separable flux of pathways for glucose and glycogen utilization studied by ^{13}C-NMR. *J. Mol. Cell. Cardiol.* **26,** 1197–1210.

Hardin, C. D., and Roberts, T. M. (1997). Differential regulation of glucose and glycogen metabolism in vascular smooth muscle by exogenous substrates. *J. Mol. Cell. Cardiol.* **29,** 1207–1216.

Hess, B., and Mikhailov, A. (1994). Self-organization in living cells. *Science* **264,** 223–224.

Hess, B., and Mikhailov, A. (1995). Microscopic self-organization in living cells: A study of time matching. *J. Theor. Biol.* **176,** 181–184.

Hitt, A. L., Cross, A. R., and Williams, R. C. (1990). Microtubule solutions display nematic liquid crystalline structure. *J. Biol. Chem.* **265,** 1639–1647.

Kaprelyants, A. S. (1988). Dynamic spatial distribution of proteins in the cell. *Trends Biochem. Sci.* **13,** 43–46.

Karkhoff-Schweizer, R., and Knull, H. R. (1987). Demonstration of tubulin-glycolytic enzyme interactions using a novel electrophoretic approach. *Biochem. Biophys. Res. Commun.* **146,** 827–831.

Kerst, A., Chmielewski, C., Livesay, C., Buxbaum, R. E., and Heidemann, S. R. (1990). Liquid crystal domains and thixotropy of filamentous actin suspensions. *Proc. Natl. Acad. Sci. U.S.A.* **87,** 4241–4245.

Lardy, H. A. (1965). On the direction of pyridine nucleotide oxidation-reduction reactions in gluconeogenesis and lipogenesis. *In* "Control of Energy Metabolism" (B. Chance, R. W. Estabrook, and J. R. Williamson, eds.), pp. 245–248. Academic Press, New York.

Luby-Phelps, K., Castle, P. E., Taylor, D. L., and Lanni, F. (1987). Hindered diffusion of inert tracer particles in the cytoplasm of mouse 3T3 cells. *Proc. Natl. Acad. Sci. U.S.A.* **84,** 4910–4913.

Luby-Phelps, K., Lanni, F., and Taylor, D. L. (1988). The submicroscopic properties of cytoplasm as a determinant of cellular function. *Annu. Rev. Biophys. Biophys. Chem.* **17,** 369–396.

Masters, C. J., Sheedy, R. J., Winzor, D. J., and Nichol, L. W. (1969). Reversible adsorption of enzymes as a possible allosteric control mechanism. *Biochem. J.* **112,** 806–808.

Mattevi, A., Obmolova, G., Schulze, E., Kalk, K. H., Westphal, A. H., de Kok A., and Hol, W. G. J. (1992). Atomic structure of the cubic core of the pyruvate dehydrogenase multienzyme complex. *Science* **255,** 1544–1550.

McConkey, E. H. (1982). Molecular evolution, intracellular organization, and the quinary structure of proteins. *Proc. Natl. Acad. Sci. U.S.A* **79,** 3236–3240.

Mendes, P., Kell, D. B., and Westerhoff, H. V. (1992). Channelling can decrease pool size. *Eur. J. Biochem.* **204,** 257–266.

Mikhailov, A. and Hess, B. (1995). Fluctuations in living cells and intracellular traffic. *J. Theor. Biol.* **176,** 181–184.

Minton, A. P. (1983). The effect of volume occupancy upon the thermodynamic activity of proteins: Some biochemical consequences. *Mol. Cell. Biochem.* **55,** 119–140.

Minton, A. P. (1992). Confinement as a determinant of macromolecular structure and reactivity. *Biophys. J.* **63,** 1090–1100.

Nelson, W. J. (1991). Cytoskeletal functions in membrane traffic in olarized epithelial cells. *Semin. Cell Biol.* **2,** 375–385.

Ovádi, J. (1988). Old pathway—new concept: Control of glycolysis by metabolite-modulated dynamic enzyme associations. *Trends Biochem. Sci.* **13,** 486–490.

Ovádi, J. (1995). "Cell Architecture and Metabolic Channeling." R. G. Landes, Austin, TX.

Pagliaro, L. (1993). Glycolysis revisited—a funny thing happened on the way to the Krebs cycle. *News Physiol. Sci.* **8,** 219–223.

Pagliaro, L., and Taylor, D. L. (1988). Aldolase exists in both the fluid and solid phases of cytoplasm. *J. Cell Biol.* **107,** 981–991; *erratum: ibid.,* p. 2463.

Pagliaro, L., and Taylor, D. L. (1992). 2-Deoxyglucose and cytochalasin D modulate aldolase mobility in living 3T3 cells. *J. Cell Biol.* **118,** 859–863.

Patel, M. S., and Roche, T. E. (1990). Molecular biology and biochemistry of pyruvate dehydrogenase complexes. *FASEB J.* **4,** 3224–3233.

Peters, R. A. (1930). Surface structure in the integration of cell activity. *Trans. Faraday Soc.* **26,** 797–809.

Rapoport, T. A. (1992). Transport of proteins across the endoplasmic reticulum membrane. *Science* **258,** 931–936

Reed, L. J. (1969). Pyruvate dehydrogenase complex. *Curr. Top. Cell. Regul.* **1,** 233–251.

Robinson, J. B., Jr., and Srere, P. A. (1985). Organization of Krebs tricarboxylic acid cycle enzymes in mitochondria. *J. Biol. Chem.* **262,** 10800–10805.

Rothman, J. E. (1994). Mechanisms of intracellular protein transport. *Nature (London)* **372,** 59–67.

Ryazanov, A. G. (1988). Organization of soluble enzymes in the cell. Relay at the surface. *FEBS Lett.* **237,** 1–3.

Schrödinger, E. (1944). "What Is Life?" Cambridge University Press, Cambridge, UK.

Small, W. C., Brodeur, R. D., Sandor, A., Fedorova, N., Li, G., Butow, R. A., and Srere, P. A. (1995). Enzymatic and metabolic studies on retrograde regulation mutants of yeast. *Biochemistry* **34,** 5569–5576.

Sperelakis, N. (1995). "Cell Physiology Source Book." Academic Press, San Diego, CA.

Srere, P. A. (1967). Enzyme concentrations in tissues. *Science* **158,** 936–937.

Srere, P. A. (1984). Why are enzymes so big? *Trends Biochem. Sci.* **9,** 387–390.

Srere, P. A. (1985). The metabolon. *Trends Biochem. Sci.* **10,** 109–110.

Srivastava, D. K., and Bernhard, S. A. (1986). Metabolite transfer via enzyme-enzyme complexes. *Science* **234,** 1081–1086.

Stryer, L. (1995). "Biochemistry," 4th ed., Chapter 35. Freeman, New York.

Swanson, J. A., and McNeil, P. L. (1987). Nuclear reassembly excludes large macromolecules. *Science* **238,** 548–550.

Volker, K. W., and Knull, H. R. (1993). Glycolytic enzyme-tubulin interactions: Role of tubulin carboxy terminals. *J. Mol. Recognition* **6,** 167–177.

Wakil, S. J., Stoops, J. K., and Joshi, V. C. (1983). Fatty acid synthesis and its regulation. *Annu. Rev. Biochem.* **52,** 537–579.

Walsh, J. L., Keith, T. J., and Knull, H. R. (1989). Glycolytic enzyme interactions with tubulin and microtubules. *Biochim. Biophys. Acta* **999,** 64–70.

Walsh, T. P., Clarke, F. M., and Masters, C. J. (1977). Modification of the kinetic parameters of aldolase on binding to the actin-containing filaments of skeletal muscle. *Biochem. J.* **165,** 165–167.

Walter, P., and Lingappa, V. R. (1986). Mechanism of protein translocation across the endoplasmic reticulum membrane. *Annu. Rev. Cell Biol.* **2,** 499–516.

Wang, J., Morris, A. J., Tolan, D. R., and Pagliaro, L. (1996). The molecular nature of the F-actin binding activity of aldolase revealed with site-directed mutants. *J. Biol. Chem.* **271,** 6861–6865.

Wang, J., Tolan, D. R., and Pagliaro L. (1997). Metabolic compartmentation in living cells: Structural association of aldolase. *Exp. Cell Res.* **237,** 445–451.

Wilson, J. E. (1978). Ambiquitous enzymes: Variation in intracellular distribution as a regulatory mechanism. *Trends Biochim. Sci.* **3,** 124–125.

Wilson, J. E. (1980). Brain hexokinase, the prototype ambiquitous enzyme. *Curr. Top. Cell. Regul.* **16,** 1–54.

Zimmerman, S. B. (1993). Macromolecular crowding effects on macromolecular interactions: Some implications for genome structure and function. *Biochem. Biophys. Acta* **1216,** 175–185.

Zimmerman, S. B., and Minton, A. P. (1993). Macromolecular crowding: Biochemical, biophysical, and physiological consequences. *Annu. Rev. Biophys. Biomol. Struct.* **22,** 27–65.

Part III

Cytoplasm and Phase Separation

Can Cytoplasm Exist without Undergoing Phase Separation?

D. E. Brooks
Department of Pathology and Laboratory Medicine and Department of Chemistry, University of British Columbia, Vancouver, Canada V6T 2B5

Studies on such systems as the lens of the eye and theoretical considerations suggest that phase separation may well occur in cytoplasm. In this chapter, several issues relevant to this question are raised. It is suggested that while the interaction between water and the macromolecules in a mixture is proving crucial to their phase separation behavior, the abnormal water that is widely observed in cytoplasm and concentrated protein solutions is unlikely to constitute a thermodynamic phase in the sense of phase separation studies. The role of fixed structures in the cytoplasm, the likelihood that the volume of separated phases would be small and subject to spreading over the fixed structures and the expectation that much of the phase volume could be occupied and dominated by properties of the interface are also discussed. Finally, some experimental approaches to studying the existence of liquid–liquid phases in cytoplasm are proposed. While there is no proof that phase separation exists in cytoplasm, application of some of the techniques outlined might well provide more positive evidence for its presence.

KEY WORDS: Phase separation, Bound water, Scanning probe microscopy, Confocal microscopy.

I. Introduction

The question posed by the title to this chapter is one of the central issues we have tried to explore in this book, the other being the possible consequences of such phase separation, which is addressed in the final chapter. As we discussed in the chapter by Johansson *et al.,* the high average concentration of macromolecules in cytoplasm suggests that phase separation could well occur. But is it inevitable? This is less clear.

II. Cytoplasm of the Lens of the Eye

In the chapter by Clark and Clark and in the published work from the Boston group (Liu *et al.,* 1995, 1996), the very interesting behavior of lens cytoplasm and some of its components is examined. Here is a case in which high protein concentrations are present but evolutionary pressure has acted to produce a system in which light is refracted and transmitted with virtually no losses due to light scattering. That is, fibrils, particulates, and regions of phase separation are absent under normal conditions. However, the referenced work shows that a relatively small decrease in temperature or a somewhat abnormal composition can result in phase separation with an attendant loss of optical clarity. Hence, under physiological conditions, the cytoplasm is a stable single phase but it exists quite close to the binodial on the phase diagram.

Because of the requirement for the lens to focus incoming light, the refractive index, hence, the total protein concentration, must be high. Certainly, macromolecular crowding must be occurring in these circumstances, yet phase separation does not occur. Therefore, crowding by itself is insufficient to produce phase separation under all conditions. In fact, it is not apparent that pure crowding effects will necessarily potentiate all kinds of phase separation in macromolecular solutions.

Phase separation occurs when the unfavorable entropy changes associated with the formation of phases (which are less "mixed" than the hypothetical single parent solution containing the combined components and volumes of the separated phases) are compensated for by the energetics of interactions among the components. These include paired interactions among all the distinguishable kinds of components in the system, particularly the solvent. Phase separation can result from repulsion (i.e., positive interaction energy) between two macromolecular components (as occurs, for instance, in dextran/poly(ethylene glycol)/water systems), from poor solubility (positive interaction energy between water and a macromolecule), or from attraction between two macromolecular species (complex coacervation). Macromolecular crowding would certainly be expected to potentiate complex coacervation, since association reactions, in general, are found to be enhanced under crowding conditions (Zimmerman and Minton, 1993). In the chapter by Johansson *et al.,* it is shown that a crowding agent that has high water solubility (i.e., an athermal interaction with solvent) lowers the critical concentrations for incompatible macromolecules if they do not mutually repel too strongly and if they differ in water solubility, as modeled in our Flory-Huggins (FH) calculations. It is also noted, however, that if the incompatibility between the macromolecules is stronger (more positive) than incompatibility of a macromolecule and the solvent, then phase separation is not affected by crowding by a soluble macromolecule.

While the FH calculation describes a somewhat different picture of the cytoplasm than does the hard particle exclusion model developed in most of the crowding literature, it has the advantage of being able to take into account the qualitative effects of repulsion or attraction between pairs of the components. A mean field model, which likewise takes into account the energetics of interactions among the three components of mixtures of two lens proteins and water, has recently been developed (Liu *et al.,* 1995, 1996). The model successfully describes the phase separation observed as a function of temperature and composition. It demonstrates that weak attraction between the proteins and a difference in protein–water interactions are required to reproduce the phase separation observed in ternary mixtures *in vitro.* Hence, as in the work described in the chapter by Johansson *et al.,* protein–water interactions are a major determinant of phase behavior.

It would be of considerable interest to examine thermodynamic models of other cytoplasmic mixtures by any of the approaches capable of describing aqueous protein mixtures. Scaled particle theory, mean field calculations, and Monte Carlo computations (Lomakin *et al.,* 1996) could all be examined. The theoretical approach may be at least as productive as an experimental examination of the same issues, due to the difficulties in demonstrating the presence of phases in cytoplasm, as discussed below.

III. Role of Bound Water

The evidence supporting the idea that water in concentrated protein solutions and in cytoplasm in particular exists in at least two distinct states is compelling, as outlined in the chapter by Garlid. There is clearly a sizeable fraction of the mass of water present that is not osmotically active, in all likelihood because it is associated in some sense with the macromolecules and surfaces present; it will be referred to here as "bound." As discussed above, water interactions are critical in predicting whether or not a macromolecular mixture will phase separate under crowded conditions. Does water bound to intracellular components, or its conjugate "free" water, constitute a thermodynamic phase in the sense of the chapter by Johansson *et al.* in this volume? There is no proof either way, but it seems to me that this is, in general, unlikely.

As Cabezas points out in his chapter, a phase is a region in which the physical properties are continuous, at most slowly varying functions of position, bounded by an interface which is characterized by having an interfacial tension. For bound water to be inactive osmotically, it is presumably strongly hydrogen bonded either to itself or to the surface of a macromolecule or structure. The layer of bound water would then be expected to follow the contours of the surface very closely. Since most proteins are irregular in their surface topography and exhibit a variety of charged (negative and positive), neutral, and hydrophobic groups to the aqueous environment, it is very unlikely, in general, that the properties of the bound water would vary continuously and smoothly over the surface. One would expect instead islands of bound water with particular average orientations that would vary with the nature of the surface groups on the macromolecule. The average amount of bound water could be quite large, but probably not continuous. Hence, it would not satisfy the above definition of a phase.

Another argument against the thermodynamic phase interpretation is the requirement for a bounding interface with a finite interfacial tension. The interface would have to delineate the boundary between the free and bound water. There would likely be a concentration gradient present in the vicinity of the interface since the bound water would not be expected to act equally well as a solvent for all small molecular species present (see chapter by Cabezas). The concentration differences on either side of the interface would result in a change in refractive index and a finite interfacial tension. The change in refractive index would produce observable light scattering if the effective diameter of the area bounded by the continuous interface was a significant fraction of the wavelength of light.

In concentrated protein solutions, the interfaces would have to interact and, if they had a significant surface tension, the interfaces would tend to fuse and try to minimize their surface area in order to minimize the free energy of the interface. Hence, in the crowded conditions of the cytoplasm, one might expect reasonably large regions of bound water to accumulate by aggregating the macromolecules to which the water was bound. These accumulations, in turn, ought to scatter significant amounts of light. At least in the lens of the eye, however, such scattering is not observed. Since the phenomenon of bound water occurs so generally in protein solutions, it seems highly unlikely that it would be uniformly abnormal in lens cytoplasm. More likely, bound water does not tend to minimize the area of its interface with free water, because it lacks a significant interfacial tension. In this case, it would not be considered a separate phase in the thermodynamic sense.

Assuming the lens components are not abnormal in water binding, the conditions that lead to phase separation *in vitro* when components of this system are examined presumably produce phases in which the proteins separate with their bound water associated with them. It is likely the hydrated species that undergo phase separation at the appropriate temperature and composition, as in any other phase separated mixture of macromolecules. In the cases in which agreement between observation and theory requires that the protein–solvent interaction energy by considered unfavorable (positive), the effective repulsion could result either from exposed hydrophobic amino acids or from bound water interacting unfavorably with free solvent, as discussed in the chapter by Garlid. If bound water does not form a thermodynamic phase in the above sense, this need not eliminate the possibility of phase separation occurring under crowded conditions in the cytoplasm.

IV. Role of Insoluble Structures

As is pointed out throughout this book, the cytoplasm contains numerous organelles and fibrillar systems which appear to be insoluble and exist as nondiffusing, solidlike structures. The total exposed area of such structures could be very high (see chapter by Luby-Phelps). This high surface area would be likely to have a significant effect on the distribution of any thermodynamic phases present in the cytoplasm because of the likelihood that one phase would exhibit a higher affinity for a given structure than the second phase of a locally phase separated region. That is, one of the phases would tend to wet or spread over a particular solidlike structure in preference to the other as a further manifestation of the difference in properties of the macromolecular components of each phase. This would occur in such a way as to minimize the free energy of the phases in contact with the structure, perhaps producing highly asymmetrical geometries if the solidlike surfaces were themselves highly asymmetric (e.g., fibrils). In the absence of such structures, multiphase systems are known to form concentric sequences of spherical phases (to minimize the area of the interface) within one another (Bungenberg de Jong, 1949). The geometric features of this type that are visible within the cytoplasm are due to the presence of membrane-bound organelles or nuclei lying within the region bounded by the plasma membrane. Liquid–liquid structures exhibiting these forms have not been reported in cytoplasm to my knowledge. They may be absent because their geometry would be determined by their wetting behavior in contact with the fixed surfaces of the cytoplasm, the geometry of the latter determining the shape of any associated liquid phases.

V. Phase and Interface Volumes

One of the interesting results, reported in the chapter by Johansson *et al.,* when the effects of a high background concentration of uninvolved macromolecule on phase separation of two macromolecules was investigated, was the prediction of the very small phase volume that resulted.

This could well be a general result since most of the individual components in cytoplasm are present in low total amount. Hence, although crowding effects can be expected in some cases to significantly lower the concentrations required to produce two phases, the total amount of material available may limit the volume of the phase to a very small fraction of the total available volume. Such phases could be difficult to detect, as pointed out in the chapter by Johansson *et al.* Moreover, when the possible thickness of the interface, as discussed in the chapter by Cabezas, is considered, the properties of these small volume phases could be somewhat different from those expected based on experience with larger volumes of phases in which the interface occupied a small fraction of the phase volume. For instance, Cabezas gives as an estimate an interface thickness of perhaps 10 molecular diameters. If a phase occupies a volume of diameter 1 μ and the largest macromolecule in the phase has a characteristic dimension of 100 Å, the interface occupies about 60% of the phase volume. Hence, the properties of such a phase would be much more dominated by the interfacial properties than would be expected at first glance. Understanding just how such phases would behave in the complex surroundings of the cytoplasm will require much further work.

VI. Experimental Approaches to Detecting Phases in Cytoplasm

While the theoretical possibility of phases appearing in cytoplasm seems very real, whether or not such phases actually exist in nature is a moot point. Experimental evidence for such phases would be invaluable. However, great difficulties attend such a demonstration because of the small volumes involved and the presence of organelles and surfaces that can be expected to complicate interpretation of observations and measurements in living cells. Nonetheless, the following avenues might be usefully explored.

A. Microelectrode Measurements

Neurophysiologists have, for many years, examined the electrical properties of intact excitable cells by recording the electrical potential difference between the inside of the cell and the external bathing medium (Katz, 1966). While the physical form of the electrodes began as fairly large, crude glass tubes, suitable for recording from the very large axon of the giant squid, more recently, much finer electrodes have been employed. It is now routine for investigators interested in sensory cells in the visual and auditory organs, for instance, to use glass micropipettes drawn down to diameters too small to observe in the optical microscope (Crawford and Fettiplace, 1980). The inside tip diameters are estimated to be a few hundred nanometers in these cases.

These very small electrodes are inserted through the cell membrane via oscillation or in small excursions controlled by piezoelectric drivers that allow fine control over the tip movement. The penetration of the membrane is detected electrically by a jump in potential (the membrane potential) or capacitance, not by visual observation. If such electrodes were used to probe the cytoplasm of a variety of cells, it is possible that the tip could sample any phases present. It is well known that the phases in aqueous macromolecular solutions frequently exhibit a potential difference of a few millivolts across the interface (Brooks *et al.,* 1985), a potential which is within the measurement range of the microelectrode systems referred to above. The capacitance of the liquid–liquid interface would probably be extremely low, however, due to its high water content. Hence, some indication of the presence of phases might be obtained if a careful series of investigations was performed using this technology.

B. Scanning Probe Microscopy

Scanning Probe Microscopy (SPM) has provided a relatively new method for examining structures at the molecular level (Wiesendanger, 1994). It has the outstanding advantage, from the biological perspective, of allowing measurements to be made with a resolution of nanometers in aqueous media. Hence, living cells, working enzyme systems, etc., can be examined. This approach, in one mode, allows the force acting on the instrument's tip to be recorded as a function of distance from a surface, which provides a characterization of the nature of the surface. The liquid–liquid interface of a phase boundary would be expected to show a characteristic force–distance response, depending on the wetting properties of the SPM tip for the phases involved. Hence, in principle, it might prove possible to distinguish liquid–liquid phase boundaries from, for instance, membrane-bound organelles or from cross-linked macromolecular fibers or surfaces on the basis of their response to the SPM tip. Clearly, much work on model systems would have to be done to allow any reasonable interpretation of such data. Also, some reproducible method for allowing the SPM tip access to the cell interior without diluting it so much that any phases present would be dissolved would have to be developed. Perhaps mechanically lysing the cell by sonication in a chamber filled with air equilibrated with an aqueous salt solution of the appropriate concentration (to avoid evaporation or dilution) could provide such access. While this approach is clearly pure speculation at this stage, the unparalleled sensitivity, resolution, and versatility of SPM suggest that the method deserves consideration with respect to the demonstration of phases in cytoplasm

C. Confocal Microscopy

The confocal microscope was designed to provide optical images in the presence of highly scattering backgrounds (Pawley, 1995). As such, it allows good observations to be made in the cytoplasm of cells. The method still suffers from the limitations of all optical instruments, namely, that structures smaller than about the wavelength of light cannot be resolved. However, the use of fluorescence reporting molecules (see the chapter by Luby-Phelps for examples) increases the effective resolution of the method, as signals from even single macromolecules can be detected and located in an image. Hence, if fluorescently labeled molecules could be utilized that either formed or partitioned strongly into cytoplasmic phases, there would be some chance of observing the geometry and behavior of such phases in a living cell. Information obtained perhaps from the *in vitro* experiments described below could be utilized to guide selection of appropriate molecular species to use as probes.

D. *In Vitro* Studies on Isolated Cytoplasmic Components

By far the best information on phase separation in cytoplasm has been obtained in the calf lens cytoplasm system described in the chapter by Clark and Clark. The approach has been powerful because it has been possible to isolate large amounts of the lens proteins and to study their phase behavior *in vitro*. This has allowed phase diagrams to be constructed and detailed theoretical models to be tested. There does not seem to be any reason why such experiments could not be carried out on cytoplasm from other organs. It should be possible, using bacterial, tissue culture, or animal sources, to isolate and fractionate cytoplasm from relatively large numbers of cells. This would, in principle, allow the composition and temperature dependence of phase separation to be examined in the laboratory, perhaps using confocal microscopy to visually examine relatively small volumes of concentrated solutions, to save on material requirements. Fluorescently labeled species isolated from the original mixtures could be examined for their potential as reporter molecules for phase separation that could be applied *in vivo*. Certainly, such experiments would be difficult, expensive, and would require considerable development of purification protocols, but the results would seem to be sufficiently valuable for our understanding of the organization of cytoplasm that they would be worth undertaking.

VII. Conclusions

It is interesting with respect to the question of whether or not phase separation occurs in cytoplasm that in the one system for which a strong evolutionary pressure exists to eliminate phase separation (the lens of the eye), a composition has evolved that *does* phase separate when the conditions are altered by only a small change in temperature or composition. It, therefore, seems legitimate to wonder if, in systems where the evolutionary pressure for transparency is *not* present, might phase separation not occur? There is presently no proof of its presence or absence in any of these systems. If, in fact, it is shown not to occur, it would suggest that further evolutionary advantage beyond transparency is gained by eliminating phase separation. What might this be? These questions cannot at present be answered. We hope, however, that by examining the issue and accumulating information relevant to it, such questions might be examined more definitively in the future.

Acknowledgments

I would like to express my appreciation to Drs. Tolstoguzov, Kopperschläger, Garlid, Cabezas, Johansson, Haynes, and particularly Harry Walter for provoking the interesting discussions and ideas that evolved while editing this book.

References

Brooks, D. E., Sharp, K. A., and Fisher, D. (1985). Theoretical aspects of partitioning. *In* "Partitioning in Aqueous Two-Phase Systems: Theory, Methods, Uses and Applications in Biotechnology" (H. Walter, D. E. Brooks, and D. Fisher, eds.), pp. 11–85. Academic Press, Orlando, FL.

Bungenberg de Jong, H. G. (1949). Morphology of coacervates. *In* "Colloid Science" (H. R. Kruyt, ed.), Vol. 2, pp. 433–480. Elsevier, New York.

Crawford, A. C., and Fettiplace, R. (1980). The frequency selectivity of auditory nerve fibres and hair cells in the cochlea of the turtle. *J. Physiol. (London)* **306,** 79–125.

Katz, B. (1966). "Nerve, Muscle and Synapse." McGraw-Hill, New York.

Liu, C., Lomakin, A., Thurston, G. M., Hayden, D., Pande, A., Pande, J., Ogun, O., Asherie, N., and Benedek, G. B. (1995). Phase separation in multicomponent aqueous-protein solutions. *J. Phys. Chem.* **99,** 454–461.

Liu, C., Asherie, N., Lomarkin, A., Pande, J., Ogun, O., and Benedek, G. B. (1996). Phase separation in aqueous solutions of lens γ-crystallins: special role of γ_s. *Proc. Natl. Acad. Sci. U.S.A.* **93,** 377–382.

Lomakin, A., Asherie, N., and Benedek, G. B. (1996). Monte Carlo study of phase separation in aqueous protein solutions. *J. Chem. Phys.* **104,** 1646–1656.

Pawley, J. B. (1995). "Handbook of Biological Confocal Microscopy." Plenum, New York.

Wiesendanger, R. (1994). "Scanning Probe Microscopy and Spectroscopy: Methods and Applications." Cambridge University Press, New York.

Zimmerman, S. B., and Minton, A. P. (1993). Macromolecular crowding: Biochemical, biophysical and physiological consequences. *Annu. Rev. Biophys. Biomol. Struct.* **22,** 27–65.

Consequences of Phase Separation in Cytoplasm

Harry Walter
Aqueous Phase Systems, Washington, DC 20008

Solutions of structurally different macromolecules, when mixed, above certain concentrations, tend to phase separate. The high concentration and diversity of proteins present in the liquid phase of cytoplasm, together with the phenomena which accompany macromolecular crowding, appear to meet the requirements for multiphase separation. The resulting cytoplasmic phase compartments, bounded by interfaces and/or other intracellular surfaces, would provide a dynamic, three-dimensional organizational structure. Based on the known physicochemical properties of aqueous phase systems and the partitioning behavior of biomaterials in them, such an organization could account for numerous phenomena observed in cytoplasm, such as its microcompartmentation. Aqueous phase separation is an attractive model for the liquid phase of cytoplasm because it comprises not only a structure (phase compartments, interfaces) for constraining biomaterials but also a mechanism (partitioning) for translocating them to other sites.

KEY WORDS: Aqueous phase separation, Aqueous phase systems, Cytoplasm, Macromolecular crowding, Microcompartmentation, Partitioning.

I. Introduction: General Thesis

Aqueous phase separation of solutions of structurally dissimilar macromolecules, above certain concentrations, is a common physical phenomenon which has, until recently (Walter and Brooks, 1995), been ignored as a possible organizational element of the three-dimensional space of the liquid phase of cytoplasm. Yet the great diversity of proteins (the typical eukaryotic cell synthesizes between 10 and 20 thousand different proteins) and their high total concentration [range of 17 to 35% (w/w)] in cytoplasm (see chapter by Luby-Phelps), together with the effects of macromolecular crowding (see chapter by Johansson *et al.*), make at least localized aqueous phase separation in cytoplasm a likely occurrence (see chapter by Brooks).

Two-polymer aqueous two-phase systems [primarily those containing dextran and poly(ethylene glycol)] (see chapters by Johansson/Walter, by Kopperschläger, and by Tjerneld/Johansson) have been widely studied and used for the separation and characterization by partitioning of macromolecules, membranes, and organelles. The phases that form are, in essence, compartmentalized solutions with different concentrations of the phase-forming polymers and with different physical properties, separated by an interface with an interfacial tension. We propose that aqueous multiphase separation in cytoplasm, enhanced by macromolecular crowding (see chapter by Johansson *et al.*), may be the basis for many observations ascribed to its "microcompartmentation," i.e., the characteristic, nonhomogeneous arrangement of biomaterials in cytoplasmic spaces not delimited by membranes (see chapters by Aw and by Ovádi/Srere; also see chapter by Luby-Phelps).

In this chapter, we review some anticipated consequences of phase separation in cytoplasm. Based on analogies with the known physicochemical properties of aqueous phase systems studied *in vitro* and on the partitioning behavior of biomaterials in them (outlined in Part I), they are examined in light of observations on the distribution and behavior of biomaterials in cytoplasm (outlined in Part II).

II. Visualizing Phase Separation in Cytoplasm

Phase separation can occur when two macromolecular species attract or repel each other in solution. When they attract, usually because of opposite charge, two phases result, one of which is rich and the other depleted with respect to both macromolecules (complex coacervation). Complex coacervation has been observed in cytoplasm from the eye lens by scattered light measurements (see chapter by Clark and Clark). Lens proteins generally form a single phase even at high crystallin protein concentrations but, under some conditions, comprising cellular components, ionic strength, pH, hydration, protein conformation, etc., attractive interactions between individual different crystallin proteins result in condensation of distinct coexisting phases.

When two macromolecular species, above certain concentrations, repel each other, a more common occurrence, and the one to be considered here, two phases result in which one phase is enriched with respect to one of the species while the other phase is enriched with respect to the second species (see chapters by Tolstoguzov and by Johansson/Walter). In principle, one can obtain as many compositionally distinct aqueous phases, in equilibrium, as one has such structurally differing, repelling macromolecules.

While macromolecular diversity in cytoplasm abounds and protein concentration is high (see chapter by Luby-Phelps), the concentration of individual macromolecular species is generally too low for phase separation to occur. Conditions of "macromolecular crowding," however, as exist in cytoplasm, cause phase separation to take place at concentrations up to orders of magnitude lower than in the absence of crowding (see chapter by Johansson *et al.*).

Protein–protein aqueous phase systems, formed *in vitro,* have been characterized (see chapter by Tolstoguzov). Thus, mixtures of some native proteins, above certain concentrations, undergo phase separation in a manner analogous to that of two polymer aqueous two-phase systems (see chapter by Johansson/Walter). Complex or aggregate formation through specific attraction or interaction between proteins tend to enhance incompatibility with and phase separation from other proteins while, conversely, dissociation of such complexes tends to increase compatibility. Adsorbed and dissolved forms of some proteins, which have different conformations, may also phase separate, while solutions of native proteins and their denatured counterparts quite generally phase separate (see chapter by Tolstoguzov).

Though characterized, protein–protein aqueous phase systems have not been used in partitioning studies. Based on general thermodynamic principles, however, as previously noted (Walter and Brooks, 1995), many dependences observed in phase systems containing neutral polymers (see chapters by Johansson/Walter, by Kopperschläger, and by Tjerneld/Johansson) are expected to be similar in systems in which proteins or proteins and nucleic acids are the phase-forming components. Phases in cytoplasm would differ in composition and concentration with respect not only to the phase-forming macromolecules, e.g., proteins, but also, due to their preference for particular phases, of partitioning proteins and/or other macromolecules, small molecules and salts (see Part I and Section III of this chapter). The magnitude of the interfacial tension between any two phase compartments, important in the partitioning of particles and some other phenomena (see below) would depend on the difference in composition and concentration of molecular components in them (see chapters by Johansson/Walter and by Cabezas).

Aqueous multiphase separation in cytoplasm would yield multiple compartments, bounded by interfaces and/or other surfaces, e.g., those of membranes or the cytoskeleton, with the resulting mesh-work extending in three dimensions. Too small to be affected by gravity, the framework of the phases would be determined by the specific or nonspecific adherence of components of some of the phases to intracellular structures, by affinity or by exclusion by an adjoining phase, forcing other phases, not in contact with the solid surface, to assume the smallest possible surface area in the available space of cytoplasm. Such events would cause the phases to become oriented in three dimensions.

The envisaged structure is dynamic (i.e., varies temporally and spatially just as the inhomogeneities contributed by the cytoskeleton do; see chapter by Luby-Phelps) since the compositionally dissimilar phases which cause biomaterials to partition differentially among them are themselves, as a consequence, subject to continual change. Even such changes are multifaceted since, quite generally, a change in a single parameter affecting phase composition, e.g., macromolecular concentration, brings with it a change in at least one additional parameter, such as the magnitude of the interfacial tension (Walter *et al.,* 1985). Localized altered phase compositions (e.g., of compatible and incompatible macromolecules) affect the structure. Phase separation of a localized homogeneous macromolecular solution could follow an increase in protein concentration (e.g., through protein synthesis), while some compositional changes in phase separated compartments (e.g., through partitioning; by dilution) would result in a homogeneous solution (see phase diagrams in chapters by Tolstoguzov and by Johansson/Walter). Chemical modification of macromolecules can also result in an increase (or decrease) in their tendency to phase separate from other macromolecules or from the same but unmodified macromolecular species (see chapters by Johansson/Walter and by Clark/Clark). Such a sequence of phase-separation, dissolution, mixing of components, and reformation of phase-separated compartments could be part of the process by which partitioning of cytoplasmic components takes place (see Section III).

The interface between phase compartments would constitute an additional major element in the organization of the liquid phase of cytoplasm. Not only is the interfacial tension, though small, adequate to hold particles greater than a few hundred angstrom in the interface and, thus, play an important role in the partitioning behavior of larger particulates (see chapters by Johansson/Walter and by Cabezas), but interface-bounded phase compartments can immobilize macromolecules, which partition into them, without the need for solid supports (see chapter by Tjerneld and Johansson). Furthermore, interfaces can retard diffusing macromolecules and affect transport processes (see chapter by Cabezas and below). In some of their properties interfaces resemble membranes and the large, calculated intracellular surface areas (see chapter by Luby-Phelps) may be further augmented by the presence of interface surfaces.

III. Visualizing Biomaterials in Phase-Separated Cytoplasm

Phase separation would provide cytplasm with an organizational structure that causes biomaterials to partition between and among the resulting phase compartments. Such partitioning would serve to guide (translocate) solutes and particulates to other locales, be they phase compartments, interfaces, or other intracellular surfaces, or other delivery systems such as cytoskeletal elements. Because proteins have a large number of both specific and non-specific binding sites (see chapter by Luby-Phelps), it is unlikely that bio-specific localization of a protein would occur solely by diffusion. In the following, we summarize aspects of the known partitioning behavior of biomaterials in aqueous phase systems (see Part I for details), together with some relevant properties of cytoplasm (see Part II).

A. Partitioning Phenomena

1. General

The partitioning of soluble materials between aqueous phases depends on their relative solubilities in them; while the distribution of particulates, such as membranes and organelles, reflects their relative affinity for the bulk phases and the interface (see chapter by Johansson/Walter). In general, soluble materials partition between the bulk phases while particulates partition between the two bulk phases and the interface and, with increasing size, between one phase and the interface (see chapter by Johansson/Walter).

The distributions of biomaterials emanate from the interaction of physical properties of the phase system employed (e.g., the relative hydrophobicity of the phases; presence or absence of a Donnan potential between them; the molecular weight of the phase-forming macromolecules) and those of the partitioned substance. The partitioning of macromolecules is influenced by both their surface properties (hydrophobicity, charge) and molecular weight, while the partitioning of larger particulates (those with surface areas exceeding 0.2 μm^2) depends predominantly on their surface properties (see chapter by Johansson/Walter for details).

Small molecules tend to partition equally between the phases (see chapter by Johansson/Walter). Unequal partitioning of small neutral molecules would follow their binding (covalently or noncovalently) to larger molecules with an unequal distribution between the phases (see chapter by Kopperschläger), while unequal distribution of small charged molecules would be due to electrostatic effects. Thus, when polyelectrolytes (proteins, nucleic acids) having different concentrations or charge densities partition to different phases, small charged molecules and salts will also partition unequally as compensation so as to maintain electroneutrality in each phase.

2. Specific Binding

A biospecific or general ligand attached, either covalently or noncovalently, to a phase-forming macromolecule is, in essence, restricted to the phase in which that macromolecule predominates. Binding reactions (specific or nonspecific) between a biomaterial (protein, nucleic acid, membrane, organelle) and such a ligand result in the biomaterial's affinity extraction into the phase containing the ligand (for examples, see chapters by Johansson/Walter and by Kopperschläger).

Dyes which intercalate certain base pairs or base sequences in nucleic acids have been used as ligands to effect selective extraction of nucleic acids (see chapters by Johansson/Walter and by Kopperschläger). Specific nucleic acids have also been extracted by hybridization with complementary oligonucleotide bound to a phase-forming macromolecule (see chapter by Kopperschläger).

Specific extraction of protein can also be effected by altering its surface by genetic engineering through, for example, the attachment of a "leader sequence" (e.g., a peptide tail) with a strong affinity for one of the phase-forming macromolecules (see chapter by Kopperschläger).

Even weak protein–protein binding can result in formation of a protein complex which partitions differently from the component proteins (see chapter by Johansson/Walter). Interactions between proteins and nucleic acids, proteins and cell particulates, as well as between nucleic acids can entail analogous partitioning results (see chapter by Johansson/Walter).

The chiral specific affinity of one of the enantiomers of a racemic mixture to a protein has been used to resolve an optically active mixture. The bound enantiomer partitions with the protein (see chapter by Johansson/Walter).

Changes in protein conformation often cause changes in the protein's partition coefficient (see chapter by Kopperschläger).

3. Concentrating Effects of Partitioning; "Bioreactors"

Binding of a partitioning protein to a phase-forming macromolecule or to a substance which itself strongly partitions into a given phase will tend to concentrate the protein in that phase. Extreme, one-sided partitioning can also be obtained when phase-forming macromolecules are present at higher concentrations and differing molecular weights (see chapter by Johansson/Walter).

In addition to providing a means for concentrating proteins in a given phase by partitioning, compartmentalization of enzymes has led to the development of bioreactors (see chapter by Tjerneld/Johansson). A biochemical reaction proceeds in the phase containing the essentially immobilized enzymes while the products, generally of smaller molecular weight, partition more equally between, or among, phases and are thus continuously removed from the reaction site. Biofeedback effects are thereby reduced or eliminated.

4. Transport

The movement of molecules from one phase compartment to an adjoining one requires that they cross the interface between them, a phenomenon which can affect molecular diffusion (see chapter by Cabezas and section III,B,2, following). The driving force for transport of a molecule across an interface depends on the difference in its chemical potential in the two phases (see chapter by Cabezas).

B. Cytoplasm: Some Relevant Observations

1. Compartmentation: Solutes and Organelles

The nonrandom distributions of organelles and macromolecules in cytoplasm and, as a consequence, that of some small solutes and ions as well (see chapters by Luby-Phelps, by Aw, and by Ovádi/Srere) indicate that cytoplasm is highly structured and organized (see chapters by Luby-Phelps, by Aw, by Ovádi/Srere, and by Pagliaro). Physical events most likely responsible for the heterogeneous distribution, without intervening membranes, of large solutes and particulates, are the interaction of these biomaterials with the cytoskeletal network, with which cytoplasm is permeated (see chapter by Luby-Phelps) and/or, as we suggest, to cytoplasmic phase separation.

Thus, the heterogeneous but defined distribution of organelles, e.g., mitochondria (see chapter by Aw), and other particulates, e.g., ribosomes (Provance *et al.,* 1993), may be mediated by their differential partitioning (see chapter by Johansson/Walter) among cytoplasmic phases; their ultimate position, under a given set of conditions, maintained by immobilization in a phase compartment bounded by interfaces and/or other surfaces.

The heterogeneous distribution of organelles within cells can create metabolite gradients, by either organellar production or consumption of small metabolites (see chapter by Aw). Partitioning of small, neutral molecules, in the absence of binding, is expected to be fairly equal among the phases (see chapter by Johansson/Walter). Thus, their continued production, or consumption, in a localized area of cytoplasm would give rise to a diffusion-mediated gradient.

The partitioning behavior of unbound, small charged molecules and ions would, as indicated earlier (see also chapter by Johansson/Walter), depend on electrostatic effects. Regional differences in ion concentration and pH, as observed in cytoplasm (see chapter by Aw; Jones, 1988), could ensue. Concentration differences of partitioned solutes in adjacent phases could give rise to gradients.

Protein partitioning can be affected by conformational changes of protein induced by binding of specific ligands, dissociation or association of individual oligomeric enzymes, or isoenzyme-specific reactions (see chapters by Johansson/Walter and by Kopperschläger). Interactions which entail such parameters in cytoplasm include those between heterologous proteins (see chapter by Ovádi/Srere).

The partitioning behavior of protein–ion (Jones, 1988), protein–small solute (Kao *et al.*, 1993), and protein–protein complexes in phase-separated cytoplasm could effect their heterogeneous distribution. Heteroenzyme associations of sequential enzymes in a metabolic pathway (see chapter by Ovádi/Srere) would be expected to result in characteristic partition coefficients (see chapter by Johansson/Walter), thus leading to compartmentation of such complexes. Channeling of intermediates (see chapter by Ovádi/Srere) and removal of final product by partitioning could follow (see chapter by Tjerneld/Johansson). Complex dissociation (as in dynamic heteroenzyme complexes; see chapter by Ovádi/Srere) would again yield the individual proteins' partition coefficients.

Specific extraction of a protein can be effected by attaching to it a "leader sequence" with affinity for a particular phase (see chapter by Kopperschläger). Intracellular targeting and trafficking of macromolecules (Pearse and Robinson, 1990; Rothman, 1994), in both secretory and endocytic pathways, often employ signal sequences on proteins which effect protein sorting, targeting, transport, and docking (see chapter by Pagliaro). Such signal sequences could display properties required to effect partitioning to a desired cytoplasmic site.

Enzymes bound to membranes or structural proteins (see chapter by Ovádi/Srere) can display altered activities and conformations; the latter may lead to phase separation from unbound enzymes of the same kind (see chapters by Tolstoguzov and by Kopperschläger). Binding of substrate to enzyme can result in modifications which affect enzyme partitioning behavior (see chapter by Kopperschläger). Analogously, ambiquitous enzymes which have soluble and particle-bound forms (see chapter by Ovádi/Srere) may phase-separate from one another in the two states.

2. Some Other Physicochemical Properties

Physical properties of cytoplasm that have received the most attention relate to the diffusibility of biomaterials including macromolecules, particulates as well as small solutes and water, and the nature of intracellular water.

Diffusion rates in cytoplasm of particulates, e.g., ribosomes and polysomes (Luby-Phelps, 1993), vesicles (see chapter by Luby-Phelps), proteins (Gershon *et al.,* 1984; Jacobson and Wojcieszyn, 1984; see chapter by Luby-Phelps), small solutes (Luxon and Weisiger, 1992; Mastro *et al.,* 1984; see chapter by Aw) and water (see chapter by Luby-Phelps) are slow. Protein diffusion, which is uncorrelated to protein size, is even slower than diffusion of uncharged macromolecules (see chapter by Luby-Phelps); while electroneutral (experimental) beads and vesicles, the diffusion of which is very restricted, appear also to be confined ("caged") for some portion of their trajectory (see chapter by Luby-Phelps).

It has been suggested that the slow diffusion of biomaterials in cytoplasm is a consequence of macromolecular crowding, viscosity, and reversible binding of diffusing biomaterials to cytoskeletal and other surfaces (see chapter by Luby-Phelps). Restriction of diffusion to one or two dimensions, e.g., along cytoskeletal elements or membranes, respectively, has also been suggested as a possible diffusional constraint (see chapter by Pagliaro). Phase separation of cytoplasm, resulting in phase compartments, requires diffusing biomaterials to cross interfaces. Diffusion of a protein would then be dependent on both its diffusion constant and its partition coefficient. Depending on the protein partition coefficient, the interface could become a retarding barrier and one not necessarily dependent on protein molecular size (see chapters by Johansson/Walter and by Cabezas).

Phase compartments could restrict diffusion by immobilizing biomaterials. The reduced Brownian motion of particulates observed in cytoplasm (see chapter by Luby-Phelps) would be expected if they are adsorbed at an interface (see chapter by Johansson/Walter).

Polymer aqueous phase systems yield solutions with different bulk viscosities (see chapter by Johansson/Walter). Different bulk viscosities would also be predicted for different phases in phase-separated cytoplasm. Bulk viscosities in cytoplasm have not been measured. Measurements with fluorescent probes reflect microviscosity which has been found, in several tissue culture cell lines, not to differ greatly from that of water (see chapter by Luby-Phelps).

The high concentration of solvated macromolecules in cytoplasm has raised the question of whether a portion of cell water is in an "ordered state" that differs from bulk water (see chapter by Luby-Phelps and by Pagliaro). Such water may also be separate from and discontinuous with bulk water (see chapter by Pagliaro). Osmotically active and inactive water, deemed "normal" and "abnormal" water, respectively, in mitochondria has been demonstrated (see chapter by Garlid). The osmotically inactive water comprises the water of hydration of proteins. If cellular water is bound, and the interior of a cell were more like a gel than a solution (see chapters by Luby-Phelps and by Garlid), the majority of proteins might never be "dissolved" in bulk water (see chapter by Garlid). The chapter by Brooks provides some further comments on the state of water in cytoplasm and its relation to our model of phase separation.

IV. Aspects of Cytoplasmic Organization

An outline of cytoplasmic characteristics and elements of cytoplasmic organization is presented in the chapter by Pagliaro. Thus, it is established that cytoplasm is heterogeneous (see chapter by Aw); that it is compartmentalized (see chapter by Aw); that it contains a high concentration of proteins [bringing into play phenomena associated with macromolecular crowding, such as the enhancement of phase separation, association of proteins, and adsorption of proteins to surfaces (see chapter by Johansson *et al.*)]; and that it is permeated by a dynamic cytoskeletal network.

Cytoplasmic phenomena are determined (as outlined in the chapter by Pagliaro) by the presence of membrane and other surfaces, intracellular targeting and trafficking of macromolecules, the formation of macromolecular complexes, and the functional compartmentation of enzymes (see chapter by Aw) and channeling (see chapter by Ovádi and Srere).

Just as the study of linear biochemical reactions does not yield an understanding of complex biological processes (Buxbaum, 1995), so a comprehensive view of the liquid phase of cytoplasm can not be obtained by cataloging characteristics of its organization. In the latter case, just as suggested for the former case by Buxbaum, "concepts at intermediate length" may prove useful. Such concepts do not necessarily provide understanding on the basis of individual molecules but may still provide a framework for an eventual understanding of the encountered complexities.

To effect the intricate, dynamic organization of cytoplasm must require the interplay of many mechanisms (see chapter by Pagliaro). Phase separation of cytoplasm is an attractive model because it provides both a three-dimensional structure (phase compartments, interfaces) to constrain cell components and a mechanism (partitioning) to steer biomaterials to a desired destination. While there may be more "specific" mechanisms for translocating biomaterials in cytoplasm than partitioning, e.g., those related to trafficking and targeting (see chapters by Pagliaro), it is not known how biomaterials initially access these or their eventual docking sites. Less specific mechanisms are a likely link to more "specific" ones. Considering the binding "promiscuity" of proteins (see chapter by Luby-Phelps), however, nonspecific mechanisms such as diffusion are unlikely to suffice to get biomaterials to their appropriate specific binding sites. Phase partitioning, which depends sensitively on a number of parameters (see chapters by Johansson/Walter and by Kopperschläger), may provide more controlled afflux to other transport mechanisms and/or binding sites. By analogy with the commonly used multiple extraction methods used in laboratories for separation and purification of materials which differ only slightly in their partition coefficients (Walter *et al.,* 1985), it is possible that selectivity of biomaterials, while low when distributed between two cytoplasmic phase compartments, increases with the number of compartments through which they partition.

V. Concluding Remarks

A structure that is compartmentalized, yet dynamic, has been suggested for the liquid phase of cytoplasm. Based on the high concentration of proteins and on the macromolecular crowding phenomena which that brings into play, phase separation of cytoplasm appears to be a likely event. Analogies between biomaterial partitioning behavior in aqueous two-phase systems and biomaterial distribution in cytoplasm reinforce this view (see Table I).

Cytoplasmic multiphase separations have, however, thus far, not been demonstrated by direct experiment. Possible reasons for this include: (a) that while certain structural compartmentations (e.g., those associated with organelles, vesicles, membranes) are readily demonstrable, other compartmentation mechanisms may be transient or easily disrupted by standard analytical techniques, i.e., be part of a "metastable organization" (see chapter by Pagliaro); (b) that phase separation occurs on a small and/or localized scale and is thus difficult to observe; (c) that, just as the eye lens has endogenous mechanisms for the regulation of phase separation (see chapter by Clark and Clark), cytoplasmic phase separation, as we visualize, is prevented from occurring because, for example, while the concentration of macromolecules is known to be high in cytoplasm, that of free macromolecules proves to be low. If this were so, cytoplasmic phase separation would be a driving force that evolution has specifically avoided.

It is our hope that a comparative reading of the chapters on properties of cytoplasm (Part II) and on partitioning phenomena (Part I) will tempt readers to devise experimental approaches to test the model.

TABLE I

Some Physical Events in Aqueous Two-Phase Systems and Analogous Observations in Cytoplasm[a]

Aqueous phase systems	Cytoplasm
Dif. partitioning across multiphases (Albertsson, 1986; see chapter by Johansson/Walter)	pH and some other (e.g., Ca^{2+}, ATP) gradients (Jones, 1988; see chapter by Aw)
Altered protein partitioning due to reaction with small molecule and conformational change (see chapter by Kopperschläger)	Altered protein distribution due to reaction with small molecule and conformational change (Jones, 1988)
Affinity partitioning: binding of ligands to a phase-forming polymer (see chapter by Kopperschläger)	Protein–ion (Jones, 1988); protein–small solute interactions (Kao *et al.,* 1993)
Protein–protein interactions (see chapters by Johansson/Walter and by Kopperschläger)	Protein–protein interactions (Jones, 1988; see chapter by Ovádi/Srere)
Fermentation in one phase and partitioning of product to other phase (see chapter by Tjerneld/Johansson)	Compartmentation of metabolism (see chapters by Luby-Phelps, by Aw and by Ovádi/Srere) and channeling of intermediates (see chapter by Ovádi/Srere)
"Steered" partitioning by attachment of "leader sequences" with desired partitioning properties (see chapter by Kopperschläger)	Destination determined by attached "sorting labels" (Pearse and Robinson, 1990; Rothman, 1994; see chapter by Pagliaro)
Diffusion across interfaces (see chapter by Cabezas)	Nonideal, slow diffusion (see chapter by Luby-Phelps) of micro- (Luxon and Weisiger, 1992; Mastro *et al.,* 1984) and macromolecules (Zimmerman and Minton, 1993; Gershon *et al.,* 1984; Jacobson and Wojcieszyn, 1984), and particulates (Luby-Phelps, 1993)
Preferential partitioning of particulates (see chapter by Johansson/Walter)	Compartmentation of organelles (e.g., mitochondria) (Jones, 1988; see chapter by Aw); ribosomes (Provance *et al.,* 1993); transport vesicles (Rothman, 1994)
Interface between phases (see chapters by Johansson/Walter and by Cabezas)	Intracellular surface areas are immense (see chapter by Luby-Phelps) interfaces may contribute to these
Exclusion of macromolecular solutes from a phase (Walter and Johansson, 1994)	Macromolecular crowding (Zimmerman and Minton, 1993; Cayley *et al.,* 1991; see chapter by Johansson *et al.*)
Preferential wetting by a phase of selected surfaces; repulsion of one phase by a surface causing another phase to adhere to that surface (see chapter by Walter)	Preferential adherence of certain proteins to cytoskeletal elements (see chapters by Luby-Phelps and by Pagliaro)
Donnan potential between phases containing certain salts (see chapter by Johansson/Walter)	Donnan potential: a likely occurrence due to unequal distribution of ions (see chapter by Aw)

[a] Adapted from Walter and Brooks (1995) by permission from Elsevier Science.

References

Albertsson, P.-Å. (1986). "Partition of Cell Particles and Macromolecules," 3rd ed. Wiley-Interscience, New York.

Buxbaum, R. E. (1995). Biological levels. *Nature* (*London*) **373,** 567–568.

Cayley, S., Lewis, B. A., Guttman, H. J., and Record, J. M. T. (1991). Characterization of the cytoplasm of *Escherichia coli* K-12 as a function of external osmolarity: Implications for protein-DNA interactions *in vivo. J. Mol. Biol.* **222,** 281–300.

Gershon, N. D., Porter, K. R., and Trus, B. L. (1984). The cytoplasmic matrix: Its volume and surface area and the diffusion of molecules through it. *Proc. Natl. Acad. Sci. U.S.A.* **82,** 5030–5034.

Jacobson, K., and Wojcieszyn, J. (1984). The translational mobility of substances within the cytoplasmic matrix. *Proc. Natl. Acad. Sci. U.S.A.* **81,** 6747–6751.

Jones, D. P., ed. (1988). "Microcompartmentation." CRC Press, Boca Raton, FL.

Kao, H. P., Abney, J. R., and Verkman, A. S. (1993). Determinants of the translational mobility of a small solute in cell cytoplasm. *J. Cell Biol.* **120,** 175–184.

Luby-Phelps, K. (1993). Effect of cytoarchitecture on the transport and localization of protein synthetic machinery. *J. Cell. Biochem.* **52,** 140–147.

Luxon, B. A., and Weisiger, R. A. (1992). A new method for quantitating intracellular transport: Application to the thyroid hormone 3,5,3′-triiodothyronine. *Am. J. Physiol.* (Gastrointestinol. Liver Physiol.) **263,** G733–G741.

Mastro, A. M., Babich, M. A., Taylor, W. D., and Keith, A. D. (1984). Diffusion of a small molecule in the cytoplasm of mammalian cells. *Proc. Natl. Acad. Sci. U.S.A.* **81,** 3414–3418.

Pearse, B. M. F., and Robinson, M. S. (1990). Clathrin, adaptors, and sorting. *Annu. Rev. Cell Biol.* **6,** 151–171.

Provance, D. W., Jr., McDowall, A., Marko, M., and Luby-Phelps, K. (1993). Cytoarchitecture of size-excluding compartments of living cells. *J. Cell Sci.* **106,** 565–578.

Rothman, J. E. (1994). Mechanisms of intracellular protein transport. *Nature* (*London*) **372,** 59–67.

Walter, H., and Brooks, D. E. (1995). Hypothesis: Phase separation in cytoplasm, due to macromolecular crowding, is the basis for microcompartmentation. *FEBS Lett.* **361,** 135–139.

Walter, H., and Johansson, G., eds. (1994). "Methods in Enzymology," Vol. 228. Academic Press, San Diego, CA.

Walter, H., Brooks, D. E., and Fisher, D., eds. (1985). "Partitioning in Aqueous Two-Phase Systems: Theory, Methods, Uses, and Applications to Biotechnology." Academic Press, Orlando, FL.

Zimmerman, S. B., and Minton, A. P. (1993). Macromolecular crowding: Biochemical, biophysical, and physiological consequences. *Annu. Rev. Biophys. Biomol. Struct.* **22,** 27–65.

INDEX

D

E

M

N

O

P